Plant–Pathogen Interactions

Annual Plant Reviews

A series for researchers and postgraduates in the plant sciences. Each volume in this series focuses on a theme of topical importance and emphasis is placed on rapid publication.

Plant–Pathogen Interactions

Edited by

NICHOLAS J. TALBOT
Professor of Molecular Genetics, School of Biological Sciences
Washington Singer Laboratories
University of Exeter
UK

Blackwell
Publishing

CRC Press

© 2004 by Blackwell Publishing Ltd

Editorial Offices:
Blackwell Publishing Ltd,
9600 Garsington Road, Oxford OX4 2DQ, UK
  Tel: +44 (0) 1865 776868
Blackwell Publishing Asia Pty Ltd,
550 Swanston Street, Carlton,
Victoria 3053, Australia
  Tel: +61 (0) 3 8359 1011

ISBN 1–4051–1433–9
ISSN 1460–1494

Published in the USA and Canada (only) by
CRC Press LLC,
2000 Corporate Blvd., N.W.,
Boca Raton, FL 33431, USA
Orders from the USA and Canada (only) to
CRC Press LLC

USA and Canada only:
ISBN 0–8493–2343–6
ISSN 1097–7570

First published 2004

Library of Congress
Cataloging-in-Publication Data:
A catalog record for this title is available from
the Library of Congress

British Library
Cataloguing-in-Publication Data:
A catalogue record for this title is available
from the British Library

Set in 10/12 pt Times
by Integra Software Services Pvt. Ltd,
Pondicherry, India
Printed and bound in Great Britain by
MPG Books Ltd, Bodmin, Cornwall

For further information on
Blackwell Publishing, visit our website:
www.blackwellpublishing.com

Contents

3 Infection with potyviruses   68
MINNA-LIISA RAJAMÄKI, TUULA MÄKI-VALKAMA,
KRISTIINA MÄKINEN and JARI P.T. VALKONEN

4 The *Ralstonia solanacearum*–plant interaction   92
CHRISTIAN BOUCHER and STÉPHANE GENIN

7 The *Ustilago maydis*–maize interaction 166
MARIA D. GARCIA-PEDRAJAS, STEVEN
J. KLOSTERMAN, DAVID L. ANDREWS and
SCOTT E. GOLD

List of Contributors

Dr David L. Andrews

Department of Plant Pathology, University of Georgia, Athens, GA, 30602-7274, USA

Dr Maike Both

Department of Biological Sciences, SAFB, Imperial College London, SW7 2AZ London, UK

Professor Christian Boucher

Laboratoire Interactions Plantes–Micro-organismes, CNRS-INRA BP 27, 31326 Castanet-Tolosan Cedex, France

Dr John P. Carr

Department of Plant Sciences, University of Cambridge, Downing Street, Cambridge CB2 3EA, UK

Dr Maria D. Garcia-Pedrajas

Department of Plant Pathology, University of Georgia, Athens, GA, 30602-7274, USA

Dr Stéphane Genin

Laboratoire Interactions Plantes–Micro-organismes, CNRS-INRA BP 27, 31326 Castanet-Tolosan Cedex, France

Professor Scott E. Gold

Department of Plant Pathology, University of Georgia, Athens, GA 30602-7274, USA

Dr Susan S. Hirano

Department of Plant Pathology, University of Wisconsin, Russell Laboratories 785B, 1630 Linden Drive, Madison, WI 53706, USA

Dr Steven J. Klosterman

Department of Plant Pathology, University of Georgia, Athens, GA, 30602-7274, USA

Dr Lei Li

Department of Botany and Plant Pathology, Purdue University, West Lafayette, IN 47907, USA

Dr Kristiina Mäkinen

Department of Applied Biology, University of Helsinki, PO Box 27, Latokartanonkaari 5-7, Helsinki FIN-00014, Finland

Dr Tuula Mäki-Valkama   Department of Applied Biology, University of
                                 Helsinki, PO Box 27, Latokartanonkaari 5-7,
                                 Helsinki FIN-00014, Finland

Dr Minna-Liisa Rajamäki   Department of Applied Biology, University of
                                 Helsinki, PO Box 27, Latokartanonkaari 5-7,
                                 Helsinki FIN-00014, Finland

Dr Kyeyong Seong   Department of Botany and Plant Pathology,
                                 Purdue University, West Lafayette, IN 47907,
                                 USA

Dr Pietro D. Spanu   Department of Biological Sciences, SAFB,
                                 Imperial College London, SW7 2AZ London,
                                 UK

Professor Nicholas J. Talbot School of Biological Sciences, University of
                                 Exeter, Washington Singer Laboratories,
                                 Exeter, EX4 4QG, UK

Dr Christen D. Upper   Department of Plant Pathology, University of
                                 Wisconsin, Russell Laboratories 785B, 1630
                                 Linden Drive, Madison, WI 53706, USA

Professor Jari P.T. Valkonen Department of Applied Biology, University of
                                 Helsinki, PO Box 27, Latokartanonkaari 5-7,
                                 Helsinki FIN-00014, Finland

Dr Pieter van West   Department of Molecular and Cell Biology,
                                 Institute of Medical Sciences, University of
                                 Aberdeen, Aberdeen, AB25 2ZD, UK

**Dr Vivianne G.A.A.   Laboratory of Plant Breeding, University of
Vleeshouwers**   Wageningen, Binnenhaven 5, 6709 PD
                                 Wageningen, The Netherlands

Dr Jin-Rong Xu   Department of Botany and Plant Pathology,
                                 Purdue University, West Lafayette, IN 47907,
                                 USA

Dr Chaoyang Xue   Department of Botany and Plant Pathology,
                                 Purdue University, West Lafayette, IN 47907,
                                 USA

Preface

Plant diseases are very destructive and still threaten virtually any crop grown on a commercial scale. They are kept in check by plant breeding strategies that have introgressed disease resistance genes into many important crops and by the deployment of costly (and often only partially effective) control measures such as antibiotics and fungicides. The capacity of the agents of plant disease – viruses, bacteria, fungi and oomycetes – to adapt to new conditions, to overcome disease resistance and to become resistant to pesticides is very great. This is because of the huge selective pressure being imposed on pathogen populations to evolve the characteristics required for further proliferation. For these reasons, understanding the biology of plant diseases is important and will be essential for the development of durable control strategies, either through the engineering of more sustainable disease resistance into crop species or by the development of new broad-spectrum chemicals that pose less of a risk to the environment.

There is, however, much more to the study of plant diseases than this economic justification. Our increased understanding of plant–pathogen interactions is also leading to a deeper insight into fundamental areas of cell and molecular biology. For example, RNA silencing, a process important in many eukaryotes, has been studied very effectively in a plant–viral interaction because of the experimental tractability of the system and the ability to carry out genetic screens with ease. Similarly, we are learning about protein degradation in cell signalling, the secretion of active eukaryotic-type signalling molecules into plant cells by bacterial pathogens and the role of cyclic AMP in cellular morphogenesis, all as a result of investigations into plant–pathogen interactions.

This volume provides an overview of our current knowledge of plant–pathogen interactions and the establishment of plant disease. The introductory chapter highlights emerging themes in plant–pathogen interactions, including our current understanding of the genetic basis of recognition of the invading pathogen by plants, the deployment of specific effector proteins by bacterial pathogens and the subversion of host metabolism and defences by biotrophic fungal pathogens. Each subsequent chapter covers a specific plant–pathogen interaction and many of the themes introduced in the first chapter re-emerge in more detail throughout the book. The volume covers two viral–plant interactions, two bacterial–plant interactions, two fungal–plant interactions, an obligate biotrophic fungus–plant interaction and finally an oomycete–plant interaction. This broad coverage should allow the reader to understand the distinct strategies employed by pathogens in their infection and colonisation of plants.

The intention of the volume is to draw together fundamental new information regarding plant infection mechanisms and host responses into a single source. The roles of molecular signals, gene regulation and the physiology of pathogenic organisms are emphasised, but the role of the prevailing environment in the conditioning of disease is also discussed. The book should be of value to researchers and professionals in plant pathology, and to postgraduates and advanced undergraduates looking for an accessible entry to the current plant pathology literature. My hope is that the book will also excite the interest of a more diverse set of plant biologists and microbiologists, who may see just how exciting the future of plant pathology research is likely to be.

Nicholas J. Talbot

1 Emerging themes in plant–pathogen interactions

Nicholas J. Talbot

1.1 Introduction

Plants are continually exposed to disease-causing agents, including viruses, bacteria, fungi, and nematodes. Understanding how plants contend with this constant onslaught and defend themselves against disease-causing organisms is of fundamental importance because of the prevalence and economic cost of crop diseases. Investigating the mechanisms by which pathogens establish disease and overcome host defences is also vital if plant diseases are to be fully understood and durable control strategies developed and deployed. This chapter will review a number of emerging hypotheses regarding the biology of plant–pathogen interactions, which have resulted from recent research. I have attempted to give a flavour of the new ideas generated from recent studies, rather than to give a completely comprehensive review of each subject. I have also tended to look at the interaction both from the viewpoint of the pathogen and that of the plant. The emphasis on pathogen biology is a recurring theme in this book that is designed to give a fresh perspective to the investigation of plant–pathogen interactions, highlighting the significance of many recent investigations regarding the establishment of plant disease.

1.2 Breaching the host cuticle

The first mechanical barrier to infection of plants is the host cuticle. This is a sufficiently impervious layer that it prevents infection by viruses and bacteria, and it is only the fungi, among the microbial pathogens, that have successfully evolved means of directly breaching the plant cuticle. To do this, many pathogenic fungi form specialised cells called appressoria at the ends of fungal germination tubes which differ greatly in form and function, but are essentially either swollen hyphal tips or completely differentiated dome-shaped cells. The biology of appressorium formation has been reviewed recently (Tucker & Talbot, 2001), and it is becoming apparent that parallels exist in the signalling pathways that bring about elaboration of these infection structures in many different pathogenic fungi, including *Magnaporthe grisea* (described in detail in Chapter 6), *Cochliobolus* species, the obligate biotrophic pathogen *Blumeria graminis*, and the dikaryotic basidiomycete *Ustilago maydis* (reviewed in Chapter 7). In each case, a cyclic AMP response pathway is required for cellular differentiation events that lead to appressorium formation. In this context, cAMP appears to accumulate in response to either external environmental cues such as nutrient status or surface signals (topology/ hydrophobicity) from the plant (Lee *et al.*, 2003).

Transmission of the cAMP signal then proceeds via cAMP-dependent protein kinase A (PKA) activity and phosphorylation of target proteins. In *M. grisea*, the *CPKA* gene, which encodes a catalytic sub-unit of protein kinase A (PKA-c), is required for pathogenicity of the fungus. Appressorium morphogenesis is delayed significantly and small non-functional appressoria are produced, which are unable to penetrate cuticle layers (Mitchell & Dean, 1995). The *M. grisea* Δ*cpkA* phenotype is very similar to the results of a targeted deletion of a PKA-c gene from *Colletotrichum trifolii*. The resulting PKA-c mutants were able to produce appressoria but these were unable to penetrate cuticles and cause disease (Yang & Dickman, 1999). In the powdery mildew fungus *B. graminis*, cAMP signalling plays a role in the initiation of appressorium development. Intracellular cAMP levels were shown to rise during conidial differentiation on barley leaf surfaces and appressorium germ tube emergence (Hall & Gurr, 1999; Hall *et al.*, 1999). Levels of cAMP then fall as the germ tube extends and at this point, application of cAMP inhibits further development. This suggests that in *B. graminis* cAMP signalling is required during the initial differentiation process leading to emergence of the specialised appressorium germ tube (after anchoring of the conidium and primary germ tube emergence). This is consistent with PKA activity, which has been measured in developing conidia and appressorium germ tubes (Hall *et al.*, 1999), and the presence of a PKA-c encoding gene in *B. graminis*, which is functionally related to *CPKA* gene from *M. grisea* (Bindslev *et al.*, 2001).

1.2.1   *MAP kinase signalling during infection-related development by fungi*

The role of MAPK signalling in regulation of infection structure formation by fungi was first reported following the isolation of the *PMK1* MAP kinase gene from *M. grisea*. *PMK1* encodes a functional homologue of *FUS3*, an MAPK gene from budding yeast involved in the response to mating pheromone (Xu & Hamer, 1996). A targeted deletion of *PMK1* generated mutants that were unable to form appressoria and were consequently non-pathogenic. *PMK1* is involved in the regulation of appressorium formation but is also necessary for invasive growth in rice tissue (Xu & Hamer, 1996). Pmk1 mutants are unable to form appressoria, although conidial germination tubes do respond to treatment with exogenous cAMP undergoing hooking and swelling. Based on these results, it has been proposed that *PMK1* operates downstream of a cAMP signal for initiation of appressorium formation, although there is no direct genetic evidence yet of a link between the pathways (Xu, 2000).

The identification of *PMK1* in *M. grisea*, and its pronounced mutant phenotype, has stimulated many investigations in a variety of other foliar pathogens. A *PMK1*-related MAPK gene was, for example, isolated from the necrotrophic fungus *Botrytis cinerea* (Zheng *et al.*, 2000). The *BMP1* gene encodes a product that is 94% identical to *M. grisea* Pmk1. To determine the role of *BMP1* in *B. cinerea*, *bmp1* deletion mutants were isolated and had both a reduced rate of vegetative growth and a loss of virulence following inoculation on host plants. Scanning electron microscopy (SEM) revealed that *bmp1* conidia germinate but germ tubes develop into an undifferentiated mycelium on the plant surface without penetration. *B. cinerea* does not produce well-defined appressoria in the same way as *M. grisea* but germ tubes do arrest tip growth and change direction in order to penetrate plant cells. It has been observed that germ tube tips usually swell slightly and can form appressorium-like structures in *B. cinerea* (Backhouse & Willetts, 1987). *BMP1* may play a role in the expression and secretion of cell wall-degrading enzymes such as the endopolygalacturonase gene reported to be involved in pathogenesis (Have *et al.*, 1998). The KSS1 MAPK pathway, for example, regulates the expression of an endopolygalacturonase gene, *PGU1*, in yeast (Madhani *et al.*, 1999). It is evident therefore that *B. cinerea* and *M. grisea* share functionally related MAPK pathways for plant infection, even though they cause completely different diseases and have different infection mechanisms.

A *PMK1*-related MAPK gene was also investigated in the southern corn leaf-blight fungus *Cochliobolus heterostrophus*. Spores of *C. heterostrophus* produce small appressoria that are distinct from the large, melanin-pigmented appressoria produced by *M. grisea* and *Colletotrichum* species (Tucker & Talbot, 2001). These small appressoria are not essential for plant penetration, which can also occur directly through the cuticle or the stomata. A MAPK, Chk1, was identified in *C. heterostrophus* and shares 90% identity with Pmk1 (Lev *et al.*, 1999). Deletion of *CHK1* led to the formation of mutants with altered colony morphology and poorly developed non-sporulating, aerial hyphae. Hyphal tips of wild-type *C. heterostrophus* respond to the presence of a glass or plastic surface by swelling to form small appressoria. As the Δ*chk1* strains did not conidiate, mycelial fragments were inoculated onto glass slides in the presence of a rich nutrient medium. Under these conditions, appressoria were not formed and instead the hyphae continued to grow but did not differentiate further (Lev *et al.*, 1999). When mycelial suspensions of Δ*chk1* mutants were inoculated into corn plants, the mutant was found to be much less virulent than an isogenic wild-type strain, although pathogenicity was not completely lost in any Δ*chk1* mutant tested. Mutation of *CHK1* causes pleiotropic phenotypes, which suggests the MAPK is involved in regulating several developmental pathways, such as the formation of appressoria, conidia, and aerial hyphae. The differences in phenotype between Δ*chk1* and Δ*pmk1* mutants from *C. heterostrophus* and *M. grisea* respectively not only emphasise the different uses of similar signalling components, even in two ascomycete foliar pathogens, but also highlight the potential commonality in the basic components of signal transduction pathways (Xu, 2000). Interestingly, although *CHK1* is most similar to *PMK1*, it shows significant similarities in mutant phenotype to a second *M. grisea* MAP kinase encoded by the *MPS1* gene (Xu *et al.*, 1998). This suggests that a single MAPK cascade in *C. heterostrophus* may be sufficient to regulate developmental processes controlled by two distinct MAPK pathways in *M. grisea*.

PMK1-related MAPK genes have now been isolated from a range of phytopatho-genic fungi including *Pyrenophora teres* and *Gaeumannomyces graminis* (Dufresne & Osbourn, 2001; Ruiz-Roldan *et al.*, 2001) and where tested by construction of targeted deletions, cause distinct effects on a range of virulence-associated functions.

1.3   Invading host tissue

The invasion of host tissue is fundamental to establishment of plant disease and the manner in which this is achieved differs greatly among the microbial pathogens and in necrotrophic and biotrophic interactions. Viruses use plasmodesmata very effectively to facilitate movement throughout plants and utilise specialised movement proteins as described in Chapter 2. It is becoming increasingly apparent that bacterial pathogens produce an array of secreted proteins which alter plant metabolism and defence mechanisms and bring about changes in host physiology that facilitate the movement and proliferation of bacteria throughout plant tissue. These proteins have been termed effectors and are described in more detail in Section 1.4.1. In fungal and oomycete pathogens, arguably the least explored aspect of pathogen-esis so far has been plant tissue colonisation, particularly in the widely studied model plant pathogen species such as *Magnaporthe grisea* and *Ustilago maydis* (Lee *et al.*, 2003; Talbot, 2003). This may be about to change, however, due to the genomic resources now available for both host and pathogen, which should allow a much more detailed analysis of plant infections process than has hitherto been possible. An example of how bioinformatic analysis and functional expression can be used in tandem to identify novel virulence factors in an oomycete pathogen, *Phytophthora infestans*, was recently reported (Torto *et al.*, 2003) and is discussed in more detail in Chapter 9.

1.4   Subverting host metabolism and defence

Biotrophic pathogens can colonise host tissue in the absence of apparent detection and stimulation of induced defences. A strategy deployed by a large number of biotrophic fungi is the development of a specialised infection structure called the haustorium, which enters plant cells and has long been thought to be a feeding structure. A recent report has provided the first experimental support for this role in the bean rust fungus *Uromyces fabae*. Haustoria formed by *U. fabae* are swollen hyphal protrusions that fill host plant cells due to invagination of the plasmalemma, creating an interface between host and pathogen. This is a highly specialised interface consisting of the haustorial membrane, an extra-haustorial matrix, and the host plasmalemma. Haustoria had been isolated from infected host tissue using conca-navalin A-affinity chromatography and were used to construct a cDNA library that was analysed to identify haustorial-specific cDNAs. Among the most abundantly expressed genes found in this study was the *HXT1* gene, encoding a hexose transporter. The *HXT1* gene product was shown to be a proton-symport driven transporter with specificity to D-glucose and D-fructose by its complementation of a yeast mutant and expression analysis in *Xenopus* oocytes (Voegele *et al.*, 2001). The Hxt1 protein was found, by immuno-localisation, to occur specifically in haustoria and was not found in any other rust fungus developmental stage. Thus, it appears that bean rust haustoria glucose is taken up specifically by the haustorium, presumably by secretion of an invertase to break down sucrose. This idea is consistent with results that have demonstrated cleavage of sucrose prior to fungal absorption during infection by wheat powdery mildew (Sutton *et al.*, 1999; Schulze-Lefert & Panstruga, 2003).

The localisation of Hxt1 was mirrored by the localisation of two amino acid transporters, AAT1 and AAT2. *AAT1* encodes a broad-spectrum amino acid proton symporter found expressed in haustoria and mycelium, and germinating spores of *U. fabae* (Struck *et al.*, 2002). *AAT2*, meanwhile, encodes an haustorial-specific amino acid transporter (Hahn *et al.*, 1997). When considered together these results have provided the first evidence that haustoria are indeed specialised feeding structures produced by biotrophic fungi. The development of new methods such as GFP tagging and epitope tagging of proteins, coupled with proteomic approaches, promises to provide the tools to dissect the interface between these specialised pathogens and their hosts. For example, it will prove particularly interesting to determine whether fungal proteins are incorporated into the host plasma membrane, and whether specific secreted effector proteins are produced by haustoria in order to sustain their existence in plant cells and suppress plant defences.

Evidence indicating that suppression of plant defence does occur during biotrophic infections has been recently provided in a number of independent studies. Biochemical analysis of the avirulence gene product Avr4, from the tomato leaf mould fungus *Cladosporium fulvum*, has provided evidence that the normal biological function of *AVR4* is to encode a chitin-binding protein that probably masks fungal chitin from host plant chitnases, which are induced during infection (van den Burg *et al.*, 2003). This protein has, however, become a pathogen-associated molecular pattern (PAMP) that has become recognised by the host plant by the product of the *Cf-4* resistance gene (see Section 1.5). The fungus has, however, countered this detection by mutation of a disulfide bridge in the Avr4 protein which prevents its detection by *Cf-4*, but still allows its instrinsic chitin-binding function (Westerink *et al.*, 2002; van den Burg *et al.*, 2003). A similar strategy to mask fungal chitin is deployed by the hemibiotrophic fungus *Colletotrichum* spp., where chitin is present on the surface of spores and hyphae at the pre-penetration stage of development but *in planta* structures contain chitosan, a deacetylation product of chitin that lacks the ability to induce host defences and is immune to plant chitinases (El Gueddari *et al.*, 2002). The conversion of chitin to chitosan limits induction of host defences because pathogen-derived chitin is known to be an elicitor of plant-defence responses. Its conversion to chitosan may be catalysed by a developmentally regulated chitin deacetylase (Deising & Siegrist, 1995). Invading pathogens also appear to have the capacity to limit host-defence mechanisms by releasing plant cell wall-derived oligogalacturonides. When an oligogalacturonide was co-inoculated with a glycoproteogalactan elicitor from germ tubes of *Puccinia gramnis* f. sp. *tritici*, causal agent of wheat stem rust, the presence of the oligogalacturonide suppressed induction of plant defence responses such as phenylalanine ammonia lyase and peroxidases (Moerschbacher *et al.*, 1999).

The mutualistic nitrogen-fixing bacterium *Bradyrhizobium japonicum* appears to have overcome plant defences at least in part due to suppression of elicitor recognition. The bacterium releases cyclic β-(1,3)-β-(1,6)-glucans which share structural similarity with a well-studied hepta-β-glucoside elicitor from the oomycete *Phytophthora sojae* (Bhagwat *et al.*, 1999). The cyclic β-glucan is a competitive inhibitor of the responses induced by the elicitor because it competes for the same binding site on a putative membrane-anchored hepta-glucan binding protein (Mithöfer, 2002).

A recent study has also highlighted how plant defence suppression can be induced by an invading pathogen in a completely unforeseen manner. The tomato leaf spot fungus *Septoria lycopersici* produces tomatinase, an extracellular enzyme that detoxifies α-tomatine, a steroidal glycoalkaloid saponin from tomato with potent anti-fungal activity. Tomatinase hydrolyses glucose from α-tomatine to give β_2-tomatine, a substantially less toxic compound to fungi. Tomatinase-deficient mutants of *S. lycopersici* cause disease on tomato, but were found to be non-pathogenic on *Nicotiana benthamiana*. Moreover, the tomatinase-deficient mutants induced elevated plant defence responses in *N. benthamiana*, and further analysis revealed that β_2-tomatine, the product of tomatinase activity, was responsible for suppressing plant defences in normal infection caused by *S. lycopersici*. Pre-infiltration of leaf tissue with β_2-tomatine allowed the tomatinase-deficient mutants of *S. lycopersici* to infect plants normally. Remarkably, the defence suppression induced by β_2-tomatine appeared to be generic. When *N. benthamiana* plants expressing the bacterial fleck resistance gene *Pto* were pre-infiltrated with β_2-tomatine and then challenged with *Agrobacterium tumefaciens* expressing the corresponding *AvrPto* avirulence gene, there was a marked reduction in the hypersensitive reaction that is normally induced by co-expression of these proteins, indicating a suppression of components of plant defence that contribute to hypersensitive cell death (Bouarab *et al.*, 2002). Consistent with this, the effect was found to be dependent on SGT1, an essential component of the signal transduction pathway involved in disease resistance (Peart *et al.*, 2002).

Viral pathogens also suppress host defences by suppressing RNA silencing. This effect, shown for example by potyviruses, protects the replicative double-stranded forms of RNA and genomic strands of RNA generated during viral infection from the host plant which would normally use its RNA silencing apparatus to target and degrade such molecules (Baulcombe, 2002). The mechanism by which this occurs is reviewed in this book in Chapter 3.

1.4.1 Effector proteins deployed by pathogenic bacteria

An emerging theme in bacterial pathogenesis that has been revealed by a combination of molecular genetics and comparative genome analysis is the presence of specific proteins that are secreted directly into host cells during the establishment of disease. Among human pathogenic bacteria there has been systematic and detailed investigation of the roles of these proteins in enteropathogenic and haemorrhagic bacteria which has revealed an unexpected degree of intervention by bacteria in host signal transduction pathways (for review see Galán, 2001). These bacterial virulence factors have been termed effector proteins because of their role in inducing responses in host cells. They are introduced into host cells using specialised secretory apparatus, and in particular the Type III secretion system. Type III secretion systems (TTSS) transport proteins that lack conventional signal peptide sequences out of bacterial cells and inject these directly into eukaryotic host cells. A typical TTSS is composed of a series of channel proteins and a basal plate complex within the outer bacterial membrane. Additionally, a projecting needle complex provides an extracellular conduit for protein transfer directly into host cells. In the human pathogen *Salmonella enterica*, the base of the needle complex is composed of three proteins: InvG, PrgH and PrgK which collectively form a cylindrical sub-structure surrounded by two rings. A single protein PrgI forms the needle structure, a protuberance that is formed only after the TTSS is itself operational, and the length of which is regulated by another protein InvJ. Proteins destined for secretion do not have defined signal peptide sequences, but secretion signals do exist which may involve mRNA, an amino-terminal domain in the polypeptide, or the presence of a cognate-chaperone protein that assists secretion (Galán, 2001). *Salmonella enterica* utilises two distinct TTSSs during pathogenesis. The SPI-1 TTSS is encoded by a group of genes within pathogenicity island-1 (SPI-1), a large cluster of co-ordinately regulated genes. The SPI-1 TTSS is required for the initial interaction of *Salmonella* with host intestinal epithelial cells and for pathogen internalisation. The SPI-2 TTSS is encoded in a second pathogenicity island and is required for systemic infection (Hensel, 2000). In *Salmonella* infections, the SPI-1 TTSS is required to introduce effector proteins into epithelial cells. These effectors appear to be injected directly into cells, and target cytoskeletal elements and signalling cascades required for internalisation of the bacterium (see Section 1.4.2).

In plant pathogens, TTSSs exist in a number of Gram-negative species including *Erwinia amylovora*, *Pseudomonas syringae*, *Xanthomonas campestris* (Cornelis & van Gijsegem, 2000), and *Ralstonia solanacearum* (see Chapter 4). These TTSSs deliver a large number of proteins to host tissue. The presence of the plant cell wall provides an extra barrier for effective protein delivery and it is not clear whether the needle complex, or Hrp pilus, found in these pathogens directly injects proteins into plant cell cytoplasm, or merely acts as a delivery system to the plant cell membrane for subsequent take-up by host cells. The Hrp pilus is produced by both *P. syringae* pv. *tomato* and *R. solanacearum* and is a filamentous structure that is up to 3 mm in length and 60–80 Å in diameter. The main structural protein in the

Hrp pilus is HrpY in *R. solanacearum* and HrpA in *P. syringae*. In both cases, this protein is absolutely required for hypersensitive response (HR) and pathogenicity, indicating that delivery of Type III effectors is fundamental to establishment of disease (van Gijsegem *et al.*, 2000; Wei *et al.*, 2000). Hrp TTSSs deliver a variety of secreted proteins termed Hrp effectors (required for HR in incompatible interactions and contributing to pathogenicity in compatible responses). Some of these proteins are avirulence gene (Avr) products, which encode PAMPs recognized by products of matching resistance genes, while others encode harpins which contribute to virulence and are also non-specific elicitors of HR. Type III effectors in plant pathogenic bacteria possess a number of features which indicate that they have evolved to alter host cell signalling and gene expression, many of them acting within plant cells and even in some cases within the cell nucleus (Cornelis & van Gijsegem, 2000).

1.4.2   Host mimicry

A new idea that has been formulated in recent years regarding disease establishment by bacterial pathogens is that bacterial effector proteins fulfil roles in the cells of their eukaryotic hosts, which mimic endogenous eukaryotic signalling components, or target host proteins in a manner that alters the activity of a host cell. This form of molecular mimicry is fascinating from an evolutionary perspective and also a means of understanding the structure–function relationships in certain eukaryotic signalling pathways.

In *S. enterica* infections, the signalling pathway which controls modulation of the actin cytoskeleton is directly targeted by bacterial Type III effectors. The Rho-family GTPases Cdc42 and Rac1 are activated due to the SopE and SopE2 effectors. These encode guanine nucleotide exchange-factor proteins, which activate Cdc42 and Rac1, stimulating actin cytoskeletal rearrangements. These are visualised in the form of membrane ruffling and subsequent filopodium formation and internalisation of *Salmonella* in a phagocytic vacuole. An additional pathway to regulate this process is regulated by the SopB effector. SopB is an inositol-3-phosphatase, which alters phosphoinositide levels in such a way that Cdc42 becomes activated, promoting actin rearrangements. At the same time, internalisation of the bacterium is facilitated by *Salmonella* actin-binding proteins that stimulate the transition of G-actin to polymerised F-actin. SipA reduces the critical concentration of actin required for stability of F-actin and enhances the rigidity of the filamentous form of actin. SipA is thought to span actin monomers, stimulating aggregation. SipC, meanwhile, has been shown to nucleate and bundle actin *in vitro*, and is also required for translocation of other Type III effectors into host cells. Collectively, these virulence factors allow *Salmonella* to become internalised in epithelial cells in the form of phagocytic vacuoles. Once this has occurred, quite remarkably *Salmonella* then reverses its effect on the cytoskeleton using the SptP effector. This is a GTPase-activating protein that exerts its effect by stimulating the intrinsic nucleotide-hydrolysing ability of G-proteins and thus switching Cdc42 and Rac1 to the inactive GDP-bound form. The consequence of this is to prevent further membrane ruffling and internalisation, once an infection is underway.

Effective host mimicry therefore facilitates bacterial entry into the host and then prevents further infection while stabilising the invading bacterium. Such exquisite control of host signalling is mirrored by a number of animal pathogens such as *Yersinia pestis* and enteropathogenic *Escherichia coli* (for reviews see Cornelis & van Gijsegem, 2000; Kenny, 2002).

In plant pathogenic species, an equally fine level of control is exerted on plant cells by invading pathogens. Proteins of the AvrBs3/Pth family from *Xanthomonas* spp., for example, contain repeated motifs that are important for the recognition of these PAMPs by corresponding plant resistance genes, and are also localised to the plant cell nucleus to exert their biological activity. AvrBs3 contains a nuclear localisation signal, which is functional and has been shown to be vital in determining the specificity of AvrBs3 interactions (van den Ackerveken *et al.*, 1996). Furthermore, AvrBs3/Pth proteins contain a domain that is structurally similar to acidic activation domains found in eukaryotic transcriptional activators. This domain has been shown, in at least three cases, to be required for avirulence. The acidic domain of the AvrXa10 protein from *Xanthomonas* has also been reported to function in *Arabidopsis* when fused into the DNA-binding domain of GAL4 (Zhu *et al.*, 1998), indicating that these proteins may be able to act as transcriptional activators during pathogenesis. The *AvrXa7* protein, another AvrBS3/Pth-type protein, has furthermore been shown to bind double-stranded DNA (Yang *et al.*, 2000). The precise intrinsic roles of bacterial avirulence gene products have proved elusive to determine, but it has been clear for some time that they contribute to bacterial virulence. Virulence functions have been reported, for example, for AvrPto and AvrRpm1 which are plasma-membrane-targeted proteins (Nimchuk *et al.*, 2000; Shan *et al.*, 2000). A selection of secreted Type III effectors produced by pathogenic bacteria is given in Table 1.1. The biology of effectors and their role in pathogenesis, as well as a discussion of the importance of environmental factors influencing bacterial pathogenesis, is presented in Chapter 5.

1.5 Perception of pathogens

Plants defend themselves using a range of strategies, including pre-formed defences such as structural barriers to infection and anti-microbial secondary metabolites, and by using induced defence reactions (for reviews see Martin *et al.*, 2003; Nimchuk *et al.*, 2003). Induced plant defences are activated when plants perceive the presence of a pathogen, either by recognising components of pathogens such as viral coat proteins or bacterial flagellin, or by responding to pathogen-induced damage to their own cells. Overlying these general forms of defence are the more sophisticated recognition processes whereby plants detect and respond to proteins produced by specific races of pathogens. The latter processes are termed gene-for-gene interactions after Flor's pioneering work on flax rust in the 1940s (Flor, 1955).

Table 1.1 Proteins trafficking through Hrp Type III secretion systems in phytopathogenic bacteria (adapted from Cornelis & van Gijsegem, 2000)[a]

Pathogen	Effector	Biochemical activity or characteristics	Similarity	Effect on host	Reference
Erwinia amylovora	HrpN		$HrpN_{Ech}$, $HrpN_{Ecc}$	HR	Wei *et al.* (1992)
	HrpW	Dual protein with the N-domain structurally similar to PopA or harpins and exhibiting HR-elicitor activity and the C-domain similar to pectate lyases	$HrpW_{Pss}$	HR	Gaudriault *et al.* (1998); Kim & Beer (1998)
	DspA/E	Similar to and functionally interchangeable with AvrE	AvrE	Virulence factor	Bogdanove & Bauer *et al.* (1998); Bogdanove & Kim *et al.* (1998); Gaudriault *et al.* (1997)
Pseudomonas syringae	HrpZ			HR	He *et al.* (1993)
	HrpW	Dual protein with the N-domain structurally similar to PopA or harpins and exhibiting HR-elicitor activity and the C-domain similar to pectate lyases, which is able to bind pectin but does not have pectinase activity	$HrpW_{Ea}$	HR	Charkowski *et al.* (1998)
	AvrRpt2			HR and resistance	Mudgett & Staskawicz (1999); Leister *et al.* (1996); McNellis *et al.* (1998)
	HrmA (HopPsyA)	Temperature- and pH-dependent secretion		HR	van Dijk *et al.* (1999)
	AvrPto	Binds to and activates the Ser/Thr kinase Pto resistance gene product; temperature- and pH-dependent secretion		HR and resistance	Scofield *et al.* (1996); Tang *et al.* (1996); van Dijk *et al.* (1999)
	HrpA	Major component of Hrp pilus		Required for secretion	Roine *et al.* (1997); Wei *et al.* (2000)
	AvrB			HR and resistance	Gopalan *et al.* (1996); Leister *et al.* (1996)
	HrmA			HR	Alfano *et al.* (1997)
	AvrPphB			HR and resistance	Stevens *et al.* (1998)
	AvrPphE			HR and resistance	Stevens *et al.* (1998)

Ralstonia solanacearum[b]	PopA			HR	Arlat et al. (1994)
	PopB	NLS, localised to plant nuclei			Guenéron et al. (2000)
	PopC	LRR protein			
	HrpY	Major component of Hrp pilus		Required for secretion	van Gijsegem et al. (2000)
X. campestris	AvrBs3	NLS, localised in plant nuclei, transcriptional activation domain		HR and resistance	van den Ackerveken et al. (1996)
	AvrB4		AvrBs3 family	HR and resistance	de Feyter et al. (1998)
	Avrb6		AvrBs3 family	HR and resistance	de Feyter et al. (1998)
	Avrb7		AvrBs3 family	HR and resistance	de Feyter et al. (1998)
	AvrBln		AvrBs3 family	HR and resistance	de Feyter et al. (1998)
	AvrB102		AvrBs3 family	HR and resistance	de Feyter et al. (1998)
	PthA		AvrBs3 family	Canker-associated symptoms in citrus, HR in other plants	Duan et al. (1999)

[a] Abbreviations: HR, hypersensitive response; NLS, nuclear localisation signals; LRR, leucine-rich repeats.

[b] For full list of *Ralstonia* effector proteins see Chapter 4.

Plants contain resistance genes (*R*-genes) which respond to proteins in pathogens encoded by avirulence genes (*Avr* genes). It is becoming increasingly apparent, however, that Avr proteins are produced by pathogens as part of their armoury of virulence factors, and that plants have subsequently developed a form of immunity that is based on perception of these proteins (Nimchuk *et al.*, 2003).

Plant resistance proteins vary greatly in form and activity, and a selection of *R*-genes and their characteristics is given in Table 1.2 (adapted from Martin *et al.*, 2003). The first resistance protein identified was Hm1 from maize, which is a reductase that detoxifies HC-toxin produced by *Cochliobolus carbonum* (Johal & Briggs, 1992). Hm1 has turned out to be a distinctive protein that appears to have evolved specifically to counter the threat caused by a single pathogen and the host-selective toxin it produces. Resistance proteins involved in perception of PAMP molecules can be broadly classified into five main groups (Martin *et al.*, 2003). The first class contains a single protein *Pto* which is responsible for perceiving the AvrPto PAMP produced by the bacterial-speck pathogen *P. syringae* pv. *tomato*. Pto is a serine-threonine kinase with an N-terminal myristylation

Table 1.2 A selection of plant disease resistance (*R*) genes. Based upon Martin *et al.* (2003)

Category	*R* gene	Plant	Pathogen	Cognate effector	Reference
Unclassified	*Hm1*	Maize	*Cochliobolus carbonum*		Johal & Briggs (1992)
	mlo	Barley	*B. graminis*		Buschges *et al.* (1997)
	RPW8	*Arabidopsis*	*Erisyphe chicoracearum*		Xiao *et al.* (2001)
	RRS1-R	*Arabidopsis*	*Ralstonia solanacearum*		Deslandes *et al.* (2002)
	HS1[pro-1]	Beet	*Heterodera schachtii*		Cai *et al.* (1997)
	Rpg1	Barley	*Puccinia graminis*		Brueggeman *et al.* (2002)
1	*Pto*	Tomato	*Pseudomonas syringae*	AvrPto, AvrPtoB	Kim *et al.* (2002); Martin *et al.* (1993); Ronald *et al.* (1992)
2	*Bs2*	Pepper	*Xanthomonas campestris*	AvrBs2	Minsavage *et al.* (1990); Tai *et al.* (1999)
	HRT[b]	*Arabidopsis*	Turnip crinkle virus	Coat protein	Cooley *et al.* (2000)
	RPM1	*Arabidopsis*	*P. syringae*	AvrRpm1, AvrB	Debener *et al.* (1991); Grant *et al.* (1995); Tamaki *et al.* (1988)
	RPP8[b]	*Arabidopsis*	*Peronospora parasitica*		McDowell *et al.* (1998)
	RPP13	*Arabidopsis*	*P. parasitica*		Bittner-Eddy *et al.* (2000)
	RPS2	*Arabidopsis*	*P. syringae*	AvrRpt2	Bent *et al.* (1994); Mindrinos *et al.* (1994); Whalen *et al.* (1991)
	RPS5	*Arabidopsis*	*P. syringae*	AvrPphB	Jenner *et al.* (1991); Warren *et al.* (1998)
	I2	Tomato	*Fusarium oxysporum*		Ori *et al.* (1997); Simons *et al.* (1998)
	Mi	Tomato	*Meloidogyne incognita*		Milligan *et al.* (1998)
	Mi	Tomato	*Macrosiphum euphorbiae*		Rossi *et al.* (1998); Vos *et al.* (1998)
	Mla	Barley	*Blumeria graminis*		Zhou *et al.* (2000)
	Pib	Rice	*Magnaporthe grisea*		Wang *et al.* (1999)
	Pi-ta	Rice	*M. grisea*	AVR-Pita	Brueggeman *et al.* 2002; Orbach *et al.* (2000)
	Gpa2[a]	Potato	*Globodera pallida*		van der Vossen *et al.* (2000)
	Hero	Potato	*G. rostochiensis, G. pallida*		Ernst *et al.* (2002)
	R1	Potato	*Phytophthora infestans*		Ballvora *et al.* (2002)
	Dm3	Lettuce	*Bremia lactucae*		Meyers *et al.* (1998)
	Rp1	Maize	*Puccinia sorghi*		Collins *et al.* (1999)
	Rx1[a]	Potato	Potato virus X	Coat protein	Bendahmane *et al.* (1995)
	Rx2	Potato	Potato virus X	Coat protein	Bendahmane *et al.* (1995, 2000); Querci *et al.* (1995)
	Sw-5	Tomato	Tomato spotted wilt virus		Brommonschenkel *et al.* (2000)
	Xa1	Rice	*X. oryzae*		Yoshimura *et al.* (1998)

3	L	Flax	*Melampsora lini*		Lawrence *et al.* (1995)
	M	Flax	*M. lini*		
	N	Tobacco	Tobacco mosaic virus	Viral helicase	
	P	Flax	*M. lini*		Dodds *et al.* (2001)
	RPS4	*Arabidopsis*	*P. syringae*	AvrRps4	Gassmann *et al.* (1999); Hinsch & Staskawicz (1996)
	RPP1	*Arabidopsis*	*P. parasitica*		Botella *et al.* (1998)
	RPP4	*Arabidopsis*	*P. parasitica*		van der Biezen *et al.* (2002)
	RPP5	*Arabidopsis*	*P. parasitica*		Parker *et al.* (1997)
4	Cf-2[c]	Tomato	*Cladosporium fulvum*	Avr2	Dixon *et al.* (1998); Luderer *et al.* (2002)
	Cf-4[d]	Tomato	*C. fulvum*	Avr4	Joosten *et al.* (1994); Thomas *et al.* (1997)
	Cf-5[c]	Tomato	*C. fulvum*	Avr5	Dixon *et al.* (1998)
	Cf-9[d]	Tomato	*C. fulvum*	Avr9	Jones *et al.* (1994)
5	Xa21	Rice	*Xanthomonas oryzae*		Song *et al.* (1995)

domain that guides it to the plasma membrane where it has been shown to interact directly with AvrPto. The second class is composed of proteins containing leucine-rich repeat sequences (LRRs), a putative nucleotide-binding site (NBS), and an N-terminal coiled coil (CC) motif (often a putative leucine zipper). The LRR domain comprises a short run of amino acids with leucine at every second or third position. This motif is repeated to form a flexible, parallel β-sheet structure that is implicated in protein–protein interactions, and which in one case, the rice blast resistance protein *Pi-ta*, has been shown to interact with the matching PAMP protein, AvrPi-ta (Jia *et al.*, 2000). The third class of disease resistance proteins also possess an LRR domain, but lack a CC domain and instead contain an N-terminal domain with similarity to the Toll (a receptor protein involved in *Drosophila* development) and interleukin-1 receptors. This has been denoted as the TIR domain. The first three classes of resistance proteins all appear to be cytoplasmically localised and this is consistent with the fact that they bestow resistance to intracellular pathogens (viruses) or bacteria that secrete intracellular proteins. The exception to this rule is *Pi-ta* that brings about resistance to the fungus *M. grisea* which indicates that its penetration into host cells may involve their rupture, or active protein secretion into plant cells. The fourth class of resistance genes all bestow resistance to the extracellular fungal pathogen *Cladosporium fulvum* which grows in the apoplast and causes tomato leaf mould. These tomato Cf proteins all possess a transmembrane domain and an extracellular LRR domain, indicating some receptor-ligand activity outside of the plant cell. The C-terminus contains a short intracellular tail. The last class of resistance protein, represented by the rice protein Xa21 bestowing resistance to *Xanthomonas oryzae*, contains an extracellular LRR and an intracellular serine-threonine kinase domain.

The predicted structures of resistance proteins, and in particular the presence of LRR domains, immediately suggested that they may operate as direct receptors to PAMP ligands produced by pathogens during infection. Such direct interactions have, however, been reported in only a very small number of cases (Pto/AvrPto and Pi-ta/AvrPi-ta). In most other cases, even exhaustive efforts have failed to show a direct interaction between R-proteins and their cognate Avr proteins. Resistance proteins do, however, clearly operate high up in a hierarchy of signalling molecules that bring about dramatic changes in host cell physiology, culminating in hypersensitive cell death, and considerable effort has been applied to determining how these signalling pathways operate, in the hope that this will shed light upon the molecular basis of hypersensitive cell death in plants, and the regulation of localised and systemic responses to pathogens. Genetic screens have been carried out to identify mutants that show defects in R-gene-mediated resistance. An example of the type of protein identified in such screens is the Rcr3 protein which is a cysteine protease, secreted by tomato cells, that is required for the function of the Cf-2 resistance protein. Its location has suggested that it may play a role in processing with the extracellular domain of the *Cf*-2 resistance protein, or perhaps the matching Avr2 protein (Kruger *et al.*, 2002). A large number of potential downstream signalling components involved in mediation of *R*-gene-mediated disease resistance have also been identified in genetic screens, including notably the *Arabidopsis EDS1* and *PAD4* genes required for resistance to *P. syringae* and the oomycete *Peronospora parasitica* mediated by TIR-NBS-LRR-type *R*-genes (Feys *et al.*, 2001). Recent reports have also suggested an important role for regulation of *R*-gene-mediated signalling by protein degradation due to character-isation of the Rar1 and Sgt1 proteins and their interaction with the SCF (Skp1, Cullin, F-box) complex and E3 ubiquitin-ligase complex, and components of the COP9 signalosome. These proteins are involved in ubiquitination and protein degradation via the 26S proteasome. Sgt1 is required for several *R*-gene-mediated responses (Peart *et al.*, 2002) and appears to physically interact with Rar1 in barley (Azevedo *et al.*, 2002). A possible role for such regulation by protein degra-dation may be the removal of negative regulators of plant resistance, thus providing a rapid means of de-repressing plant resistance responses. Alternatively, it may be that *R*-gene products themselves are targets for degradation following signal transduction. The extensive signalling apparatus identified to operate in association with *R*-gene signalling by the Pto protein in tomato has recently been reviewed (Pedley & Martin, 2003) and, together with other recent reviews (Martin *et al.*, 2003; Nimchuk *et al.*, 2003), provides considerable insight into the manner in which *R*-gene signalling is regulated.

1.5.1   Recognition in gene-for-gene interactions and the guard hypothesis

The lack of experimental evidence for a direct interaction between R-proteins and cognate AVR proteins has led to a number of investigations to identify additional proteins that interact with either proteins individually and that might therefore be likely to contribute to larger recognition complexes. A yeast two-hybrid screen using the *P. syringae* effector protein AvrB identified an *Arabidopsis* protein Rin4 that also interacts with the AvrRpm1 effector and was found to be required for *RPM1*-mediated disease resistance. Rin4 is a membrane-associated protein, as are the effectors AvrB and AvrRpm1 and the Rpm1 disease resistance protein (Mackey *et al.*, 2002). RIN4 can bind with either effector and can become phosphorylated in the process. Mutations in Rin4 cause loss of *RPM1* function and also appear to affect Rpm1 stability, while interestingly, reductions in *Rin4* expression also lead to constitutive defence responses. Complete null mutations in *RIN4* are lethal, but this can be suppressed by mutations in another disease resistance gene *RPS2*, suggesting that the cause of constitutive defence responses in *rin4* mutants is due to the activation of the RPS2 protein and its downstream signalling pathway (Mackey *et al.*, 2003). Significantly, Rin4 is also the target of the AvrRpt2 effector which brings about degradation of the Rin4 protein in a manner that is independent of the presence of the Rps2 protein. Taken together, the reports indicate that Rin4 is the target of a number of Type III virulence effectors and that in the case of avrRpt2, the resistance response is triggered by degradation of Rin4 leading to activation (or de-repression) of *RPS2* (Mackey *et al.*, 2003; Axtell and Staskawicz, 2003).

These results, and other examples of multiprotein complexes such as the Pto/AvrPto interaction which requires a CC-NBS-LRR protein Prf (Scofield *et al.*, 1996), have led to formulation of the guard hypothesis which proposes that bacterial effectors target proteins involved in regulation of host defences in order to suppress these defences during pathogen invasion. Plants have subsequently evolved *guard* proteins that prevent such targeting, or react to PAMP–host target interactions, stimulating hypersentive cell death to prevent plant disease from becoming established. There are a number of permutations of this hypothesis and for a full discussion of these the reader is referred to Martin *et al.* (2003) and Nimchuk *et al.* (2003).

1.6   Genome-level analysis of pathogens

The rapid increases in the efficiency of DNA sequencing have provided whole genome sequences for a variety of plant pathogens, starting with the viruses, followed by several phytopathogenic bacteria, and finally a growing number of fungal pathogens. This has provided unparalleled opportunities for investigating the gene inventories of pathogens and comparing these to related saprotrophic species, identifying the differences in genome organisation that occur in pathogens, and investigating evolution of the ability to cause plant disease among very diverse species.

The first phytopathogenic bacterium to have its genome fully sequenced was the citrus pathogen *Xylella fastidiosa* (Simpson *et al.*, 2000). This report has been followed by completion of the genome sequences of the crown gall pathogen *Agrobacterium tumefaciens* (Goodner *et al.*, 2001) and the wilt pathogen *Ralstonia solanacearum* (Salanoubat *et al.*, 2002), as well as the mutualistic plant-associated bacteria such as *Mesorhizobium loti* and *Sinorhizobium meliloti* (Kaneko *et al.*, 2000; Galibert *et al.*, 2001). Sequencing initiatives are also in place for a number of other phytopathogenic bacteria and fungi, as shown in Table 1.3. Results of comparative genome analysis between these species have shown a considerable degree of gene-order conservation among taxonomically related species such as the *Xanthomonas* species (Table 1.3) and the Rhizobiaceae, and far less gene-order conservation, although considerable similarity in gene inventories exists, among *Xylella* and *Ralstonia*, for example (van Sluys *et al.*, 2002).

One of the most striking features of bacterial pathogen genomes is the occurrence of distinct regions of the genome that exhibit a different GC content and codon utilisation pattern from the rest of the sequence. The genome islands are regions that appear to have been acquired by horizontal gene transfer, where entire genetic regions, often containing genes and regulons specific to a given ecological adaptation, are transferred between bacterial species via transposon-mediated exchanges or as plasmids, prophages, or integrons. In bacterial pathogens, these regions are often the site of entire TTSSs and associated effector protein-encoding genes and are termed pathogenicity islands, indicating that rapid evolution of virulence-associated characteristics can occur in bacteria (for review see Hacker and Kaper, 2000). Five regions of altered codon usage exist, for example, in the *Ralstonia* genome, and include TTSS-dependent effectors (see Chapter 4), and similar

Table 1.3 A selection of genome sequencing projects for phytopathogenic bacteria and fungi

Organism	Host plant	Disease	Reference	Web-site
Xylella fastidiosa	Citrus	Citrus variegated chlorosis	Simpson *et al.* (2000)	http://aeg.lbi.unicamp.br/xf/
Ralstonia solanacearum	Tomato/ potato	Vascular wilt	Salanoubat *et al.* (2002)	http://sequence.toulouse.inra.fr/ R.solanacearum
Pseudomonas syringae pv. tomato DC3000	Tomato	Bacterial speck		www.tigr.org
Agrobacterium tumefaciens	Tobacco	Crown gall	Goodner *et al.* (2001)	www.agrobacterium.org
Xanthomonas campestris	Crucifers	Black rot		http://cancer.lbi.ic.unicamp. br/ xanthomonas/
Xylella fastidiosa Temecula 1	Grapevine	Pierce's disease		http://aeg.lbi.unicamp.br/xf-grape/
Leifsonia xyli	Sugarcane	Ratoon stunting disease		http:/ /aeg.lbi.unicamp.br/
Magnaporthe grisea (F) *Ustilago maydis* (F)	Rice	Blast disease		http://www-genome.wi.mit.edu/ annotation/fungi/magnaporthe/ http://www-genome.wi.mit.edu/ annotation/fungi/ustilago_maydis/ index.html

F = phytopathogenic fungus.

regions exist in *Xanthomonas* and *Pseudomonas* species (van Sluys *et al.*, 2002). Pathogenicity islands also contain genes involved in toxin biosynthesis (syringomycin production, for example, in *Xanthomonas*) and genes encoding other virulence-associated proteins such as haemolysin and haemagglutinin, as well as Hrp and Avr effectors.

Comparative genome analysis has proved revealing in identifying particular adaptations to the host environment. *Xylella*, for instance, is a citrus pathogen which invades the xylem of its citrus host, where it exists on simple sugars present in plant cells and apoplast or released by degradation of host cell walls. Its genome mirrors this environmental adaptation, as the bacterium lacks all enzymes associated with lipid metabolism and fatty acid β-oxidation (Simpson *et al.*, 2000). Other pathogenic bacteria in contrast, such as *Ralstonia* and *Agrobacterium*, can survive in soil and thus display considerably more metabolic versatility in order to contend with glucose-deficient environments.

Global genome-level analysis has also provided much more detail concerning the secretory systems displayed by pathogenic bacteria. In addition to the TTSS mechanism described above, pathogenic bacteria in fact possess up to five distinct types of secretory apparatus. Type I secretion is responsible for toxin secretion of haemolysisns and rhizobiocin, and involves ABC transporters – ATP-dependent pumps for exporting these products. Type II secretion involves the standard *sec*-dependent secretion of peptides containing a signal peptide and is used in phytopathogens such as *Xanthomonas*, for example, to secrete cell wall-degrading enzymes like pectinases and cellulases. Type IV secretion involves a pilus-like structure that targets bacterial proteins into host cells in a manner reminiscent of TTSSs and the most well-studied example among bacterial pathogens is the *virB* operon from *A. tumefaciens*, which is responsible for T-DNA transfer to plant cell cytoplasm (Kado, 2000). Both *Xylella* and *Xanthomonas* also possess a putative Type V secretion system in the form of surface adhesin proteins that are autotransported proteins found to be involved in cellular adhesion in a number of human pathogens (Henderson *et al.*, 1998).

When gene conservation among bacterial pathogens is considered, it is striking that most of the highly conserved orthologous genes that occur in phytopathogens, but not free-living related Gram-negative species, are involved in membrane or cell wall biogenesis and function. This highlights how specificity and successful virulence is, not surprisingly, centred around the point of interface between pathogen and host.

1.6.1   Fungal and oomycete phytopathogen genomics

Phytopathogenic fungi are now the targets for a number of genome sequencing projects and the larger genomes of these organisms – typically 30–50 Mb for an ascomycete pathogen and up to 250 Mb for an oomycete – are now well within the sequencing capabilities of the dedicated genome centres around the world. A draft sequence of the genome of the rice blast pathogen *M. grisea* has already appeared and has revealed a number of surprises, including the large number of putative genes (up to 12 000 predicted ORFs (Open Reading Frames)) in a 40 Mb genome and the high frequency of predicted secreted proteins (Mitchell *et al.*, 2003). The availability of the full genome of the related saprotrophic fungus *Neurospora crassa* (Galagan *et al.*, 2003) is of great utility in providing an excellent resource for future comparative genome analysis. Sequencing projects are also underway for *Ustilago maydis*, *Fusarium graminearum*, and the oomycetes *Phytophthora infestans* and *P. sojae*.

1.7 The future

Considerable progress has been made in understanding the regulation of virulence in bacterial and fungal pathogens, the suppression of host defences by viruses, and the response of incompatible host plants to infection, in recent years. In all areas, there exists at least a basic framework to describe the signalling apparatus that mediates the outcome of plant–pathogen interactions, either compatibility or incompatibility. The framework is, however, in most cases still extremely basic, and in many instances only a small number of the components of pathways are known. The application of genome-level analysis will allow for the first time the orchestrated actions of host and pathogen genes and proteins to be investigated in a holistic manner. Transcriptional profiling has already revealed, for example, that around 10% of *Arabidopsis* genes alter their expression level during infection by the fungus *Alternaria brassisicola* (Schenck *et al.*, 2000). This form of analysis will allow the battleground of successful pathogen invasion to be investigated in great detail, and the downstream targets of many of the signalling pathways, such as the PMK1-type MAP kinases in fungal pathogens, and the *R*-gene-mediated defence response pathways in hosts, to be identified. In addition, high-throughput methods for analysis of proteins by mass spectrometry and peptide mass finger-printing will identify the proteome-level changes accompanying pathogenesis. Along with this flood of new information, however, there is a need for more systematic functional analysis of plant–pathogen interactions, in particular those that are still most poorly understood – the diseases caused by fungi and oomycetes. A key question here, for example, is do effector proteins exist that are specifically targeted to plant cells in the same way as those of bacterial pathogens? A novel informatics-driven screen for such proteins in the oomycete *P. infestans* (Torto *et al.*, 2003) highlights the type of study required, but in more genetically tractable fungal pathogens, a systematic effector screen, such as that carried out in bacterial pathogens (Guttman *et al.*, 2002), is likely to provide the answer to this question.

The use of genetically tractable models is also vital for future investigation of plant–pathogen interactions as witnessed by the explosion of new information on the various interactions of pathogens with *Arabidopsis* in recent years. The existence of activation-tagged lines, insertional mutants, gene chips, and an organised infrastructure of expertise and resources will allow rapid progress to continue. The virus-induced gene silencing system (VIGS) used in *N. benthamiana* has also proved very revealing in identifying signalling components involved in plant defence. In VIGS, a viral vector carries a gene fragment to a host cell where it is targeted to its corresponding transcript, providing an elegant means of silencing a particular component of a signalling pathway (see Peart *et al.*, 2002 for an example). High-throughput VIGS screens offer an unparalleled opportunity to identify host signalling components in plant–pathogen interactions. Overall, the genetic and biochemical methods currently available and being developed, in tandem with the wealth of genomic information being generated, indicate that there has been no better time in which to study plant–pathogen interactions.

References

Alfano, J.R., Kim, H.S., Delaney, T.P. & Collmer, A. (1997) Evidence that the *Pseudomonas syringae* pv. syringae *hrp*-linked *hrmA* gene encodes an Avr-like protein that acts in an hrp-dependent manner within tobacco cells. *Mol. Plant Microbe Interact.*, **10**, 580–588.

Arlat, M., van Gijsegem, F., Huet, J.C., Pernollet, J.C. & Boucher, C.A. (1994) PopA1, a protein which induces a hypersensitivity-like response on specific *Petunia* genotypes, is secreted via the HRP pathway of *Pseudomonas solanacearum*. *EMBO J.*, **13**, 543–553.

Axtell, M.J. & Staskawicz, B.J. (2003) Initiation of RPS2-specified disease resistance in *Arabidopsis* is coupled to AvrRpt2-directed elimination of RIN4. *Cell*, **112**, 369–377.

Azevedo, C., Sadanandom, A., Kitagawa, K., Freialdenhoven, A., Shriasu, K. & Schulze-Lefert, P. (2002) The RAR1 interactor SGT1, an essential component of *R* gene triggered disease resistance. *Science*, **295**, 2073–2076.

Backhouse, D. & Willetts, H.J. (1987) Development and structure of infection cushions of *Botrytis cinerea*. *Trans. Brit. Mycol. Soc.*, **89**, 89–95.

Ballvora, A., Ercolano, M.R., Weiss, J., Meksem, K., Bormann, C.A. *et al.* (2002) The *R1* gene for potato resistance to late blight (*Phytophthora infestans*) belongs to the leucine zipper/NBS/LRR class of plant resistance genes. *Plant J.*, **30**, 361–371.

Baulcombe, D. (2002) Viral suppression of systemic silencing. *Trends Microbiol.*, **10**, 306–308.

Bendahmane, A., Kohn, B.A., Dedi, C. & Baulcombe, D.C. (1995) The coat protein of potato virus X is a strain-specific elicitor of Rx1-mediated virus resistance in potato. *Plant J.*, **8**, 933–941.

Bendahmane, A., Querci, M., Kanyuka, K. & Baulcombe, D.C. (2000) *Agrobacterium* transient expression system as a tool for the isolation of disease resistance genes: application to the *Rx2* locus in potato. *Plant J.*, **21**, 73–81.

Bent, A.F., Kunkel, B.N., Dahlbeck, D., Brown, K.L., Schmidt, R. *et al.* (1994) *RPS2* of *Arabidopsis thaliana*: a leucine-rich repeat class of plant disease resistance genes. *Science*, **265**, 1856–1860.

Bhagwat, A.A., Mithöfer, A., Pfeffer, P.E., Kraus, C., Spickers, N. *et al.* (1999) Further studies of the role of cyclic b-glucans in symbiosis. An *ndvC* mutant of *Bradyrhizobium japonicum* synthesises cyclodecakis-(1-3)-b-glucosyl. *Plant Physiol.*, **119**, 1057–1064.

Bindslev, L., Kershaw, M.J., Talbot, N.J. & Oliver, R.P. (2001) Complementation of the cAMP-dependent protein kinase A mutant of *Magnaporthe grisea* by *Blumeria graminis pka-c* gene: functional genetic analysis of an obligate biotrophic fungus. *Mol. Plant Microbe Interact.*, **14**, 1368–1375.

Bittner-Eddy, P.D., Crute, I.R., Holub, E.B. & Beynon, J.L. (2000) *RPP13* is a simple locus in *Arabidopsis thaliana* for alleles that specify downy mildew resistance to different avirulence determinants in *Peronospora parasitica*. *Plant J.*, **21**, 177–188.

Bogdanove, A.J., Bauer, D.W. & Beer, S.V. (1998) *Erwinia amylovora* secretes DspE, a pathogenicity factor and functional AvrE homolog, through the hrp (Type III secretion) pathway. *J. Bacteriol.*, **180**, 2244–2247.

Bogdanove, A.J., Kim, J.F., Wei, Z.M., Kolchinsky, P. *et al.* (1998) Homology and functional similarity of an *hrp*-linked pathogenicity locus, *dspEF*, of *Erwinia amylovora* and the avirulence locus *avrE* of *Pseudomonas syringae* pathovar tomato. *Proc. Natl. Acad. Sci. USA*, **95**, 1325–1330.

Botella, M.A., Parker, J.E., Frost, L.N., Bittner-Eddy, P.D., Beynon, J.L. *et al.* (1998) Three genes of the *Arabidopsis RPP1* complex resistance locus recognize distinct *Peronospora parasitica* avirulence determinants. *Plant Cell*, **10**, 1847–1860.

Bouarab, K., Melton, R., Peart, J., Baulcombe, D. & Osbourn, A. (2002) A saponin-detoxifying enzyme mediates suppression of plant defences. *Nature*, **418**, 889–892.

Brommonschenkel, S.H., Frary, A. & Tanksley, S.D. (2000) The broad-spectrum tospovirus resistance gene *Sw-5* of tomato is a homolog of the root-knot nematode resistance gene *Mi*. *Mol. Plant Microbe Interact.*, **13**, 1130–1138.

Brueggeman, R., Rostoks, N., Kudrna, D., Kilian, A., Han, F. *et al.* (2002) The barley stem rust-resistance gene *Rpg1* is a novel disease-resistance gene with homology to receptor kinases. *Proc. Natl. Acad. Sci. USA*, **99**, 9328–9333.

Buschges, R., Hollricher, K., Panstruga, R., Simons, G., Wolter, M. *et al.* (1997) The barley *Mlo* gene: a novel control element of plant pathogen resistance. *Cell*, **88**, 695–705.

Cai, D., Kleine, M., Kifle, S., Harloff, H.J., Sandal, N.N. *et al.* (1997) Positional cloning of a gene for nematode resistance in sugar beet. *Science*, **275**, 832–834.

Charkowski, A.O., Alfano, J.R., Preston, G., Yuan, J., He, S.Y. *et al.* (1998) The *Pseudomonas syringae* pv. tomato HrpW protein has domains similar to harpins and pectate lyases and can elicit the plant hypersensitive response and bind to pectate. *J. Bacteriol.*, **180**, 5211–5217.

Collins, N., Drake, J., Ayliffe, M., Sun, Q., Ellis, J. *et al.* (1999) Molecular characterization of the maize *Rp1-D* rust resistance haplotype and its mutants. *Plant Cell*, **11**, 1365–1376.

Cooley, M.B., Pathirana, S., Wu, H.J., Kachroo, P. & Klessig, D.F. (2000) Members of the *Arabidopsis HRT/RPP8* family of resistance genes confer resistance to both viral and oomycete pathogens. *Plant Cell*, **12**, 663–676.

Cornelis, G.R. & van Gijsegem, F. (2000) Assembly and function of Type III secretory systems. *Annu. Rev. Microbiol.*, **54**, 735–774.

Debener, T., Lehnackers, H., Arnold, M. & Dangl, J.L. (1991) Identification and molecular mapping of a single *Arabidopsis thaliana* locus determining resistance to a phytopathogenic *Pseudomonas syringae* isolate. *Plant J.*, **1**, 289–302.

de Feyter, R., McFadden, H. & Dennis, L. (1998) Five avirulence genes from *Xanthomonas campestris* pv. malvacearum cause genotype-specific cell death when expressed transiently in cotton. *Mol. Plant Microbe Interact.*, **11**, 698–701.

Deising, H. & Siegrist, J. (1995) Chitin deacetylase activity of the bean rust fungus *Uromyces viciae fabae* is controlled by fungal morphogenesis. *FEMS Microbiol. Lett.*, **127**, 207–211.

Deslandes, L., Olivier, J., Theulieres, F., Hirsch, J., Feng, D.X. *et al.* (2002) Resistance to *Ralstonia solanacearum* in *Arabidopsis thaliana* is conferred by the recessive *RRS1-R* gene, a member of a novel family of resistance genes. *Proc. Natl. Acad. Sci. USA*, **99**, 2404–2409.

Dixon, M.S., Hatzixanthis, K., Jones, D.A., Harrison, K. & Jones, J.D. (1998) The tomato *Cf-5* disease-resistance gene and six homologs show pronounced allelic variation in leucine-rich repeat copy number. *Plant Cell*, **10**, 1915–1925.

Dodds, P., Lawrence, G. & Ellis, J. (2001) Six amino acid changes confined to the leucine-rich repeat beta-strand/beta-turn motif determine the difference between the P and P2 rust resistance specificities in flax. *Plant Cell*, **13**, 163–178.

Duan, Y.P., Castaneda, A., Zhao, G., Erdos, G. & Gabriel, D.W. (1999) Expression of a single, host-specific, bacterial pathogenicity gene in plant cells elicits division, enlargement, and cell death. *Mol. Plant Microbe Interact.*, **12**, 556–560.

Dufresne, M. & Osbourn, A.E. (2001) Definition of tissue-specific and general requirements for plant infection in a phytopathogenic fungus. *Mol. Plant Microbe Interact.*, **14**, 300–307.

El Gueddari, N.E., Rauchhaus, U., Moerschbacher, B.M. & Deising, H.B. (2002) Developmentally regulated conversion of surface-exposed chitin to chitosan in cell walls of plant pathogenic fungi. *New Phytol.*, **156**, 103–112.

Ernst, K., Kumar, A., Kriseleit, D., Kloos, D.U., Phillips, M.S. & Ganal, M.W. (2002) The broad-spectrum potato cyst nematode resistance gene (Hero) from tomato is the only member of a large gene family of NBS-LLR genes with an unusual amino acid repeat in the LRR region. *Plant J.*, **31**, 127–136.

Feys, B.J., Moisan, L.J., Newman, M.A. & Parker, J.E. (2001) Direct interaction between the *Arabidopsis* disease resistance signaling proteins EDS1 and PAD4. *EMBO J.*, **20**, 5400–5411.

Flor, H.H. (1955) Host–parasite interactions in flax – its genetics and other implications. *Phytopathology*, **45**, 680–685.

Galagan, J.E., Calvo, S.E., Borkovich, K.A., Selker, E.U., Read, N.D., Jaffe, D., FitzHigh, W. *et al.* (2003) The genome sequence of the filamentous fungus *Neurospora crassa. Nature*, **422**, 859–868.

Galán, J.E. (2001) *Salmonella* interactions with host cells: Type III secretion at work. *Annu. Rev. Cell Dev. Biol.*, **17**, 53–86.

Galibert, F., Finan, T.M., Long, S.R., Puhler, A., Abola, P. *et al.* (2001) The composite genome of the legume symbiont *Sinorhizobium meliloti. Science*, **293**, 668–672.

Gassmann, W., Hinsch, M.E. & Staskawicz, B.J. (1999) The *Arabidopsis RPS4* bacterial-resistance gene is a member of the TIR-NBS-LRR family of disease-resistance genes. *Plant J.*, **20**, 265–277.

Gaudriault, S., Malandrin, L., Paulin, J.P. & Barny M.A. (1997) DspA, an essential pathogenicity factor of *Erwinia amylovora* showing homology with AvrE of *Pseudomonas syringae*, is secreted via the Hrp secretion pathway in a DspB-dependent way. *Mol. Microbiol.*, **26**, 1057–1069.

Gaudriault, S., Brisset, M.N. & Barny, M.A. (1998) HrpW of *Erwinia amylovora*, a new Hrp-secreted protein. *FEBS Lett.*, **428**, 224–228.

Goodner, B., Hinkle, G., Gattung, S., Miller, N., Blanchard, M. *et al.* (2001) Genome sequence of the plant pathogen and biotechnology agent *Agrobacterium tumefaciens C58. Science*, **294**, 2323–2328.

Gopalan, S., Bauer, D.W., Alfano, J.R., Loniello, A.O. *et al.* (1996) Expression of the *Pseudomonas syringae* avirulence protein AvrB in plant cells alleviates its dependence on the hypersensitive response and pathogenicity (Hrp) secretion system in eliciting genotype-specific hypersensitive cell death. *Plant Cell*, **8**, 1095–1105.

Grant, M.R., Godiard, L., Straube, E., Ashfield, T., Lewald, J. *et al.* (1995) Structure of the *Arabidopsis RPM1* gene enabling dual specificity disease resistance. *Science*, **269**, 843–846.

Guenéron, M., Timmers, A.C.J., Boucher, C. & Arlat, M. (2000) Two novel proteins, PopB, which has functional nuclear localization signals, and PopC, which has a large leucine-rich repeat domain, are secreted through the Hrp-secretion apparatus of *Ralstonia solanacearum. Mol. Microbiol.*, **36**, 261–277.

Guttman, D.S., Vinatzer, B.A., Sarkar, S.F., Ranall, M.V., Kettler, G. & Greenberg, J.T. (2002) A functional screen for the Type III (Hrp) secretome of the plat pathogen *Pseudomonas syringae. Science*, **295**, 1722–1726.

Hacker, J. & Kaper, J.B. (2000) Pathogenicity islands and the evolution of microbes. *Annu. Rev. Microbiol.*, **54**, 641–679.

Hahn, M., Neef, U., Struck, C., Gottfert, M. & Mendgen, K. (1997) A putative amino acid transporter is specifically expressed in haustoria of the rust fungus *Uromyces fabae. Mol. Plant Microbe Interact.*, **10**, 438–445.

Hall, A.A. & Gurr, S.J. (1999) Initiation of appressorial germ tube differentiation and appressorial hooking: distinct morphological events regulated by cAMP signalling in *Blumeria graminis* f. sp *hordei. Physiol. Mol. Plant Pathol.*, **56**, 39–46.

Hall, A.A., Bindsle, L., Rouster, J., Rasmussen, S.W., Oliver, R.P. & Gurr, S.J. (1999) Involvement of cAMP and protein kinase A in conidial differentiation by *Erysiphe graminis* f. sp. *hordei. Mol. Plant Microbe Interact.*, **12**, 960–968.

Have, A., Mulder, W., Visser, J. & van Kan, J.A.L. (1998) The endopolygalacturonase gene *Bcpg1* is required for full virulence of *Botrytis cinerea. Mol. Plant Microbe Interact.*, **11**, 1009–1016.

He, S.Y., Huang, H.C. & Collmer, A. (1993) *Pseudomonas syringae* pv. *syringae harpin Pss*: a protein that is secreted via the Hrp pathway and elicits the hypersensitive response in plants. *Cell*, **73**, 1255–1266.

Henderson, I.R., Navarro-Garcia, F. & Nataro, J.P. (1998) The great escape: structure and function of the autotransporter proteins. *Trends Microbiol.*, **6**, 370–378.

Hensel, M. (2000) *Salmonella* pathogenicity island 2. *Mol. Microbiol.*, **36**, 1015–1023.

Hinsch, M. & Staskawicz, B. (1996) Identification of a new *Arabidopsis* disease-resistance locus, *RPS4*, and cloning of the corresponding avirulence gene, *avrRps4*, from *Pseudomonas syringae* pv. *pisi. Mol. Plant Microbe Interact.*, **9**, 55–61.

Jenner, C., Hitchin, E., Mansfield, J., Walters, K., Betteridge, P. *et al.* (1991) Gene-for gene interactions between *Pseudomonas syringae* pv. *phaseolicola* and *Phaseolus. Mol. Plant Microbe Interact.*, **4**, 553–562.

Jia, Y., McAdams, S.A., Bryan, G.T., Hershey, H.P. & Valent, B. (2000) Direct interaction of resistance gene and avirulence gene products confers rice blast resistance. *EMBO J.*, **19**, 4004–4014.

Johal, G.S. & Briggs, S.P. (1992) Reductase activity encoded by the *HM1* disease resistance gene in maize. *Science*, **258**, 958–987.

Jones, D.A., Thomas, C.M., Hammond-Kosack, K.E., Balint-Kurti, P.J. & Jones, J.D.G. (1994) Isolation of the tomato *Cf-9* gene for resistance to *Cladosporium fulvum* by transposon tagging. *Science*, **266**, 789–793.

Joosten, M.H.A.J., Cozijnsen, T.J. & de Wit, P.J.G.M. (1994) Host resistance to a fungal tomato pathogen lost by a single base-pair change in an avirulence gene. *Nature*, **367**, 384–386.

Kado, C.I. (2000) The role of T-pilus in horizontal gene transfer and tumorigenesis. *Curr. Opin. Microbiol.*, **3**, 643–648.

Kaneko, T., Nakamura, Y., Sato, S., Asamizu, E., Kato, T. *et al.* (2000) Complete genome structure of the nitrogen-fixing symbiotic bacterium *Mesorhizobium loti. DNA Res.*, **7**, 331–338.

Kim, J.F. & Beer, S.V. (1998) HrpW of *Erwinia amylovora*, a new harpin that contains a domain homologous to pectase lyases of a distinct class. *J. Bacteriol.*, **180**, 5203–5210.

Kim, Y.J., Lin, N.C. & Martin, G.B. (2002) Two distinct *Pseudomonas* effector proteins interact with the Pto kinase and activate plant immunity. *Cell*, **109**, 589–598.

Kenny, B. (2002) Enteropathogenic *Escherichia coli* (EPEC) – a crafty, subversive little bug. *Microbiology*, **148**, 1967–1978.

Kruger, J., Thomas, C.M., Golstein, C., Dion, M.S., Smoker, M. *et al.* (2002) A tomato cysteine protease required for *Cf-2*-dependent disease resistance and suppression of autonecrosis. *Science*, **296**, 744–747.

Lawrence, G.J., Finnegan, E.J., Ayliffe, M.A. & Ellis, J.G. (1995) The *L6* gene for flax rust resistance is related to the *Arabidopsis* bacterial resistance gene *RPS2* and tobacco viral resistance gene *N. Plant Cell*, **7**, 1195–1206.

Lee, N., D'Souza, C.A. & Kronstad, J.W. (2003) Of smuts, blasts, mildews, and blights: cAMP signaling in phytopathogenic fungi. *Annu. Rev. Phytopathol.*, **41**, 399–427.

Leister, R.T., Ausubel, F.M. & Katagiri, F. (1996) Molecular recognition of pathogen attack occurs inside of plant cells in plant disease resistance specified by the Arabidopsis genes RPS2 and RPM1. *Proc. Natl. Acad. Sci. USA*, **93**, 15497–15502.

Lev, S., Sharon, A., Hadar, R., Ma, H. & Horwitz, B.A. (1999) A mitogen-activated protein kinase of the corn leaf pathogen *Cochliobolus heterostrophus* is involved in conidiation, appressorium formation and pathogenicity: diverse roles for mitogen-activated protein kinase homologs in foliar pathogens. *Proc. Natl. Acad. Sci. USA*, **96**, 13542–13547.

Luderer, R., de Kock, M.J.D., Dees, R.H.L., de Wit, P.J.G.M. & Joosten, M.H.A.J. (2002) Functional analysis of cysteine residues of ECP elicitor proteins of the fungal tomato pathogen *Cladosporium fulvum. Mol. Plant Pathol.*, **3**, 91–95.

Mackey, D., Holt, B.F., Wiig, A. & Dangl, J.L. (2002) RIN4 interacts with *Pseudomonas syringae* Type III effector molecules and is required for RPM1-mediated resistance in *Arabidopsis. Cell*, **108**, 743–754.

Mackey, D., Belkhadir, Y., Alonso, J.M., Ecker, J.R. & Dangl, J.L. (2003) Arabidopsis RIN4 is a target of Type III virulence effector AvrRpt2 and modulates RPS2-mediated resistance. *Cell*, **112**, 379–389.

Madhani, H.D., Galitski, T. & Fink, G.R. (1999) Effectors of a developmental mitogen-activated protein kinase cascade revealed by expression signatures of signalling mutants. *Proc. Natl. Acad. Sci. USA*, **96**, 12530–12535.

Martin, G.B., Brommonschenkel, S.H., Chunwongse, J., Frary, A., Ganal, M.W. *et al.* (1993) Map-based cloning of a protein kinase gene conferring disease resistance in tomato. *Science*, **262**, 1432–1436.

Martin, G.B., Bogdanove, A.J. & Sessa, G. (2003) Understanding the functions of plant disease resistance proteins. *Annu. Rev. Plant Biol.*, **54**, 23–61.

McDowell, J.M., Dhandaydham, M., Long, T.A., Aarts, M.G., Goff, S. *et al.* (1998) Intragenic recombination and diversifying selection contribute to the evolution of downy mildew resistance at the *RPP8* locus of *Arabidopsis. Plant Cell*, **10**, 1861–1874.

McNellis, T.W., Mudgett, M.B., Li, K., Aoyama, T., Horvath, D. *et al.* (1998) Glucocorticoid-inducible expression of a bacterial avirulence gene in transgenic Arabidopsis induces hypersensitive cell death. *Plant J.*, **14**, 247–257.

Meyers, B.C., Shen, K.A., Rohani, P., Gaut, B.S. & Michelmore, R.W. (1998) Receptor-like genes in the major resistance locus of lettuce are subject to divergent selection. *Plant Cell*, **10**, 1833–1846.

Milligan, S.B., Bodeau, J., Yaghoobi, J., Kaloshian, I., Zabel, P. & Williamson, V.M. (1998) The root knot nematode resistance gene *Mi* from tomato is a member of the leucine zipper, nucleotide binding, leucine-rich repeat family of plant genes. *Plant Cell*, **10**, 1307–1319.

Mindrinos, M., Katagiri, F., Yu, G.-L. & Ausubel, F.M. (1994) The *A. thaliana* disease resistance gene *RPS2* encodes a protein containing a nucleotide-binding site and leucine-rich repeats. *Cell*, **78**, 1089–1099.

Minsavage, G.V., Dahlbeck, D., Whalen, M.C., Kearney, B., Bonas, U. *et al.* (1990) Gene-for-gene relationships specifying disease resistance in *Xanthomonas campestris* pv. *vesicatoria*-pepper interactions. *Mol. Plant Microbe Interact.*, **3**, 41–47.

Mitchell, T.K. & Dean, R.A. (1995) The cAMP-dependent protein kinase catalytic subunit is required for appressorium formation and pathogenesis by the rice blast fungus *Magnaporthe grisea. Plant Cell*, **7**, 1869–1878.

Mitchell, T.K., Thon, M.R., Jeong, J.S., Brwon, D., Deng, J. & Dean, R.A. (2003) The rice blast patho-gystem as a case study for the development of new tools and raw materials from genome analysis of fungal plant pathogens. *New Phytol.*, **159**, 53–61.

Mithöfer, A. (2002) Suppression of plant defence in rhizobia-legume symbiosis. *Trends Plant Sci.*, **7**, 440–444.

Moerschbacher, B.M., Mierau, M., Graesner, B., Noll, U. & Mort, A.J. (1999) Small oligomers of glacturonic acid are endogenous suppressors of disease resistance in wheat leaves. *J. Exp. Bot.*, **50**, 605–612.

Mudgett, M.B. & Staskawicz, B.J. (1999) Characterization of the *Pseudomonas syringae* pv. tomato AvrRpt2 protein: demonstration of secretion and processing during bacterial pathogenesis. *Mol. Microbiol.*, **32**, 927–41.

Nimchuk, Z., Marois, E., Kjemtrup, S., Leister, R.T., Katagiri, F. & Dangl, J.L. (2000) Eukaryotic fatty acid acylation drives plasma membrane targeting and enhances function of several Type III effector proteins from *Pseudomonas syringae. Cell*, **101**, 353–363.

Nimchuk, Z., Eulgem, T., Holt III, B.F. & Dangl, J.L. (2003) Recognition and response in the plant immune system. *Annu. Rev. Genet.*, **37**, 579–609.

Orbach, M.J., Farrall, L., Sweigard, J.A., Chumley, F.G. & Valent, B. (2000) A telomeric avirulence gene determines efficacy for the rice blast resistance gene *Pi-ta. Plant Cell*, **12**, 2019–2032.

Ori, N., Eshed, Y., Paran, I., Presting, G., Aviv, D. *et al.* (1997) The *12C* family from the wilt disease resistance locus *12* belongs to the nucleotide binding leucine-rich repeat superfamily of plant resistance genes. *Plant Cell*, **9**, 521–532.

Parker, J.E., Coleman, M.J., Dean, C. & Jones, J.D.G. (1997) The *Arabidopsis* downy mildew resistance gene *RPP5* shares similarity to the Toll and Interleukin-1 receptors with N and L6. *Plant Cell*, **9**, 879–894.

Peart, J.R., Lu, R., Sadanandom, A., Malcuit, I., Moffett, P. *et al.* (2002) Ubiquitin ligase-associated protein SGT1 is required for host and non-host disease resistance in plants. *Proc. Natl. Acad. Sci. USA*, **99**, 10865–10869.

Pedley, K.F. & Martin, G.B. (2003) Molecular basis of PTO-mediated resistance to bacterial speck disease in tomato. *Annu. Rev. Phytopathol.*, **41**, 215–243.

Querci, M., Baulcombe, D.C., Goldbach, R.W. & Salazar, L.F. (1995) Analysis of the resistance-breaking determinants of potato virus X (PVX) strain HB on different potato genotypes expressing extreme resistance to PVX. *Phytopathology*, **85**, 1003–1010.

Roine, E., Wei, W.S., Yuan, J., NurmiahoLassila, E.L. *et al.* (1997) Hrp pilus: an *hrp*-dependent bacterial surface appendage produced by *Pseudomonas syringae* pv. tomato DC3000. *Proc. Natl. Acad. Sci. USA*, **94**, 3459–3464.

Ronald, P.C., Salmeron, J.M., Carland, F.M. & Staskawicz, B.J. (1992) The cloned avirulence gene *avrPto* induces disease resistance in tomato cultivars containing the *Pto* resistance gene. *J. Bacteriol.*, **174**, 1604–1611.

Rossi, M., Goggin, F.L., Milligan, S.B., Kaloshian, I., Ullman, D.E. & Williamson, V.M. (1998) The nematode resistance gene *Mi* of tomato confers resistance against the potato aphid. *Proc. Natl. Acad. Sci. USA*, **95**, 9750–9754.

Ruiz-Roldan, M.C., Maier, F.J. & Schafer, W. (2001) *PTK1*, a mitogen-activated protein kinase gene is required for conidiation, appressorium formation, and pathogenicity of *Pyenophora teres* on barley. *Mol. Plant Microbe Interact.*, **14**, 116–125.

Salanoubat, M., Genin, S., Artiguenave, F., Gouzy, J., Mangenot, S., Arlat, M., Billault, A., Brottier, P., Camus, J.C., Cattolico, L., Chandler, M., Choisne, N., Claudel-Renard, C., Cunnac, S., Demange, N., Gaspin, C., Lavie, M., Moisan, A., Robert, C., Saurin, W., Schiex, T., Siguier, P., Thebault, P., Whalen, M., Wincker, P., Levy, M., Weissenbach, J. & Boucher, C.A. (2002) Genome sequence of the plant pathogen *Ralstonia solanacearum*. *Nature*, **415**, 497–502.

Schenck, P.M., Kazan, K., Wilson, I., Anderson, J.P., Richmond, T. *et al.* (2000) Coordinated plant defense responses in *Arabidopsis* revealed by microarray analysis. *Proc. Natl. Acad. Sci. USA*, **97**, 11655–11660.

Schulze-Lefert, P. & Panstruga, R. (2003) Establishment of biotrophy by parasitic fungi and reprogramming of host cells for disease resistance. *Annu. Rev. Phytopathol.*, **41**, 641–667.

Scofield, S.R., Tobias, C.M., Rathjen, J.P., Chang, J.H., Lavelle, D.T. *et al.* (1996) Molecular basis of gene-for-gene specificity in bacterial speck disease of tomato. *Science*, **274**, 2063–2065.

Shan, L., Thara, V.K., Martin, G.B., Zhou, J.M. & Tang, X. (2000) The *Pseudomonas* AvrPto protein is differentially recognized by tomato and is localized to the plant plasma membrane. *Plant Cell*, **12**, 2323–2328.

Simons, G., Groenendijk, J., Wijbrandi, J., Reijans, M., Groenen, J. *et al.* (1998) Dissection of the *Fusarium I2* gene cluster in tomato reveals six homologs and one active gene copy. *Plant Cell*, **10**, 1055–1068.

Simpson, A.J., Reinach, F.C., Arruda, P., Abreu, F.A., Acencio, M. *et al.* (2000) The genome sequence of the plant pathogen *Xylella fastidiosa*. *Nature*, **406**, 151–157.

Song, W.Y., Wang, G.-L., Chen, L.-L., Kim, H.-S., Pi, L.-Y. *et al.* (1995) A receptor kinase-like protein encoded by the rice disease resistance gene, *Xa21*. *Science*, **270**, 1804–1806.

Stevens, C., Bennett, M.A., Athanassopoulos, E., Tsianis, G., Taylor, J.D. & Mansfield, J.W. (1998) Sequence Variations in alleles of the avirrulence gene avrPphE.R2 from *Pseudomonas syringae* pv. phaseolicola lead to loss of recognition of the AvrPphE protein within bean cells and a gain of cultivar-specific virulence. *Mol. Microbiol.*, **29**, 167–177.

Struck, C., Ernst, M. & Hahn, M. (2002) Characterization of a developmentally regulated amino acid transporter (AAT1p) of the rust fungus *Uromyces fabae*. *Mol. Plant Pathol.*, **3**, 23–30.

Sutton, P.N., Henry, M.J. & Hall, J.L. (1999) Glucose, and not sucrose, is transported from wheat to wheat powdery mildew. *Planta*, **208**, 426–430.

Tai, T.H., Dahlbeck, D., Clark, E.T., Gajiwala, P., Pasion, R. *et al.* (1999) Expression of the *Bs2* pepper gene confers resistance to bacterial spot disease in tomato. *Proc. Natl. Acad. Sci. USA*, **96**, 14153–14158.

Talbot, N.J. (2003) On the trail of a cereal killer: investigating the biology of *Magnaporthe grisea*. *Annu. Rev. Microbiol.*, **57**, 177–202.

Tamaki, S., Dahlbeck, D., Staskawicz, B. & Keen, N.T. (1988) Characterisation and expression of two avirulence genes cloned from *Pseudomonas syringae* pv. *glycinea*. *J. Bacteriol.*, **170**, 4846–4854.

Tang, X.Y., Frederick, R.D., Zhou, J.M., Halterman, D.A., Jia, Y.L. & Martin, G.B. (1996) Initiation of plant disease resistance by physical interaction of AvrPto and Pto kinase. *Science*, **274**, 2060–2063.

Thomas, C.M., Jones, D.A., Parniske, M., Harrison, K., Balint-Kurti, P.J. *et al.* (1997) Characterization of the tomato *Cf-4* gene for resistance to *Cladosporium fulvum* identifies sequences that determine recognitional specificity in Cf-4 and Cf-9. *Plant Cell*, **9**, 2209–2224.

Torto, T.A., Li, S., Styer, A., Huitema, E., Testa, A., Gow, N.A.R., van West, P. & Kamoun, S. (2003) EST mining and functional expression assays identify extra-cellular signal proteins from the plant pathogen *Phytophthora*. *Genome Res.* (in press).

Tucker, S.L. & Talbot, N.J. (2001) Surface attachment and pre-penetration stage development by plant pathogenic fungi. *Annu. Rev. Phytopathol.*, **39**, 385–417.

van den Ackerveken, G., Marois, E. & Bonas, U. (1996) Recognition of the bacterial avirulence protein AvrBs3 occurs inside the host plant cell. *Cell*, **87**, 1307–1316.

van den Burg, H.A., Westerink, N., Francoijs, K.J., Roth, R., Woesterenk, E., Boeren, S., de Wit, P.J.G.M., Joosten, M.H.A.J. & Vervort, J. (2003) Natural disulfide bond-disrupted mutants of AVR4 of the tomato pathogen *Cladosporium fulvum* are sensitive to proteolysis, cicrumvent Cf-4-mediated resistance, but retain their chitin-binding ability. *J. Biol. Chem.*, **278**, 27340–27346.

Van der Biezen, E.A., Freddie, C.T., Kahn, K., Parker, J.E. & Jones, J.D. (2002) *Arabidopsis RPP4* is a member of the *RPP5* multigene family of TIR-NB-LRR genes and confers downy mildew resistance through multiple signalling components. *Plant J.*, **29**, 439–451.

van der Vossen, E.A., van der Voort, J.N., Kanyuka, K., Bendahmane, A., Sandbrink, H. *et al.* (2000) Homologues of a single resistance-gene cluster in potato confer resistance to distinct pathogens: a virus and a nematode. *Plant J.*, **23**, 567–576.

van Dijk, K., Fouts, D.E., Rehm, A.H., Hill, A.R., Collmer, A. *et al.* (1999) The Avr (effector) proteins HrmA (HopPsyA) and AvrPto are secreted in culture from *Pseudomonas syringae* pathovars via the Hrp (Type III) protein secretion system in a temperature- and pH-sensitive manner. *J. Bacteriol.*, **181**, 4790–4797.

van Gijsegem, F., Vasse, J., Camus, J.C., Marenda, M. & Boucher, C. (2000) *Ralstonia solanacearum* produces *hrp*-dependent pili that are required for PopA secretion but not for attachment of bacteria to plant cells. *Mol. Microbiol.*, **36**, 249–260.

van Sluys, M.A., Monteiro-Vitorello, C.B., Camargo, L.E.A., Menck, C.F.M., da Silva, A.C.R., Ferro, J.A., Oliveira, M.C., Setubal, J.C., Kitajima, J.P. & Simpson, A.J. (2002) Comparative genomic analysis of plant-associated bacteria. *Annu. Rev. Phytopathol.*, **40**, 169–189.

Voegele, R.T., Struck, C., Hahn, M. & Mendgen, K. (2001) The role of haustoria in sugar supply during infection of broad bean by the rust fungus *Uromyces fabae*. *Proc. Natl. Acad. Sci. USA*, **98**, 8133–8138.

Vos, P., Simons, G., Jesse, T., Wijbrandi, J., Heinen, L. *et al.* (1998) The tomato *Mi-1* gene confers resistance to both root-knot nematodes and potato aphids. *Nat. Biotechnol.*, **16**, 1365–1369.

Wang, Z.X., Yano, M., Yamanouchi, U., Iwamoto, M., Monna, L. *et al.* (1999) The *Pib* gene for rice blast resistance belongs to the nucleotide binding and leucine-rich repeat class of plant disease resistance genes. *Plant J.*, **19**, 55–64.

Warren, R.F., Henk, A., Mowery, P., Holub, E. & Innes, R.W. (1998) A mutation within the leucine-rich repeat domain of the *Arabidopsis* disease resistance gene *RPS5* partially suppresses multiple bacterial and downy mildew resistance genes. *Plant Cell*, **10**, 1439–1452.

Wei, Z.M., Laby, R.J., Zumoff, C.H., Bauer, D.W., He S.Y., Collmer, A. & Beer, S.V. (1992) Harpin, elicitor of the hypersensitive response produced by the plant pathogen *Erwinia amylovora*. *Science*, **257**, 85–88.

Wei, W.S., Plovanich-Jones, A., Deng, W.L., Jin, Q.L., Collmer, A. *et al.* (2000) The gene encoding for the Hrp pilus structural protein is required for Type III secretion of Hrp and Avr proteins in *Pseudomonas syringae* pv. *tomato*. *Proc. Natl. Acad. Sci. USA*, **97**, 2247–2252.

Westerink, N., Roth, R., van den Burg, H.A., De Wit, P.J.G.M. & Joosten, M.H.A.J. (2002) The AVR4 elicitor protein of *Cladosporium fulvum* binds to fungal components with high affinity. *Mol. Plant Microbe Interact.*, **15**, 1219–1227.

Whalen, M.C., Innes, R.W., Bent, A.F. & Staskawicz, B.J. (1991) Identification of *Pseudomonas syringae* pathogens of *Arabidopsis* and a bacterial locus determining avirulence on both *Arabidopsis* and soybean. *Plant Cell*, **3**, 49–59.

Xiao, S., Ellwood, S., Calis, O., Patrick, E., Li, T. *et al.* (2001) Broad-spectrum mildew resistance in *Arabidopsis thaliana* mediated by *RPW8*. *Science*, **291**, 118–120.

Xu, J.R. (2000) MAP kinases in fungal pathogens. *Fungal Genet. Biol.*, **31**, 137–152.

Xu, J.-R. & Hamer, J.E. (1996) MAP kinase and cAMP signalling regulate infection structure formation and pathogenic growth in the rice blast fungus *Magnaporthe grisea*. *Genes Dev.*, **10**, 2696–2706.

Xu, J.-R., Staiger, C.J. & Hamer, J.E. (1998) Inactivation of the mitogen-activated protein kinase Mps1 from the rice blast fungus prevents penetration of host cells but allows activation of plant defence responses. *Proc. Natl. Acad. Sci. USA*, **95**, 12713–12718.

Yang, Z. & Dickman, M.B. (1999) *Colletotrichum trifolii* mutants disrupted in the catalytic subunit of cAMP-dependent protein kinase are non-pathogenic. *Mol. Plant Microbe Interact.*, **12**, 1753–1790.

Yang, B., Zhu, W., Johnson, L.B. & White, F.F. (2000) The virulence factor AvrXa7 of *Xanthomonas oryzae* pv. oryzae is a Type III-secretion pathway-dependent nuclear-localized double-stranded DNA-binding protein. *Proc. Natl. Acad. Sci. USA*, **97**, 9807–9812.

Yoshimura, S., Yamanouchi, U., Katayose, Y., Toki, S., Wang, Z.X. *et al.* (1998) Expression of *Xa1*, a bacterial blight-resistance gene in rice, is induced by bacterial inoculation. *Proc. Natl. Acad. Sci. USA*, **95**, 1663–1668.

Zheng, L., Campbell, M., Murphy, J., Lam, S. & Xu, J.-R. (2000) The *BMP1* gene is essential for pathogenicity in the gray mold fungus *Botrytis cinerea*. *Mol. Plant Microbe Interact.*, **13**, 724–732.

Zhou, J.-M., Trifa, Y., Silva, H., Pontier, D., Lam, E. *et al.* (2000) NPR1 differentially interacts with members of the TGA/OBF family of transcription factors that bind an element of the *PR-1* gene required for induction by salicylic acid. *Mol. Plant Microbe Interact.*, **13**, 191–202.

Zhu, W.G., Yang, B., Chittoor, J.M., Johnson, L.B. & White, F.F. (1998) AvrXa10 contains an acidic transcriptional activation domain in the functionally conserved C terminus. *Mol. Plant Microbe Interact.*, **11**, 824–832.

2 Tobacco mosaic virus

John Peter Carr

2.1 Introduction

Tobacco mosaic virus (TMV) occupies a rather special position in scientific history, since it was the first virus to be defined as such (Beijerinck, 1898, reviewed by Zaitlin 1998). Subsequently, TMV was used extensively as a model for the exploration of many important fundamental concepts in modern biology. These included some of the first work on the role of RNA as a carrier of genetic information and the elucidation of the genetic code (see Hull, 2002 and earlier editions). Studies of the properties of the TMV virion also contributed significantly to some of the earliest investigations of the mechanisms governing the assembly and self-assembly of viruses, and other biological macromolecular structures (Butler, 1999; Klug, 1999).

Within the plant sciences, studies of TMV have contributed significantly to the understanding of plant–microbe interaction in general and in particular to our current knowledge of induced-resistance phenomena such as systemic acquired resistance (SAR). Efforts to understand the TMV–host interaction also led directly to the development of the first pathogen-resistant transgenic plants (Powell-Abel *et al.*, 1986). For in-depth reviews of the significance of TMV research in the history of biology, the reader is referred to the article by Harrison and Wilson (1999) and its companion reviews published in the same issue of the *Philosophical Transactions of The Royal Society of London*. In addition, Scholthof *et al.* (1999) have collected together key papers from the first century of TMV research.

Since the history of TMV, and the contributions of TMV research to the development of virology and molecular biology, has been ably and comprehensively covered by others, this chapter will focus predominantly on recent developments in research on TMV and closely related viruses. In particular, I will attempt to review our current knowledge of TMV genome structure, replication, and movement through the host. Subsequently, the effects of TMV on its hosts, both susceptible and resistant, will be discussed with particular emphasis on recent progress in identifying host or viral gene products that appear to play roles in the virus–plant interaction. In the final section of this chapter, the potential of TMV as a biotechnological tool as well as other potential future research directions will be discussed.

2.2   *Tobacco mosaic virus*: virion and genome structure

2.2.1   *Genome structure and the taxonomy of TMV and related viruses*

TMV is the type species of the genus *Tobamovirus* (Zaitlin, 2000; Gibbs, 1999) which belongs to the alphavirus-like supergroup of positive sense, single-stranded RNA viruses (Haseloff *et al.*, 1984; Gibbs, 1999; Zaitlin, 2000). The tobamoviral genome is monopartite, consisting of a capped, single-stranded RNA molecule of approximately 6300 nucleotides [6395 in the case of the common (also called *vulgare* or U1) strain of TMV: Goelet *et al.*, 1982]. As with all viruses, classification is constantly changing, as new sequence information becomes available. Currently, the genus can be divided into three subgroups (Melcher, 2003). Subgroup 1 includes TMV U1 and *Tomato mosaic virus* (ToMV), subgroup 2 includes *Cucumber green mottle mosaic virus* (CGMMV), and subgroup 3 includes the crucifer-infecting tobamoviruses (Melcher, 2003). Okada (1999) further divided subgroup 1 into subgroups 1a (including TMV U1 and ToMV) and subgroup 1b (including *Pepper mild mottle virus*) based on genome organization. The *typical* tobamovirus genome map in Fig. 2.1 is based on the subgroup 1a organization.

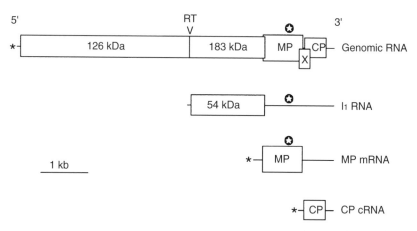

Figure 2.1 Tobamoviral genome organization. The map represents the organization of the genome for a subgroup 1a tobamovirus such as the common or U1 strain of TMV. Horizontal lines represent RNA and boxes represent open reading frames (ORFs) for known or putative viral proteins. The 183 kDa replicase protein is synthesized by read-through of a leaky termination codon at the end of the 126 kDa replicase protein ORF (position indicated by RT). These proteins are synthesized by translation of the genomic RNA. The movement protein (MP) and coat protein (CP) gene sequences are synthesized by translation of subgenomic mRNAs. An additional subgenomic RNA (I_1) encoding a putative 54-kDa protein is produced during infection but it is not known if this RNA is a functional mRNA *in vivo*. The position of an additional ORF for the putative X protein (Morozov *et al.*, 1993) is indicated on the map of the genomic RNA. 5'-Cap structures are indicated by * and the position of the origin of assembly sequence (OAS) is indicated by a star. It should be noted that the OAS is located in the ORF for the CP in subgroup 2 tobamoviruses (Okada, 1999).

Viral genomes are modular in nature – i.e., entire sections of viral genome can arise from recombination between viral genomes or between host and viral genomes (Haseloff *et al.*, 1984). This, together with the lack of a fossil record, makes it difficult to trace the ancestry of the tobamoviruses with exactitude. Nevertheless, based on comparisons of gene and genome sequences between tobamoviruses infecting various groups of plants, Gibbs (1999) has inferred that the *proto-tobamoviruses* probably first arose around 120–140 million years ago, around the same time that angiosperms appeared. Further back than this, there may have been even earlier progenitor(s) of modern tobamoviruses. This idea is based on the isolation and partial sequencing of a tobamo-like virus infecting the alga *Chara australis* (Gibbs, 1999).

2.2.2   The TMV virion

The virus particle or virion of TMV is considered to be the archetypal example of a rod-shaped virus (Klug, 1999). The TMV virion is a rigid rod 300 nm long, 18 nm in diameter, and with a central hole of radius 2 nm (Butler, 1999; Culver, 2002). The TMV genomic RNA (6395 nt) is encapsidated by 2130 coat or capsid protein (CP) subunits in a helical arrangement and the RNA forms a helix within the coat located at a radius of 4 nm from the major axis of the virion (Butler, 1999; Culver, 2002). By mass, the RNA constitutes approximately 5% of the virion and the CP 95%. For further up-to-date information on the bio- and physico-chemical properties of the CP and the virion, the reader is referred to the recent review by Culver (2002).

2.2.3   Virion assembly

A remarkable property of the TMV virion is its ability to self-assemble *in vitro* from a mixture of purified viral RNA and protein (reviewed by Butler, 1999; Klug, 1999; Culver, 2002). Under various conditions of the solution, CP can be induced to form a range of aggregation states. One of these, a cylindrical structure consisting of two layers of 17 CP subunits called the 20S disk, is the form that is able to interact with the viral RNA and participate in the assembly process. Incorporation of the TMV RNA into the virion is initiated at the origin of assembly (OAS), a stem-loop hairpin structure formed by the sequence between nucleotides 5420 and 5546 in the TMV *vulgare* sequence (Zimmern and Wilson, 1976; Zimmern, 1977; Goelet *et al.*, 1982). In this strain of TMV and in most tobamoviruses, the OAS is located within the open reading frame (ORF) encoding the protein responsible for viral cell-to-cell movement (see Section 2.4.1). Interestingly, it was later demonstrated that the incorporation of the TMV OAS sequence into virtually any recombinant RNA sequence can allow that RNA to be encapsidated into virus-like particles both *in vitro* (Sleat *et al.*, 1986) and *in vivo* (Gallie *et al.*, 1987). During assembly, the interaction with the OAS causes the disk to convert to a helical form. As subsequent disks are recruited, the nascent virion grows in a bi-directional fashion with both the free ends of the TMV RNA being drawn *up* through the central hole as the RNA complexes with the CP subunits (Butler, 1999; Klug, 1999; Culver, 2002).

Does TMV assembly occur *in vivo* spontaneously without any need for host factors, much as it occurs *in vitro*? At present, this is not known for certain but a clue suggesting that the assembly process *in vivo* may not be identical to that seen *in vitro* comes from a comparison of protein composition of reconstituted virions with those purified from infected plants. Zaitlin and colleagues found that TMV from infected plants, unlike reconstituted virus, contains an additional protein component, which they called *H-protein* (Asselin & Zaitlin, 1978), that was subsequently shown to be a monoubiquitinylated CP subunit (Dunigan *et al.*, 1988). Ubiquitin is a highly conserved 76-amino-acid eukaryotic protein that is covalently attached to proteins as monomers or lysine-linked chains of polyubiquitin (Weissman, 2001). One copy of H-protein is found on each virion at the end corresponding to the 3' terminus of the viral RNA (Dunigan *et al.*, 1988).

Conjugation with ubiquitin can mark a protein down for destruction, particularly if the target protein is polyubiquitinated. However, monoubiquitination is associated with regulation of protein activity or the compartmentalization of proteins in animals and plants (Bachmair *et al.*, 2001; Weissman, 2001). Speculatively, ubiquitination plays some role in either the *in vivo* assembly of virions or in their localization in the infected cell. However, the importance, if any, of H-protein remains unknown (Zaitlin, 1999). TMV was the first virus that was found to be ubiquitinated, although subsequently others have been found (Dunigan *et al.*, 1988; Hazelwood & Zaitlin, 1990).

In vitro, the disassembly of TMV can be induced by a wide range of denaturing chemicals or solution conditions. Studies of disassembly *in vitro* contributed to our understanding of virus structure and to early efforts to map the TMV genome. For example, controlled disassembly of TMV rods using alkaline solutions can generate a family of partially stripped virus particles that, after treatment with nucleases, can be used to obtain discrete portions of the viral genome (for example, Wilson *et al.*, 1978). However, these chemical treatments do not accurately mimic the *in vivo* mechanism of uncoating and will not be dealt with in detail here.

2.2.4   *Proteins encoded by the TMV genome*

The U1 type common or *vulgare* strain of TMV was fully sequenced over twenty years ago (Goelet *et al.*, 1982) which means that TMV research has been in the post-genomic era for rather a long time. Nevertheless, there are still unanswered questions and even controversy regarding the function(s) of certain TMV gene sequences. However, there is general agreement that at least four proteins are encoded by TMV. These are the 126- and 183-kDa proteins, the movement protein (MP: Section 2.4.1) and the CP (Fig. 2.1).

The 126- and 183-kDa proteins are encoded by the 5'-proximal ORF and are translated directly from the genomic RNA. The 183-kDa protein is synthesized as the result of occasional read-through of the termination codon of the 126-kDa protein sequence (Pelham, 1978; Skuzeski *et al.*, 1991). In translational read-through, certain tRNAs (termed suppressor tRNAs) mis-read a termination codon as a codon specifying an amino acid. In the case of TMV, a UAG codon is suppressed by certain tyrosine-specific tRNAs (Beier *et al.*, 1984). The 126- and 183-kDa proteins are components of the viral replication complex (replicase: Section 2.3).

The 30-kDa MP and the CP (17.5 kDa) are encoded by ORFs downstream of the 126/183-kDa ORF (Fig. 2.1). These proteins cannot be synthesized by direct translation of the full-length genomic RNA. This is because, in the majority of cases, the protein synthesis machinery of eukaryotic cells is not able to re-initiate protein synthesis following termination. Thus, for most polycistronic mRNAs, only the ORF closest to the 5' cap is normally translated in a eukaryotic system (Kozak, 2002). In order to evade this constraint, the synthesis of the TMV CP and MP occurs by translation of two subgenomic mRNAs, which are generated during viral infection (Section 2.3.1) (Hunter *et al.*, 1976; Palukaitis *et al.*, 1983; Sulzinski *et al.*, 1985).

In addition to the subgenomic RNAs utilized for the synthesis of MP and CP, an additional subgenomic RNA, called the I_1 RNA, has been detected in TMV-infected tobacco tissue (Beachy & Zaitlin, 1977; Sulzinski *et al.*, 1985). The I_1 RNA contains an ORF for a 54-kDa protein with a predicted amino acid sequence that is identical to that of the read-through region of the 183-kDa replicase protein (Sulzinski *et al.*, 1985). Can the I_1 RNA function as an mRNA? Certainly, the I_1 RNA does occur in association with polyribosomes, and 54-kDa protein can be synthesized from I_1 RNA or synthetic mRNA *in vitro* (Sulzinski *et al.*, 1985; Carr *et al.*, 1992). However, the 54-kDa protein has never been successfully detected *in vivo*, whether in TMV-infected plants, in transgenic plants engineered to constitutively express the 54-kDa ORF (Golemboski *et al.*, 1990; Zaitlin, 1999), or in preparations of the TMV replicase (Osman & Buck, 1997: Section 2.3). Furthermore, Okada and colleagues were unable to detect either the I_1 RNA or 54-kDa protein in TMV-infected tobacco protoplasts, suggesting that the 54-kDa protein may not be essential for replication of the virus (see Okada, 1999 and references therein).

In an attempt to determine definitively whether or not the 54-kDa protein has a biological function, Zaitlin and colleagues constructed transgenic plants expressing the 54-kDa ORF (Golemboski *et al.*, 1990). The original objective of this work was to determine if these plants would support replication of TMV mutants lacking the 54-kDa ORF (Zaitlin, 1999). In the event, this work was not carried out because the plants showed extreme resistance to the replication of the wild-type virus (Golemboski *et al.*, 1990; Carr & Zaitlin, 1991). Thus, it has still not been determined whether or not the 54-kDa protein really does occur during TMV infection. However, the work resulted in the development of a highly effective approach to plant protection (Palukaitis & Zaitlin, 1997) and contributed to the debate over the various mechanisms (protein-mediated *versus* RNA-mediated) that may explain pathogen-derived resistance (Carr & Zaitlin, 1993; Baulcombe, 1996; Marano & Baulcombe, 1998) (Section 2.5.2.4).

Another potential gene product of TMV is p4, identified by Atabekov and associates (Morozov *et al.*, 1993). This putative viral gene product is a positively charged 4-kDa polypeptide that is encoded by an ORF (ORF X) that overlaps the 3′ end of the ORF for the MP and the 5′ end for the CP gene (Morozov *et al.*, 1993). ORF X was identified in both the TMV U1 and the crucifer strain (cr-TMV: Dorokhov *et al.*, 1994). A later report indicated that it is also present in ToMV and that the p4 encoded by this virus can, when synthesized by *in vitro* translation, bind with the elongation factor EF-1-α (Fedorkin *et al.*, 1995). However, the biological significance of p4 and its biochemical properties remain unclear.

2.2.5   Untranslated and regulatory RNA sequences within the TMV genome

In addition to protein coding information, the viral RNA also carries information in the form of *cis*-acting regulatory sequences. In addition to the OAS discussed above, these include sequences that control the translation of viral proteins and/or the synthesis of full-length plus- and minus-sense viral RNAs, as well as transcription of subgenomic RNAs (Grdzelishvili *et al.*, 2000: Section 2.3). Perhaps the most striking structural feature of the genomic RNAs of tobamoviruses and certain other virus groups is the possession of a 3′ untranslated region (UTR) comprising a tRNA-like structure and three associated upstream pseudoknot domains (Okada, 1999). Curiously, the 3′ tRNA-like structure of TMV RNA can be aminoacylated with histidine *in vitro* and it does appear that the histidinylation is required for the interaction of the 3′ UTR with a host protein, the elongation factor eEF1A (Zeenko *et al.*, 2002). The 3′ tRNA-like structure and the pseudoknot structures play roles in the maintenance of RNA stability, in the initiation of translation of viral RNA into protein, and as a promoter for synthesis of negative-sense RNA by the viral replicase (see Section 2.2).

Successful translational initiation of most host mRNAs requires an association between the cap structure in the 5′ UTR and the 3′ poly A tail. This association is mediated by the poly A-binding protein and a complex of initiation factors (Gallie, 1996). In the case of TMV RNA, which is not polyadenylated, the role of the 3′ poly A tail in initiation and binding of translation factors is performed by sequences within the 3′ UTR (Leathers *et al.*, 1993; Gallie, 1996, 2002; Zeenko *et al.*, 2002). The 5′ UTR of TMV is similar to most host mRNAs in that it possesses a cap structure (in the case of TMV: m^7GpppG). However, it also possesses a 68-nucleotide leader sequence (called Ω) containing a 25-nucleotide poly (CAA) region responsible for enhancement of translational initiation (Gallie, 1996, 2002). Recent work indicates that enhancement of translation by Ω is mediated by binding of the heat shock protein HSP101 and the initiation factor eIF4F (Gallie, 2002).

2.2.6   The establishment of TMV infection

Unlike many other plant viruses that are vectored from host to host by insects or nematodes, TMV has no known vector organism. Instead, it is mechanically transmitted. That is, the virus can enter the plant only when mild abrasion of the plant surface leads to the wounding of host cells. For successful inoculation, the wounding must be sufficient to allow the virus to enter through breaches in the cell wall and plasmalemma, but not so severe as to lead to irreparable damage that would kill the inoculated cell outright.

For infection of the primary inoculated cell to occur, the viral RNA must be released from the virion. The uncoating process is driven by the effects of entry into the cell environment and interactions between components of the virus and host factors. The lower pH and calcium ion concentrations found in the plant cell compared with the extracellular environment are thought to cause a loss of the hydrogen and calcium ions that stabilize the RNA–CP interaction (Culver, 2002). The Ω leader sequence, which contains no G residues, is thought to interact relatively loosely with CP subunits, so the weakening of the CP–RNA interaction induced by the loss of hydrogen and calcium ions will be more pronounced in this region of the TMV RNA (Shaw, 1999; Culver, 2002). This may explain how the 5′ terminal region of TMV RNA becomes uncovered and exposed to the translational machinery of the inoculated cell.

Subsequently, further uncoating of the viral RNA occurs as the host ribosomes displace CP subunits during translation of the 126- and 183-kDa replicase proteins: a process known as co-translational disassembly (Wilson, 1984). Wilson first discovered co-translational disassembly when he found that *in vitro* translation systems could be programmed with preparations of TMV that had been pre-incubated briefly at pH 8 to release a small number of CP subunits from the 5′ end of the viral RNA (Wilson, 1984). Electron microscopy showed that these virus-programmed cell-free translation systems contained striposomes: complexes of partly stripped TMV virions and ribosomes (Wilson, 1984). Later on, striposomes were identified in TMV-inoculated plant cells showing that co-translational disassembly does occur *in vivo* (Wu et al., 1994).

Co-translational disassembly can explain only the uncoating of the TMV RNA from the 5′ cap as far as the stop codon of the 183-kDa protein gene sequence. Since translation cannot re-initiate internally on the genomic RNA, co-translational disassembly can allow only the synthesis of the 126- and 183-kDa proteins. It cannot expose the remaining viral sequences, including the ORFs for the MP and CP, in the virus particle. Yet, the viral RNA is entirely stripped of CP within 20 minutes following inoculation (Wu et al., 1994; Shaw, 1999). What mechanism completes the release of viral genomic RNA from the particle? Wilson (1984) had speculated early on that since 126- and 183-kDa proteins are components of the viral replicase complex (Section 2.3), the replicase might participate in some way in the completion of the stripping process. But at that time it was not technically feasible to test the idea. Eventually, Wu and Shaw (1997) obtained evidence that appears to vindicate Wilson's co-replicational hypothesis using a sensitive coupled reverse-transcription PCR (RT-PCR) assay to detect the first appearance of minus-strand viral RNA and mutant versions of TMV RNA lacking functional 126/183-kDa sequences. They showed that in synchronously infected tobacco protoplasts the initiation of synthesis of the minus-strand, which is the intermediate in replication and the first product of viral replicase activity (Section 2.3), occurs simultaneously with the removal of CP from the 3′ end of the viral RNA (Wu & Shaw, 1997). When the protoplasts were inoculated with encapsidated RNA for the mutant lacking functional replicase activity, there was no uncoating of the 3′ end of these RNAs and no minus-strand synthesis was detected (Wu & Shaw, 1997).

In summary, the establishment of infection with TMV begins with the loss of a small number of CP subunits to expose the 5′ cap structure and the 5′ UTR including the Ω leader sequence. Translation will initiate on this exposed RNA and the host ribosomes strip the viral RNA as far as the terminator of the 183-kDa coding region. Once sufficient 126- and 183-kDa proteins have accumulated, they associate with host factors to form an active replicase complex (Section 2.3.2). This will be able to interact with 3′ viral RNA sequences to initiate transcription of the viral minus-sense RNA and to simultaneously remove the remaining CP subunits from the positive-sense viral genomic RNA. From this point, full-scale viral replication and gene expression will be unimpeded.

2.3   TMV replication and the synthesis of subgenomic mRNAs

2.3.1   The process of replication and subgenomic RNA synthesis

In eukaryotes infected with positive-sense RNA viruses, the synthesis of new viral RNA is carried out by an RNA-dependent RNA polymerase (RdRp) that comprises part of a replicase complex containing virus- and host-encoded proteins, and which occurs in the cytoplasm of infected cells (Buck, 1996). For TMV, the full cycle of replication, i.e. the regeneration of progeny RNA molecules identical to the original parental molecule, requires the replicase complex to possess helicase and methyl-transferase activities, as well as RdRp activity.

First, the replicase complex binds with the 3′ UTR of the genomic RNA, and the RdRp component of the replicase complex uses the full-length genomic RNA as a template to synthesize a complementary full-length minus-strand viral RNA. The resulting dsRNA molecule (replicative form or RF RNA: Buck, 1999) is unwound by the helicase activity of the RdRp, releasing the template and the single-stranded minus-sense RNA. If RNA synthesis by the RdRp is initiated at the 5′ end of the minus-strand RNA, it serves as a template for synthesis of new full-length genomic RNA. However, RNA synthesis can also be initiated at internal promoters on the minus-strand to yield the subgenomic mRNAs (Grdzelishvili et al., 2000). The amount of minus-strand RNA synthesized during a TMV infection is much less than the amount of plus-sense RNA synthesized, indicating that each full-length minus-strand RNA can be used multiple times as a template for genomic or sub-genomic TMV RNA (Aoki & Takebe, 1975; Buck, 1999). New genomic and subgenomic RNAs are released from minus-sense templates by helicase activity. Subsequently, the methyltransferase domain of the replicase complex caps new molecules of the genomic RNA, the subgenomic CP mRNA (Keith & Frankel-Conrat, 1975; Zimmern, 1975), and probably the subgenomic MP mRNA (Grdzelishvili *et al.*, 2000).

2.3.2 The composition of the TMV replicase complex and its location in the infected cell

The TMV replicase complex contains two virus-encoded subunits that are required for efficient RNA replication, the 126- and 183-kDa proteins (Ishikawa *et al.*, 1986, 1991a; Osman & Buck, 1996, 1997). Within these proteins, three functional domains have been identified (Haseloff *et al.*, 1984; Hodgman, 1988; Koonin & Dolja, 1993). The two proteins share an N-terminal methyltransferase-like domain that has been shown to have guanylyltransferase activity (Dunigan & Zaitlin, 1990), and a second domain that has sequence similarity to many viral and cellular helicases (Gorbalenya *et al.*, 1988). Recently, this second domain was confirmed to be an ATP-dependent helicase capable of unwinding double-stranded RNA molecules (Goregaoker and Culver, 2003). The read-through region of the 183-kDa protein contains a third domain containing amino acid sequence motifs common to many nucleic acid polymerases, including the Mg^{2+}-binding GDD motif (Argos, 1988). This polymerase-like domain is also present in the sequence of the putative 54-kDa protein although this protein, even if it exists *in vivo* at all, is not found in the TMV replicase complex (Osman & Buck, 1996, 1997).

Watanabe *et al.* (1999) used antibodies specific for either the RNA polymerase (anti-P) or the methyltransferase (anti-M), domains of the TMV 126/183-kDa proteins to carry out co-immunoprecipitation experiments on a replicase preparation from tobacco infected with TMV strain OM. They found that anti-M precipitated both the 126- and the 183-kDa proteins, with the 126-kDa protein being in excess of the 183-kDa protein, while anti-P precipitated similar amounts of 126- and 183-kDa proteins (Watanabe *et al.*, 1999). This suggested that the 183-kDa protein occurs in the viral replicase complex predominantly as a heterodimer with the 126-kDa protein (Watanabe *et al.*, 1999). The idea has now been confirmed, and specific regions within the helicase domain and in the intervening region between the methyltransferase and helicase domains have been identified as being required for heterodimerization and for full replicase activity (Goregaoker *et al.*, 2001; Goregaoker & Culver, 2003).

The TMV replicase complex is associated with the cytoplasmic face of one or more types of intracellular membrane (Osman & Buck, 1996). Several groups have sought to determine how exactly the replicase complex associates with intracellular membranes, if only certain membrane types are involved, or how the association between the replicase complex and the membrane may change over time during the course of the replication cycle. In the 1980s, immuno-electron-microscopic (EM) studies indicated that a large amount of 126/183-kDa protein is associated with virus-induced inclusion bodies (X-bodies) that occur in the cytoplasm of TMV-infected cells, suggesting that these are sites of replication (Hills *et al.*, 1987; Saito *et al.*, 1987). On account of much earlier work which indicated that the

X-bodies contain endoplasmic reticulum (ER) (Esau & Cronshaw, 1967), this led to the suggestion that the ER is the membrane to which the TMV replicase is anchored (Buck, 1999).

The results of some recent studies of the distribution of TMV proteins fused with the green fluorescent protein (GFP) seem to support the idea that the replicase does localize to the ER. Heinlein *et al.* (1998) used confocal laser-scanning microscopy to study MP localization in *Nicotiana benthamiana* plants and tobacco protoplasts infected with a strain of TMV modified to express an MP–GFP fusion protein. Their results led them to propose that the MP may mediate an association of replicase with the ER network, giving rise to the so-called replication factories on the membrane surface (Heinlein *et al.*, 1998), although in subsequent studies this idea has been challenged (Boyko *et al.*, 2000a: Section 2.4). But, using the biolistic method of transient gene expression, Dos Reis Figueira *et al.* (2002) have shown that, at least in onion epidermis cells, the 126-kDa protein (tagged with GFP) can bind with the ER membranes in the absence of MP or any other viral proteins. The ability to bind with, and form aggregations on, the ER membrane was controlled by two regions of the 126-kDa protein sequence, one of which, curiously, is an apparently cryptic nuclear localization sequence (NLS) (Dos Reis Figuera *et al.*, 2002). Interestingly, although there is no indication that any significant amounts of 126-kDa protein accumulate in the host cell nucleus, it does have a biological function since a TMV mutant lacking the NLS sequence in the 126-kDa protein cannot replicate (Dos Reis Figuera *et al.*, 2002).

There is now very strong evidence indicating that several host-coded proteins participate in the localization of the TMV replicase complex to host cell membranes. Ishikawa and colleagues screened mutant *Arabidopsis thaliana* populations for plants unable to support the replication of tobamoviruses (Ishikawa *et al.*, 1991b, 1993; Ohshima *et al.*, 1998). The subsequent characterization of some of these *tom* (*tobamovirus multiplication*) mutants has revealed the existence of several membrane-spanning host proteins that are required by tobamoviruses (but not other virus types) for efficient replication (Yamanaka *et al.*, 2000; Hagiwara *et al.*, 2003; Tsujimoto *et al.*, 2003).

Up to now, the Ishikawa group has reported the detailed characterization of four *Arabidopsis* proteins that appear to function in TMV replication: TOM1, TOM3, TOM2A, and TOM2B (Yamanaka *et al.*, 2000, 2002; Hagiwara *et al.*, 2003; Tsujimoto *et al.*, 2003). An additional factor, *ttm1*, can influence the replication of tobamoviruses in *tom2* mutants but has not yet been characterized in detail (Tsujimoto *et al.*, 2003). The TOM1 and TOM3 proteins are homologous, seven-pass transmembrane proteins, both of which can interact with the helicase domain common to the TMV 126- or 183-kDa proteins in the yeast two-hybrid system (Yamanaka *et al.*, 2002). TOM2A and TOM2B are two distinct factors, one of which, TOM2A, has been shown to be a four-pass transmembrane protein able to self-interact and interact with TOM1 (Tsujimoto *et al.*, 2003). Homologues of TOM1 and TOM2A have been found in tobacco, which have been named NtTOM1 and NtTOM2A respectively (Hagiwara *et al.*, 2003).

Confocal microscopic examination of the distribution of GFP-tagged TOM1 and TOM2A in transgenic tobacco cell cultures and in transiently transformed *Arabidopsis* epidermal cells confirmed that these proteins localize to intracellular membranes (Hagiwara *et al.*, 2003). However, further more detailed imaging studies indicated that the two proteins can associate with tonoplast (i.e. vacuolar) membranes (Hagiwara *et al.*, 2003). Cell fractionation analysis using buoyant-density centrifugation indicated that TOM2A and NtTOM2A occurred exclusively in fractions enriched in tonoplast membranes, while TOM1 and NtTOM1 were found both in these fractions and in fractions enriched in membranes derived from several different organelles, possibly including the ER (Hagiwara *et al.*, 2003). In TMV-infected protoplasts, it was found that although the TMV-encoded 126- and 183-kDa replicase proteins did not co-purify exclusively with NtTOM1 and NtTOM2A, RdRp activity was detected only in fractions in which NtTOM1 was present (Hagiwara *et al.*, 2003). The results show that TOM1 and its homologues are required for the RdRp activity of the TMV replicase and that the tonoplast membrane is an important site for viral replication.

At least one function of TOM1 (and its homologues) would seem to act as a physical bridge between the TMV-encoded replicase proteins and host membranes. Two separate lines of evidence support the idea. First, mutations in the helicase domain of the 126/183-kDa protein sequence abolish the interaction of this domain with TOM1 in the yeast two-hybrid system (Yamanaka *et al.*, 2000). Second, when these sequences are deleted from a 126-kDa–GFP fusion protein, it will not localize to cell membranes in biolistically transformed onion epidermis cells (Dos Reis Figuera *et al.*, 2002). This is at least suggestive that proteins similar to TOM1 and capable of participating in support of tobamoviral replication may be very widely distributed in the plant kingdom, even in plants that, like onion, are not normal hosts for TMV. Together with the discovery of NtTOM1 and NtTOM2A in tobacco, this result is consistent with the idea that tobamoviruses exploit homologous host factors in different hosts to anchor the replicase complex to the membrane (Dos Reis Figuera *et al.*, 2002; Hagiwara *et al.*, 2003). It would be interesting to know what normal function(s) these proteins perform in uninfected cells.

Various host proteins have been found to be present in replicase preparations purified from TMV-infected tissue (Osman & Buck, 1996, 1997; Watanabe *et al.*, 1999). Of these, a 56-kDa protein identified by Osman and Buck (1997) in replicase preparations from TMV-infected tomato appeared to be of particular interest. This protein cross-reacted with antibodies raised against the GCD10 subunit of the yeast translation initiation factor eIF-3. The viral 126- and 183-kDa proteins, plus the host encoded 56-kDa protein, could be purified using either antibodies against the TMV-L 126-kDa protein or antibodies against the yeast GCD10 protein (Osman & Buck, 1997). Two lines of evidence indicated that the 56-kDa host protein was a component of the replicase. The first was that antibodies against the GCD10 protein inhibited the synthesis of TMV genomic RNA and dsRNA by purified TMV replicase. The second was that inhibition of viral RNA synthesis by the antibody was prevented by competition with the purified yeast GCD10 protein

(Osman & Buck, 1997). Further work using the yeast two-hybrid system showed that the authentic yeast GCD10 protein was capable of interacting with the methyl-transferase domain shared by the 126- and 183-kDa TMV replicase proteins (Taylor & Carr, 2000).

Although the evidence for a GCD10-related host factor in TMV replication may seem persuasive, there is a need for caution. Following the sequencing of the *Arabidopsis* genome, searches for sequences encoding GCD10-like proteins have been unsuccessful and therefore there is to date no evidence that a protein with homology to GCD10 occurs in plants (unpublished data). Presumably, the 56-kDa protein, whatever its true identity, has enough similarity to GCD10 in terms of its secondary and tertiary structure, to be able to cross-react with anti-GCD10 anti-bodies. Until further information on its identity becomes available, the biological relevance of the presence of the 56-kDa protein in replicase preparations or of the interaction of authentic GCD10 proteins with the TMV replicase proteins' methyl-transferase domain will remain difficult to assess.

2.4 Movement of TMV within the host

Higher plants possess a unique transport and communication network based on the symplast, i.e. the interconnected nature of almost all of the cells in the plant body. This complex network is formed at the local level by plasmodesmata, bridges of cytoplasm that penetrate through the cell walls separating neighboring cells. They may be formed during the process of cell division (primary plasmodesmata) or produced *de novo* at later stages of development (secondary plasmodesmata) (Blackman & Overall, 2001). Symplastic communication and transport can occur over longer distances between tissues and organs through the phloem tissues, the main conduit for transport of soluble carbohydrate, which are connected via plas-modesmata to the surrounding tissues (Nelson & van Bel, 1997; Blackman & Overall, 2001; van Bel, 2003). In order to reach all parts of a plant, viruses must be able to exploit this complex network of plasmodesmata and phloem elements. Typically, plant virus movement is, for the purposes of discussion, divided into two phases: local, cell-to-cell movement, and long distance, or systemic, movement.

2.4.1 Cell-to-cell movement of tobamoviruses

2.4.1.1 Plasmodesmata and the viral movement protein
We now know that plasmodesmata are intercellular transport routes not only for exchange of low molecular weight substances but also for macromolecules (Blackman & Overall, 2001; Heinlein, 2002). These include mRNAs and proteins, including transcription factors (Blackman & Overall, 2001; Heinlein, 2002; Roberts & Oparka, 2003). Thus, cells can influence gene expression, protein synthesis, and larger-scale developmental processes in their neighbors (reviewed in detail by Heinlein, 2002). Furthermore, inducing alterations in the permeability of

plasmodesmata can have major physiological effects, for example on the distribution of carbohydrate stores between photosynthetic and storage organs (Lucas *et al.*, 1993; Olesinski *et al.*, 1996).

Plasmodesmata are now seen as dynamic structures that alter during development. For example, plasmodesmata linking mesophyll cells of immature sink leaves are permeable to molecules of up to 60 kDa (i.e. the size exclusion limit (SEL) is 60 kDa). In contrast, plasmodesmata in mature tissues may have an SEL of 1 kDa or less (Wolf *et al.*, 1989; Oparka *et al.*, 1999; Roberts *et al.*, 2001). In addition, different cell types exhibit varying SEL values with trichome cells having values of around 7 kDa (Waigmann & Zambryski, 1995). The SEL values of plasmodesmata in different cell types are thought to be controlled by the plasmodesmal structure (single channels or branched) and by the amount of callose laid down around them in the cell wall (Oparka *et al.*, 1999; Bucher *et al.*, 2001; Roberts & Oparka, 2003). Fifteen years ago, we knew little of this and much of what is now known about the properties and roles of plasmodesmata has resulted directly or indirectly from studies of plant virus movement and in particular from work on the movement of TMV.

The early pioneering experiments in which fluorescent dyes coupled to dextrans of known size were microinjected into tobacco mesophyll cells indicated that the SEL for these cells' plasmodesmata was less than 1 kDa. This is too small to allow the free passage of either a TMV particle or its RNA (Wolf *et al.*, 1989). The question was how could TMV overcome this apparently insurmountable roadblock? Previous data obtained with movement-defective TMV mutants indicated that the 30-kDa protein encoded by the virus was the putative MP required for cell-to-cell movement (Leonard & Zaitlin, 1982; Ohno *et al.*, 1983). When the 30-kDa protein gene was expressed constitutively in transgenic tobacco plants it was able to enhance the local spread of a movement-defective TMV mutant, demonstrating that it was indeed the MP (Deom *et al.*, 1987). Furthermore, microinjection studies showed that the SEL of plasmodesmata between the leaf mesophyll cells of the *MP*-transgenic plants was increased to 10 kDa or greater (Wolf *et al.*, 1989).

Meanwhile, a separate line of investigation by Citovsky and colleagues showed that the TMV MP can complex with and unfold single-stranded nucleic acids (Citovsky *et al.*, 1990, 1992). Taken together with the *MP*-transgenic plant studies, this suggested that the mechanism of action of the TMV MP involves unfolding of TMV RNA and formation of a ribonucleoprotein complex that can pass through plasmodesmata that have been modified by the MP. Subsequent mutational analysis of the *MP* gene showed that there were distinct domains within the MP which controlled its ability to accumulate within plasmodesmata and to associate with ssRNA (Berna *et al.*, 1991; Lapidot *et al.*, 1993; Kahn *et al.*, 1998).

The mode of action of the TMV MP suggested that it, and similar MPs from other viruses, could either be mimicking host factors that move from cell to cell, or could have been derived from progenitors that had been 'stolen' from plants. Support for this idea came first from the discovery that the maize KNOTTED 1 transcription factor can move itself and its mRNA through plasmodesmata (Lucas *et al.*, 1995). Since then, more host proteins and mRNAs capable of moving between cells have been found and it is likely that this process is an important determinant in plant development (see reviews by Blackman & Overall, 2001; Heinlein, 2002; Wu *et al.*, 2002; Roberts & Oparka, 2003). Recently, it was shown that an insect-infecting virus, *Flock house virus*, which does not normally infect plants, is able to propagate systemically through transgenic *Nicotiana benthamiana* plants expressing either the TMV MP or the MP of another plant virus, *Red clover necrotic mosaic virus* (Dasgupta *et al.*, 2001). These results lend further credence to the idea that viruses might be able to acquire MP-type proteins either from the host or from other viruses in mixed infections.

2.4.1.2 Interactions of the TMV MP with host cell ultrastructure

During the 1990s, viral vectors expressing GFP or MP–GFP fusion proteins were developed (Oparka *et al.*, 1996). These GFP-expressing viruses made it possible to examine plant viral movement non-destructively and in real time. Combined with the superior imaging offered by confocal laser-scanning microscopy, viruses expressing GFP or similar fluorescent proteins like DsRed (Toth *et al.*, 2002) have provided insights into the plant–virus relationship at the subcellular, cellular, and whole tissue levels.

When inoculation sites of TMV expressing an MP–GFP fusion (TMVMP-GFP) are monitored over time in tobacco leaves, it was seen that the MP GFP accumulates predominantly at the leading edge of the infection site and is absent from the center (Oparka *et al.*, 1997; Boyko *et al.*, 2000a). This is because accumulation of the MP is regulated by turnover via ubiquitin-dependent proteolysis (Reichel & Beachy, 2000). This tight control of MP accumulation means that only the plasmodesmata at the leading edge of the infection site have increased SEL values (Oparka *et al.*, 1997).

Imaging of MP-GFP using epifluorescence and confocal laser microscopy confirmed immuno-EM studies showing that the TMV MP accumulates within the plasmodesmata (Ding *et al.*, 1992; Heinlein *et al.*, 1998). These methods also demonstrated that the MP associates with the ER (Heinlein *et al.*, 1995, 1998) and with components of the cytoskeleton: microtubules (Heinlein *et al.*, 1995; McLean *et al.*, 1995), and possibly with F-actin (McLean *et al.*, 1995). Biochemical studies showed that the MP-GFP accumulation sites on the ER were also sites of replicase accumulation (Heinlein *et al.*, 1998; Reichel & Beachy, 1998; Más & Beachy, 1999). When confocal microscopy was used to observe the localization of MP-GFP in individual tobacco protoplasts, it was seen that it changed over the course of infection (Más & Beachy, 1998, 1999). Initially, GFP fluorescence occurred in ER surrounding the nucleus but as infection proceeded it became predominantly associated with larger structures at the cell periphery, possibly the sites at which plasmodesmata had existed (Más & Beachy, 1998, 1999). These and other studies pointed to a model in which the MP associates with viral replicase complexes on the ER before the nascent viral RNA is conducted along microtubules or ER, or a combination of both, to the plasmodesmata for trafficking into an adjacent cell (Reichel *et al.*, 1999).

Although the model is attractive and appears to mesh together virus replication and movement in a compelling way, certain aspects of this model have been called into question. Boyko *et al.* (2000a) found that a mutant form of TMV MP lacking 55 amino acids in the C terminus was still able to promote viral cell-to-cell movement but a GFP fusion protein derived from this mutant MP did not localize strongly with ER-derived bodies. The same group noted that several tobamovirus MP sequences shared a motif that they contended was similar to a motif found in the protein tubulin, which promotes tubulin–tubulin interactions (Boyko *et al.*, 2000b). Point mutations in the putative tubulin-binding sequences of tobamovirus MPs appeared to compromise the ability of the MPs to associate with microtubules and to promote viral cell-to-cell movement (Boyko *et al.*, 2000b). Further analysis of MP mutants appeared to confirm the view that association of the MP with microtubules was needed for viral cell-to-cell movement (Boyko *et al.*, 2002). Thus, the work of this group suggests that association with microtubules is required for MP-mediated viral cell-to-cell movement but that the association with the ER is dispensable.

However, this is by no means the final word about the relationship between the intracellular localization of TMV MP and how it functions in intercellular virus movement. Toth *et al.* (2002) generated a variant of TMV.GFP (a TMV-derived vector expressing free GFP) which showed enhanced cell-to-cell movement due to mutations within the MP gene sequence. Subsequent work showed that this novel MP, MP^{R3}, was highly efficient at gating plasmodesmata, and more effective at promoting viral movement possibly because it was less subject to turnover than wild-type MP (Gillespie *et al.*, 2002). A DsRed-MP^{R3} fusion protein associated with the host cell ER but did not show the same pattern of association with the microtubule network that was seen with a wild-type MP-DsRed fusion protein (Gillespie *et al.*, 2002). Also, pharmacological and RNAi-based methods that disrupted the microtubule network of the host cells did not prevent plasmodesmal gating by MP^{R3} or movement of the TMV MP^{R3} variant (Gillespie *et al.*, 2002). These workers concluded that the association of TMV MP with the microtubule network is not required for viral cell-to-cell movement. Furthermore, they propose that the transfer of MP from the ER to the microtubule network is a prelude to ubiquitin-mediated breakdown of MP (Gillespie *et al.*, 2002). This would suggest that association of the MP with the microtubules provides a negative regulation of plasmodesmal gating and viral cell-to-cell movement.

Evidently, the nature and the relative importance of the interactions of the TMV MP with the ER are still not fully understood. New models to explain the relationship between the intracellular localization of the MP with its role in intercellular transport of TMV RNA will be difficult to construct until apparent contradictions between the results of Boyko *et al.* (2002) and Gillespie *et al.* (2002) are resolved experimentally.

2.4.1.3   Interactions of the TMV MP with host proteins

Although there are uncertainties as to the nature and function(s) of the association of the TMV MP with the ER and cytoskeleton, it appears certain that its association with plasmodesmata is essential for the promotion of viral cell-to-cell movement.

It is presumed that the MP interacts with one or more protein components of the plasmodesmata and it is likely that these proteins may be the targets of several viruses' MPs. This conclusion can be drawn since expression of a mutant TMV MP that can bind with plasmodesmata, but which cannot induce gating, interferes not only with the movement of TMV but also with the movement of several other viruses (Cooper et al., 1995).

Two groups have identified a ubiquitous cell-wall-associated enzyme, pectin methylesterase (PME: Micheli, 2001), as a potential plasmodesmal receptor for the TMV MP (Dorokhov et al., 1999; Chen et al., 2000). The TMV MP, as well as the MPs from two other viruses, was found to bind with PME, and evidence that PME has potential RNA-binding activity supports the idea that this enzyme is a receptor for the TMV-MP RNA complex (Dorokhov et al., 1999; Chen et al., 2000). Other proteins that appear to be able to interact with TMV include two phloem-sap proteins from melon (Shalitin & Wolf, 2000). Such proteins might plausibly be involved in viral entry into or exit out of the phloem. Interactions between tobamoviral MPs and putative transcriptional factors in far-western assays have been demonstrated but the biological significance of these interactions is not yet clear (Matsushita et al., 2002).

The MPs of tobamoviruses can occur in phosphorylated forms and it has been hypothesized that this may affect their function (Watanabe et al., 1992; Haley et al., 1995). Phosphorylation has been reported at various serine and threonine residue sites in the MPs of TMV and ToMV (Haley et al., 1995; Kawakami et al., 1999; Waigmann et al., 2000; Karger et al., 2003). Protein kinases capable of phosphorylating tobamoviral MPs have been found in protein fractions derived from the cell wall (Citovsky, 1999) and ER (Karger et al., 2003). Recently, it was shown that a protein kinase from tobacco that is similar to casein kinase 2 can complex with and phosphorylate the ToMV MP in vitro (Matsushita et al., 2003).

Phosphorylation of tobamoviral MPs has been proposed to be important for intracellular localization and stability of the MP and for its function in the gating of plasmodesmata (Beachy & Heinlein, 2000; Waigmann et al., 2000; Kawakami et al., 2003). When the Ser258, Thr261, and Ser265 residues of TMV MP were replaced with negatively charged amino acids (to mimic the effect of phosphorylation), the recombinant MPs were not able to gate plasmodesmata although they could still interact with PME (Waigmann et al., 2000). This suggests that phosphorylation may negatively regulate plasmodesmal gating. An additional function of phosphorylation of the MP, which was suggested by Karpova and colleagues, was that MP limits the availability of TMV RNA for translation, and that this effect may be negatively regulated by phosphorylation of MP (Karpova et al., 1997, 1999).

However, the effect of MP phosphorylation on plasmodesmal gating was seen only in tobacco, not in Nicotiana benthamiana (Waigmann et al., 2000). Furthermore, no phosphorylation of the TMV MP was detected at all in Arabidopsis thaliana (Hughes et al., 1995). Thus, phosphorylation of tobamoviral MPs is a host-dependent phenomenon and it is not clear whether or not it is biologically significant in all hosts.

2.4.1.4   Are other virus-encoded proteins involved in cell-to-cell movement?
The MP is absolutely required for the cell-to-cell movement of tobamoviruses (for example, see Wright *et al.*, 2000). Nevertheless, the properties of other viral proteins may influence the rate of movement. Hirashima and Watanabe (2001) found that a recombinant virus (TMV-UR-hel) possessing a TMV strain U1 sequence in which the helicase domain of the 126-kDa protein gene had been replaced with that from TMV strain R was unable to move from cell to cell. However, this was not a secondary effect of impaired viral replication since the recombinant virus and the parental viruses replicate to similar levels in protoplasts (Hirashima & Watanabe, 2001). The results suggest that the 126-kDa protein may have a role in cell-to-cell movement.

For some plant viruses, e.g. *cucumber mosaic virus* (CMV), the CP as well as the MP are required for cell-to-cell movement (Canto *et al.*, 1997) but the CP of TMV is not required for cell-to-cell movement (Dawson *et al.*, 1988). This does not completely rule out a role for the TMV CP in regulating cell-to-cell movement, and in a recent paper Bendahmane *et al.* (2002) characterized a mutant CP that inhibited production of MP and virus movement. This effect possibly resulted from sequestration of the MP-subgenomic mRNA (which includes the OAS sequence) by the mutant CP. However, Bendahmane *et al.* (2002) suggest that even in wild-type viral infections the CP may indirectly affect the rate of movement by regulating the synthesis of MP.

2.4.2   Systemic movement of tobamoviruses

2.4.2.1   Virus movement and the phloem
For TMV and other tobamoviruses, as well as for most plant viruses, the phloem tissue provides the most rapid route for spread out of the initially inoculated tissue and for propagation throughout the plant. Once it enters the phloem, TMV spreads in a similar pattern to that of photosynthetically fixed carbon compounds, moving from mature (source) tissues to younger, actively growing tissues (sinks) (reviewed by Leisner & Turgeon, 1993; Nelson & van Bel, 1997). This is illustrated in Fig. 2.2, which shows the pattern of systemic movement of TMV.GFP in *Nicotiana benthamiana*. However, the ability of the virus to move systemically also depends to various extents upon the properties of the MP, CP, and the replicase proteins.

2.4.2.2   The movement protein
MP is required by tobamoviruses to move outwards from the initial point of infection to reach and enter the phloem tissue. If an MP cannot function in a particular host species, then those plants will not become systemically infected and this is one way that MP can influence the host range of tobamoviruses (Fenczik *et al.*, 1995; Deom *et al.*, 1997). From analysis of the responses of 14 ecotypes of *Arabidopsis* to the U1 strain of TMV, it was concluded that the speed of cell-to-cell movement is also important in determining whether or not a plant will be susceptible to systemic disease (Dardick *et al.*, 2000). However, once the virus has entered the phloem, the role of the MP becomes less clear.

Figure 2.2 Movement of a *Tobacco mosaic virus* vector expressing the green fluorescent protein. Panel A shows the appearance, under ultraviolet illumination, of a *Nicotiana benthamiana* plant one week after inoculation with a TMV vector carrying the gene for the green fluorescent protein (TMV.GFP). A brightly fluorescing TMV.GFP infection focus on one of the inoculated leaves is indicated by an arrow. Panel B shows the same plant three weeks after inoculation photographed under ultraviolet illumination. TMV.GFP has moved preferentially into younger tissues, primarily the newly emerging leaves at the top of the plant, as well as immature (carbon sink) leaf tissues lower down. The appearance of the same plant under normal illumination is shown in panel C.

Gera *et al.* (1995) carried out experiments with a TMV deletion mutant lacking the *MP* gene (TMV.ΔMP). TMV.ΔMP could move cell-to-cell in transgenic tobacco plants engineered to express the MP coding sequence under the control of the *Cauliflower mosaic virus* (CaMV) 35S promoter. Experiments were carried out in which sections of *MP*-transgenic and non-transgenic tobacco were grafted together. It was found that TMV.ΔMP from an infection established in *MP*-transgenic rootstock was able to move through the phloem of a grafted section of non-transgenic tobacco and into the *MP*-transgenic tissue grafted above (Gera *et al.*, 1995). Subsequently, another group who used independently generated TMV.ΔMP and *MP*-transgenic tobacco carried out similar experiments (Arce-Johnson *et al.*, 1997). In contrast to the results of Gera *et al.* (1995), they found that their TMV.ΔMP construct did not move through the phloem of non-transgenic tissue grafted onto an *MP*-transgenic rootstock (Arce-Johnson *et al.*, 1997). Curiously, although the TMV.ΔMP did move through *MP*-transgenic tobacco in which the *MP* gene expression was driven by the CaMV 35S promoter, the mutant virus did not move through the phloem of transgenic tobacco tissue in which *MP* expression was driven by the phloem-specific promoter, *rolC* from *Agrobacterium* (Arce-Johnson *et al.*, 1997).

It is difficult to understand why the two groups obtained contradictory results, although conceivably there may have been sequence differences between the two TMV.ΔMP strains. However, both sets of results are consistent with the idea that the MP is required for loading of the virus into the phloem, while it is not clear whether the MP is required for translocation of the virus once it has arrived in the phloem.

Whether or not MP plays a role within the phloem tissue, additional factors influence the success or failure of a systemic tobamovirus infection. Arce-Johnson *et al.* (2003) showed that in several ecotypes of *Arabidopsis*, a crucifer-infecting strain of TMV (TMV-Cg) has a faster rate of systemic movement than the type strain, TMV U1. However, a chimeric virus in which the *MP* gene in TMV U1 was replaced with the TMV-Cg *MP* gene did not have an increased rate of systemic movement in *Arabidopsis* or a decreased rate in tobacco (Arce-Johnson *et al.*, 2003). Both of the viral strains and the chimeric virus moved cell-to-cell with similar efficiency in the inoculated leaves of *Arabidopsis*. The results indicate that the MP does not, at least by itself, control differences in the rate of systemic movement in different host plants (Arce-Johnson *et al.*, 2003).

2.4.2.3 The coat protein and systemic movement

The CP is required for movement of tobamoviruses through the phloem. TMV mutants that lack a *CP* gene (TMV.ΔCP) or encode a CP that is not competent for assembly into virus particles can enter the vein parenchyma cells but cannot load into the sieve elements and be propagated through the rest of the plant (Siegel *et al.*, 1962; Hilf & Dawson, 1993; Ding *et al.*, 1996). In most cases, since they can move only using the plasmodesmata, TMV.ΔCP mutants are restricted to the initially inoculated tissue. The only exception to this occurs in extremely slow growing hosts such as *Nicotiana sylvestris*, in which the relatively slow rate of spreading of the virus by cell-to-cell movement exceeds even slower the rate of plant growth and can allow the virus to reach other leaves (Hilf & Dawson, 1993).

Since viral mutants that encode CPs which are not competent for assembly into virus particles cannot move in the phloem, this suggests that tobamoviruses travel in the phloem in the form of virions. To test this idea Simon-Buela and Garcia-Arenal (1999) took advantage of the ease with which phloem sap can be collected from cucumber plants to analyze the nature of the viral ribonucleoprotein complex in plants infected with the tobamovirus CGMMV. Using electron microscopy they found that CGMMV particles did occur in the phloem sap. Furthermore, using an RNase A sensitivity assay, they showed that the majority of viral RNA in the phloem sap was protected in ribonucleoprotein complexes with the same properties as that of virions (Simon-Buela & Garcia-Arenal, 1999). These results indicate that, at least in cucurbit hosts, tobamovirus virions are the predominant form taken by these viruses during systemic movement in the phloem in natural infections.

Interestingly, TMV RNA does not *have* to move in the phloem in the form of an authentic TMV virion. Spitsin *et al.* (1999) replaced the *CP* gene of TMV with the *CP* gene from *Alfalfa mosaic virus* (AlMV), a virus that unlike TMV forms spherical or bacilliform particles rather than rods. The hybrid RNA replicated and was encapsidated *in vivo* by the AlMV CP and moved systemically in tobacco, *N. benthamiana*, and spinach (Spitsin *et al.*, 1999). Interestingly, spinach is a host for AlMV but not for TMV, indicating that the ability of a viral CP to effect the systemic movement of a virus is an important determinant of host range (Spitsin *et al.*, 1999).

However, tobamoviral RNA can be transported in a form other than a virion. Umbraviruses, such as *Groundnut rosette virus* (GRV), *Pea enation mosaic virus-2*, and *Tobacco mottle virus*, do not encode conventional CPs but some of the functions of a CP are carried out by the product of the umbraviral ORF3 (Ryabov *et al.*, 2001). If the *CP* gene is removed from TMV and replaced with the *ORF3* sequence from any of these umbraviruses, the resulting chimeras can move systemically in the phloem (Taliansky *et al.*, 2003). Although TMV ORF3 chimeras cannot form TMV virions, it has been shown that the GRV-derived ORF3 protein, for example, forms a protective ribonucleoprotein complex with the viral RNA that protects it and mediates its transport in the phloem (Taliansky *et al.*, 2003). However, it should be noted that the ability of the ORF3 proteins to mediate the movement of TMV RNA is host specific, taking place in *N. benthamiana* and *N. clevelandii*, but not in tobacco (Ryabov *et al.*, 2001).

All of these results taken together indicate that tobamoviral RNAs must be transported in the phloem protected in the form of a ribonucleoprotein complex, which, in natural, unmixed infections, is most likely to be the virion.

2.4.2.4   The replicase proteins, virus replication, and systemic movement

Mutations in the sequences encoding the 126- and 183-kDa replicase proteins can influence the efficiency of TMV transport through the phloem into sink tissue (Ding *et al.*, 1995; Derrick *et al.*, 1997). Similarly, the properties of the replicase proteins of other plant viruses, such as *Brome mosaic virus* (Traynor *et al.*, 1991) and CMV (Carr *et al.*, 1994; Gal-On *et al.*, 1994; Canto & Palukaitis, 2001), can influence systemic virus movement. For these other viral replicase proteins it appears that their effects on movement are distinct from their role in virus replication. But is this true also for TMV and other tobamoviruses? The general consensus is that tobamoviruses do not need to replicate in the phloem in order to be translocated. The grafting experiments of Gera *et al.* (1995) and the sap analysis experiments of Simon-Buela and Garcia-Arenal (1999) are consistent with the idea that tobamovirus virions move passively in the phloem. In support of this idea, Susi *et al.* (1999) showed that chilling a host plant stem to 4°C did not prevent systemic movement of TMV, even though at this temperature replication of the virus was completely inhibited.

One study contradicts this consensus. Arce-Johnson *et al.* (1997) found that TMV could not move through a stem graft derived from a *54-kDa*-transgenic plant (Section 2.5.2.4). These transgenic plants express strong pathogen-derived resistance to TMV (Golemboski *et al.*, 1990). There was evidence that in the original lines of the *54-kDa*-transgenic plant made by Golemboski *et al.* (1990), resistance was caused by an almost complete inhibition of TMV replication (Carr & Zaitlin, 1991). This led Arce-Johnson *et al.* (1997) to argue that TMV replication was required for phloem transport of TMV to occur.

However, Arce-Johnson *et al.* (1997) used their own independently transformed line of tobacco and not that of Golemboski *et al.* (1990). There is evidence that resistance mediated by expression of the TMV or *Pepper mild mottle virus 54-kDa* sequence in transgenic plants can in many cases be attributed to RNA silencing, or a homology-dependent RNA degradation mechanism (Tenllado *et al.*, 1995, 1996; Marano & Baulcombe, 1998) (see Section 2.5.2.4). Thus, the inability of TMV to move through the intergraft (Arce-Johnson *et al.*, 1997) may not have been due to the inhibition of replication. But if the resistance was due to RNA silencing, it suggests either, that RNA in virions is susceptible to degradation by this mechanism, or that a systemic silencing signal had passed up into the non-transgenic graft from the *54-kDa*-transgenic intergraft.

2.4.2.5 The role of the host in the systemic movement of tobamoviruses

Detailed microscopic examination of the movement of TMV.GFP in *N. benthamiana* has revealed that the entry and exit of tobamoviruses into and out of the phloem is a highly ordered process (Cheng *et al.*, 2000). As mentioned earlier, systemic virus movement tends to follow the trend of photosynthate translocation, i.e. from carbon source to carbon sink tissues. In inoculated source leaves, TMV.GFP enters the major and minor leaf veins by three days post-inoculation. Subsequent export of the virus occurs via the external (abaxial) phloem of the petiole and stem but somehow, during the movement of the virus in the stem, there is a transfer of virus to the internal (adaxial) phloem. It is these phloem elements which, apparently, serve as the sole conduits of the virus into the sink leaves, and virus exit into the surrounding tissues occurs from the major veins (Cheng *et al.*, 2000).

For tobamoviruses, there is less known about the specific host genes that regulate systemic virus movement than for certain other systems, most notably the potyviruses (see Chapter 3). Nevertheless, some progress has been made in the last five years or so. Citovsky and colleagues (Lartey *et al.*, 1998) isolated an *Arabidopsis* mutant, *vsm1*, in which systemic movement of two tobamoviruses, *Turnip vein clearing virus* (TVCV) and ToMV, is inhibited. The wild-type *Vsm1* gene product is thought to facilitate the loading of virus into the phloem of the inoculated leaves (Lartey *et al.*, 1998). Recently, during a study of cadmium ion-induced resistance to TVCV in tobacco the same group identified a host gene for a glycine-rich protein (cdiGRP). CdiGRP appears to regulate the accumulation of callose in the phloem tissue, which in turn affects the unloading of the virus in sink tissues (Ueki & Citovsky, 2002). This indicates that callose plays a role in both phases of virus movement, systemic as well as local.

2.5 Host reactions to TMV

2.5.1 The 'susceptible' host

TMV is considered to have a host range of intermediate size (Schneider & Roossinck, 2000), being able to infect around 199 plant species in 30 families (Zaitlin, 2000). This host range size is presumably typical of most tobamoviruses. In hosts considered to be susceptible to infection, the virus is able to spread outwards from the initial point of infection, spread systemically and, usually, cause disease symptoms.

In tobamovirus-infected plants, symptoms are most usually apparent on the non-inoculated leaves and take the form of vein clearing and chlorosis (yellowing) on immature sink leaves and mosaic on new leaves that develop after infection has become established (Fig. 2.3). The directly inoculated leaves do not usually display any obvious symptoms apart from chlorosis around the primary infection sites (which are only visible in plants exposed to high light intensity or inoculated with severe TMV strains). Iodine staining may also reveal the accumulation or depletion of starch in primary infected cells. Younger leaves may also be distorted and display *bubbling* of the leaf lamina, and depending on how young the plants were when infected, plants may be stunted. The mechanisms by which TMV causes distortion or stunting of plants are currently not understood. Although tobamovirus infection has a wide range of effects on host metabolism, they are often difficult to relate directly to symptomology (Dawson, 1999). Similarly, symptom severity cannot be directly related to virus load since there is no simple correlation between the titre of virus and the severity of the disease (Dawson, 1999). For example, the mutant TMV strain YSI/1 (Garcia-Arenal *et al.*, 1984) induces more severe symptoms than its parent, the U1 strain of TMV, but accumulates to lower levels in tobacco and *Arabidopsis* (Handford, 2000). Conversely, TMV U1 and the *masked* strain of TMV reach similar titres but the symptoms induced by the masked strain are far milder (Derrick *et al.*, 1997). In this case, the mildness of the

Figure 2.3 Symptoms induced on tobacco plants by TMV. Panel A shows the symptoms induced on a plant of a susceptible variety of tobacco infected with TMV strain U1. The plant was inoculated on one lower leaf approximately three weeks before the photograph was taken. Virus has spread systemically and induced symptoms including vein clearing (most clear on the large leaf on the lower right of the photograph), leaf distortion, and mosaic (most clearly seen on the younger upper leaves). Panel B shows the symptoms induced on a directly inoculated leaf of a plant from a resistant tobacco variety carrying the *N* resistance gene. Plant cells in the primary infection foci undergo programmed cell death (the hypersensitive response) about two days after inoculation. Virus does not spread beyond a small zone of tissues around these lesions and the virus cannot spread to non-inoculated leaves of the same plant which therefore remain healthy (Panel C).

symptoms may depend on the rate at which the masked strain enters the phloem, which in turn is dependent upon the properties of the virus' 126-kDa protein (Derrick *et al.*, 1997).

2.5.1.1   Chlorosis and vein clearing in systemically infected plants

Chlorosis is caused by tobamovirus infection, at least in part due to the uptake of viral CP into the chloroplasts, where it associates with the thylakoid membranes and disrupts photosystem II (PSII) (Reinero & Beachy, 1986, 1989). The disruption of PSII leads to increased levels of reactive oxygen species (ROS) in the chloroplast, leading to the destruction of chlorophyll and other pigments (Lehto *et al.*, 2003). Although some genomic length TMV RNA has been detected in chloroplasts from infected plants, this is apparently not used as a template for protein synthesis, so the CP must be imported from the cytosol (Schoelz & Zaitlin, 1989). However, the CP does not possess a conventional N-terminal plastid import sequence and cannot exploit any of the known mechanisms responsible for the uptake of host nuclear-encoded chloroplast proteins (Banerjee & Zaitlin, 1992; Banerjee *et al.*, 1995). *In vitro* import experiments using isolated chloroplasts showed that CP can be taken up into the chloroplasts and can localize to the thylakoid membranes in an ATP-independent manner, if it is provided in the form of 20S disks (Banerjee & Zaitlin, 1992; Banerjee *et al.*, 1995). The mechanism for this is unknown but is thought to involve a hydrophobic interaction between the disk face and the chloroplast mem-branes (Banerjee & Zaitlin, 1992; Banerjee *et al.*, 1995). The severity of chlorosis seen in TMV-infected plants is related to the amount of CP that accumulates in the chloroplast. Thus, the TMV *flavum* strain and the YSI/1 mutant of TMV strain U1 both induce severe chlorosis and eventually necrosis in tobacco, and both have CPs that accumulate to high levels in chloroplasts (Banerjee & Zaitlin, 1992; Banerjee *et al.*, 1995; Lehto *et al.*, 2003). There is evidence that damage to PSII may in some way help increase the level of virus accumulation (Abbinck *et al.*, 2002).

The symptom of vein clearing, an apparent chlorosis of the cells surrounding the vasculature, is probably not primarily due to damage to the chloroplasts of those cells. Rather, it is most likely due to a change in the optical properties of these cells, perhaps due to cell expansion, causing them to become more translucent than equivalent cells in uninfected leaves (Dawson, 1999).

2.5.1.2   Mosaic symptoms in systemically infected plants: an effect of localized RNA silencing?

Leaves with mosaic or mottling symptoms have sectors of dark green (green islands) and light green or yellow (chlorotic) tissue (Fig. 2.3). Detailed examination of mosaic leaves by Matthews and colleagues (reviewed by Dawson, 1999) showed that cells in a chlorotic sector containing an abundant amount of virus can be imme-diately adjacent to a virus-free cell in a green sector. The mosaic pattern comes about because the green island and chlorotic sectors represent clones of cells that arose from cells that were uninfected or infected with the virus (Dawson, 1999). However, how do the green islands remain free of infection when, in principle, the virus could invade via plasmodesmata linking the green island cells with infected cells? It is worth remembering that the SEL values for plasmodesmata linking cells in immature leaf tissues are relatively high (Blackman & Overall, 2001), which should facilitate cell-to-cell viral movement.

It now appears likely that the virus-free nature of green island tissue is due to the localized induction of RNA silencing. RNA silencing (also known as RNA interference) is a homology-dependent gene-silencing phenomenon that occurs in a wide range of eukaryotic organisms and has attracted considerable attention over the last few years. In tissues expressing RNA silencing, foreign, overexpressed, or aberrant RNA molecules are targeted for destruction in a sequence-specific manner (Grant, 1999; Baulcombe, 2001). Work with *Arabidopsis* mutants defective in RNA silencing showed that their susceptibility to viral infection was increased (Morrain *et al.*, 2000). These, and other findings, have lent support to the theory that RNA silencing may have evolved as an intrinsic mechanism to protect plants from virus infection (reviewed by Waterhouse *et al.*, 1999, 2001; Voinnet, 2001).

The production of short (21–25 nucleotide) dsRNA molecules is a characteristic feature in cells or tissues displaying RNA silencing (Hamilton & Baulcombe, 1999; Waterhouse *et al.*, 2001). These short RNAs, now known as short interfering RNAs (siRNAs: Voinnet, 2001), are produced by an enzyme termed dicer (a member of the RNase III family of nucleases). Dicer specifically cleaves dsRNA, producing small nucleotides which then serve as guide sequences that target an associated nuclease complex to degrade specific mRNA (Bernstein *et al.*, 2001; Di Serio *et al.*, 2001; Lipardi *et al.*, 2001). The substrate for dicer is dsRNA which is produced by the action of either viral or host RdRp enzymes which are thought to play a key role in the initiation of RNA silencing (reviewed by Ahlquist, 2002).

Recently, Chen and colleagues demonstrated that accumulation and activity of a host-encoded RdRp, NtRdRp1, are increased in tobacco during a systemic infection with TMV (Xie *et al.*, 2001). Production of NtRdRp1 was also stimulated by the resistance-inducing chemical salicylic acid (SA) (Section 2.5.2.1). When the expression of the *NtRdRp1* gene was inhibited in transgenic tobacco plants harboring anti-sense constructs, infection with TMV resulted in a greater yield of virus and greater chlorosis in infected leaves. However, they did not produce green islands on the systemically infected leaves (Xie *et al.*, 2001). The results offer very strong support for the idea that an RNA-silencing mechanism protects virus-free green islands. They also show that ostensibly 'susceptible' plants do offer at least a basal resistance to viral infection. What is puzzling, perhaps, is why only some sectors of tissue are able to protect themselves in this way. How does TMV avoid inducing widespread, uniform, and complete silencing against itself?

As part of an examination of the ability of several viruses to suppress virus-induced gene silencing in *N. benthamiana* it was found that TMV is able to suppress RNA silencing, although this effect is seen preferentially in the veins (Voinnet *et al.*, 1999). To explain the ability of TMV to evade or suppress RNA silencing in other tissues, these authors have speculated that TMV may employ other, as yet unknown, counter-silencing strategies and that TMV may interfere with the systemic induction of RNA silencing (Voinnet *et al.*, 1999). It is not known at present what viral gene product(s) are involved in the suppression of silencing by tobamoviruses.

Recently, another instance of localized silencing of TMV was demonstrated in developing lateral roots of *N. benthamiana* (Valentine *et al.*, 2002). In plants infected with TMV.GFP, newly emerging lateral roots initially display intense GFP fluorescence, but this is subsequently silenced. This effect appears to be triggered by a signal from the root meristem (Valentine *et al.*, 2002). It remains unclear whether this effect is directed against viral infection or whether it is an intrinsic part of the process of tissue differentiation during which, perhaps, there is activation of RNA silencing to eradicate unwanted host cell transcripts.

2.5.2 Resistance to tobamoviruses

2.5.2.1 The N resistance gene
Several genes conferring resistance to tobamoviruses have been identified in various hosts and some of these have proven to be commercially important in crop protection (Okada, 1999). Probably the most intensively studied of these resistance genes is the *N* gene, which has been introgressed into several cultivars of tobacco from *N. glutinosa* (reviewed by Dunigan *et al.*, 1987). This dominant resistance gene confers the ability to respond to nearly all tobamoviruses with a hypersensitive response (HR), a form of programmed cell death (Fig. 2.3). Following the HR, virus is prevented from spreading out to the rest of the plant, although it should be noted that cells at the periphery of the HR lesions still contain live virus for several days following the appearance of the lesion (Weststeijn, 1981). This can be seen directly when using TMV.GFP to infect *NN* genotype tobacco plants (Wright *et al.*, 2000; Murphy *et al.*, 2001).

Induction of the HR by TMV (or by other pathogens) can result in the induction of SAR (Ross, 1961a,b) and an enhanced degree of resistance to a broad range of pathogens, not just against viruses. The induction of SAR is regulated by a signal transduction pathway that includes SA as an important component (Malamy *et al.*, 1990; Métraux *et al.*, 1990). SA induces the activation of pathogenesis-related (PR) proteins, which contribute to resistance to fungal and bacterial pathogens (reviewed by Carr & Klessig, 1989). SA also triggers resistance to viruses, including TMV, by inducing the inhibition of virus movement and replication, and possibly by increasing the levels of host RdRp (Chivasa *et al.*, 1997; Chivasa & Carr, 1998; Naylor *et al.*, 1998; Murphy & Carr, 2002; Gilliland *et al.*, 2003). Pharmacological and genetic evidence obtained from tobacco and *Arabidopsis* indicates that the induction of resistance against viruses by SA involves a signaling pathway that is distinct from that leading to *PR* gene activation (Chivasa *et al.*, 1997; Wong *et al.*, 2002; Gilliland *et al.*, 2003). For a more detailed description of this area, the reader is referred to the reviews by Murphy *et al.* (1999, 2001) and Gilliland *et al.* (2004). Recent results from Canto and Palukaitis (2002) also indicate that *N* gene-mediated resistance may involve SA-independent mechanisms.

The *N* gene was isolated from tobacco by transposon tagging (Whitham *et al.*, 1994). The gene encodes a full-length protein with Toll, Interleukin receptor-like (TIR), nucleotide binding site (NBS), and leucine-rich repeat (LRR) domains. The TIR and NBS domains are thought to be involved in signaling while the LRR domain is presumed to be involved in ligand binding (Erickson *et al.*, 1999a). Due to alternative splicing, the *N* gene can give rise to transcripts encoding the full-length N protein (132 kDa) or a truncated protein of 75 kDa, called N^{tr} that includes the TIR and NBS domains and only a portion of the LRR domain. Both gene products appear to be necessary for resistance to TMV (Dinesh-Kumar & Baker, 2000; Marathe *et al.*, 2002).

The elicitor for the *N* gene, i.e. the viral gene product which triggers the *N* gene-mediated HR, is a 50-kDa sequence within the helicase domain of the 126-kDa replicase protein (Padgett *et al.*, 1997; Abbink *et al.*, 1998; Erickson *et al.*, 1999b). However, it is considered unlikely that the N or N^{tr} proteins interact directly with the 126-kDa protein (Marathe *et al.*, 2002). Based on what is now known about other resistance gene products, it is likely that the products of the *N* gene guard a cellular target of the 126-kDa protein.

2.5.2.2 *Resistance genes that are elicited by the tobamoviral CP*

The *N′* gene is a dominant resistance gene from *N. sylvestris* that conditions a HR triggered by infection with some TMV strains. Biochemical studies which were later vindicated by molecular biological approaches showed that the elicitor for *N′* gene was the viral CP (Fraser, 1983; Knorr & Dawson, 1988; Culver and Dawson, 1989; Culver *et al.*, 1994). Subsequent site-directed mutagenesis experiments defined the face of the CP that participates in recognition by the host, and together with other studies suggested that recognition of the elicitor CP occurs at the level of either monomers or small CP aggregates (Taraporewala & Culver, 1996, 1997; Culver, 2002).

Tobamoviral CPs are also the elicitors of the HR in pepper harboring the *L* resistance genes, some of which can be overcome by some strains of the tobamovirus pepper mild mottle virus with appropriate mutations in their *CP* genes (Berzal-Herranz *et al.*, 1995; de la Cruz *et al.*, 1997; Dardick *et al.*, 1999). The CP also governs the elicitation of the tobamovirus-induced HR in eggplant (or aubergine; *Solanum melongena*) (Dardick & Culver, 1997; Dardick *et al.*, 1999).

2.5.2.3 *Genes for resistance to tobamoviruses in tomato*

Three genes conferring resistance to ToMV have been characterized: *Tm-1*, *Tm-2*, and *Tm-2²* (Fraser, 1985; Hull, 2002). *Tm-2* and *Tm-2²* are dominant genes that confer the ability to express a HR against ToMV. The *Tm-1* gene is incompletely dominant: homozygous plants support very little or no virus multiplication and heterozygous plants support an intermediate level. All of these genes have been used to protect commercial cultivars of tomato, although each can be overcome by various virus strains that have evolved. The characterization of resistance-breaking mutants has aided our understanding of the resistance phenotypes.

Investigation of a *Tm-1* resistance-breaking mutant, ToMV Lta1, showed that mutations mapped to the 126-kDa replicase protein sequence (Meshi *et al.*, 1988). The results suggest that either the *Tm-1* gene encodes an inhibitor of replication or it encodes a defective host factor needed for 126-kDa protein function. The MP can act as the elicitor for both the *Tm-2* and the *Tm-2^2* genes (Meshi *et al.*, 1989; Weber *et al.*, 1993; Weber & Pfitzner, 1998). Deployment of these genes alone or in combination is now widespread in commercial tomato cultivars and has superseded cross-protection in the protection of tomato from ToMV infection (Fraser, 1998).

2.5.2.4 *Genetically engineered resistance to tobamoviruses*
Cross-protection is the protection of a plant from a severe virus strain by intentional inoculation with a closely related, but mild strain of the virus. It is used currently in the protection of a variety of crops and at one time was used to protect tomato from tobamovirus infection (Fraser, 1998). One of the suggested mechanisms of action for cross-protection between tobamovirus strains is interference with uncoating of the severe strain by the CP from the mild strain (Fraser, 1998). To further explore this idea, Beachy and co-workers carried out their ground-breaking experiments on transgenic tobacco expressing the TMV *CP* gene (Powell Abel *et al.*, 1986; Nelson *et al.*, 1987). Their work demonstrated for the first time that pathogen-derived resistance, that is the expression of a pathogen gene in the host, could protect the host from pathogen infection and that genetic engineering was a practical method-ology for the generation of new disease-resistant lines of plants.

Since that time, pathogen-derived resistance has been employed to generate transgenic resistance to a wide range of other viruses. In many cases, genes other than the CP gene have been used to engender resistance and in many of these cases, resistance has resulted from induction of RNA silencing rather than from interference with the processes of the virus life cycle. For more information on this area, and its controversies, the reader is referred to specialized reviews on the topic (for example Baulcombe, 1996; Palukaitis & Zaitlin, 1997; van den Boogaart *et al.*, 1998; Beachy, 1999).

Isolation of natural resistance genes has raised the possibility that these genes could be moved between different species or genera of plants. The *N* gene, has been successfully transferred into tomato (Whitham *et al.*, 1996) and *N. benthamiana* independently by the laboratories of Dinesh-Kumar (Liu *et al.*, 2002) and Baulcombe (Bendahmane *et al.*, 1999). In these transgenic plants, the response to TMV infection is either slower (in tomato: Whitham *et al.*, 1996), or in the case of some of the *N*-transgenic *N. benthamiana* lines the resistance does not work well against wild-type TMV (Peart *et al.*, 2002). There are no published reports in the literature of successful transfer of *N* gene-mediated resistance to non-Solanaceous plants. These results are compatible with what has been observed for cross-species transfer of several other resistance genes in that, with some notable exceptions, these genes are not universally compatible with plant resistance signaling systems (see Xiao *et al.*, 2003 and references therein). Thus, for the successful transfer of some resistance genes, like *N*, the resistance gene will need to be modified to be compatible with the recipient plant's defensive signaling system.

2.6 Future directions for TMV research: making an old foe into a new friend?

TMV will no doubt continue to be used for the foreseeable future as a model virus for the investigation of plant–pathogen interactions and, as the preceding text makes clear at a number of points, there are still many unresolved questions concerning the mechanistic details of tobamoviral replication and movement. Thus, for some time to come work on TMV and related viruses will continue to contribute to fundamental studies in plant virology and biology in general. However, what new areas of research outside of basic biology are likely to profit from studies of the tobamoviruses? I suggest that there are two main areas of technology in which these viruses may have a significant impact in the short to medium term.

TMV was one of the first plant viruses to be utilized as a vector for the expression of foreign or modified proteins or RNA sequences in plants (recently reviewed by Pogue *et al.*, 2002). As mentioned earlier in this review its use as a carrier of GFP has contributed to a revolution in our understanding of virus–plant interactions and plant cell biology. However, in general TMV has not been a useful vector for the targeted knockout of gene expression by virus-induced gene silencing (VIGS; Baulcombe, 1999), except in very few cases in which silencing has been patchy (for example see Hiriart *et al.*, 2003). Tobamovirus-mediated VIGS may be facilitated by the use of a two-component system as described by Gossele *et al.* (2002), in which TMV strain U2 was used to support replication of a modified satellite tobacco mosaic virus (STMV) carrying the target gene sequence, the so-called satellite virus-induced silencing system (SVISS).

However, the ability of TMV to largely avoid inducing silencing against itself is a major advantage for production of useful protein products in plants on a commercial scale (Turpen, 1999; Pogue *et al.*, 2002). Plant viral vectors in general have a number of advantages over other systems in that they are easier and cheaper to use and manipulate than mammalian tissue culture or transgenic animals and are faster to produce than transgenic plants (Pogue *et al.*, 2002).

Current work with TMV-based vectors includes large-scale field production of pharmaceuticals (Turpen, 1999; Pogue *et al.*, 2002). One company, the Large Scale Biology Corporation, has expressed several cytokines, interferons, antimicrobial enzymes, vaccine epitopes, and other biologically active peptides from non-plant sources in tobacco infected with viral vectors (Pogue *et al.*, 2002). But it is the area of orphan drugs, i.e. drugs required by such a small number of patients (or by patients lacking sufficient spending power) for the drug to be economical to produce by conventional means, that the use of plant viral vectors comes into its own (Turpen, 1999; Pogue *et al.*, 2002). An example of this type of small-scale production by *in planta* synthesis using TMV vectors is the manufacture of Gal A. This is used in enzyme-replacement therapy for patients suffering from Fabry's disease, an inherited lysosomal enzyme disorder (Pogue *et al.*, 2002). An even more ambitious idea is the use of TMV vectors for *individualized* therapy, i.e. tailored to the treatment of a single patient. A human disease for which this approach is highly appropriate is non-Hodgkin's lymphoma. This is a cancer in which a single clone of malignant B cells proliferates in the patient. These cells express a surface immunoglobulin that is unique both to the tumor and to the patient. Genes encoding these unique tumor cell markers can be cloned into a TMV vector and can be expressed and correctly processed *in planta* (McCormick *et al.*, 1999). Once purified, this product can be used to provoke the production of anti-idiotypic antibodies in animals and is currently under investigation as an immunotherapy for human patients with non-Hodgkin's lymphoma (McCormick *et al.*, 1999; Grill, 2003).

Nanotechnology is, broadly speaking, the name given to precision engineering and chemistry carried out on the smallest scale, i.e. down as far as the scale of individual molecules and atoms. Virus particles are, of course, natural nanotechnological devices and have in some cases inspired ideas in the field or become useful tools or models for research and development. The first virus to be used as a nanotechnological chemical reaction vessel was a virus with isometric virions, *Cowpea chlorotic mottle virus*, which were used for the nanofabrication of paratungstate crystals of defined size and shape (Douglas & Young, 1998). Recently, it was shown that manipulation of TMV particle structure through engineering of the CP gene or manipulation of the assembly process *in vitro* can permit controlled deposition on TMV-derived rod structures of nanoparticles of platinum, gold, or silver compounds (Dujardin *et al.*, 2003). As has happened in several instances in the past, TMV research is contributing to the birth of a new area of science.

Acknowledgments

I wish to thank Dr Keith Johnstone and Ms Clare Scovell for generously allowing me to use their images of plants infected with TMV and TMV.GFP respectively, and Dr Alex Murphy for critically reading sections of this review. Research from this laboratory described in this review has been funded by a variety of sources including the Biotechnological and Biological Sciences Research Council, The Leverhulme Trust, The Gatsby Technical Education Project, and The Land Settlement Association.

References

Abbink, T.E.M., Tjernberg, P.A., Bol, J.F. & Linthorst, H.J.M. (1998) Tobacco mosaic virus helicase domain induces necrosis in *N* gene-carrying tobacco in the absence of virus replication. *Mol. Plant Microbe Interact.*, **11**, 1242–1246.

Abbink, T.E.M., Peart, J.R., Mos, T.N.M., Baulcombe, D.C., Bol, J.F. & Linthorst, H.J.M. (2002) Silencing of a gene encoding a protein component of the oxygen-evolving complex of photosystem II enhances virus replication in plants. *Virology*, **295**, 307–319.

Ahlquist, P. (2002) RNA-dependent RNA polymerases, viruses, and RNA silencing. *Science*, **296**, 1270–1273.

Aoki, S. & Takebe, I. (1975) Replication of tobacco mosaic virus RNA in tobacco mesophyll protoplasts *in vitro*. *Virology*, **65**, 343–345.

Arce-Johnson, P., Reimann-Philipp, U., Padgett, H.S., Rivera Bustamante, R. & Beachy, R.N. (1997) Requirement of the movement protein for long distance spread of tobacco mosaic virus in grafted plants. *Mol. Plant Microbe Interact.*, **10**, 691–699.

Arce-Johnson, P., Medina, C., Padgett, H.S., Huanca, W. & Espinoza, C. (2003) Analysis of local and systemic spread of the crucifer-infecting TMV-Cg virus in tobacco and several *Arabidopsis thaliana* ecotypes. *Funct. Plant Biol.*, **30**, 401–408.

Argos, P. (1988) A sequence motif in many polymerases. *Nucleic Acids Res.*, **16**, 9909–9916.

Asselin, A. & Zaitlin, M. (1978) Characterization of a second protein associated with virions of tobacco mosaic virus. *Virology*, **91**, 173–181.

Bachmair, A., Novatchkova, M., Potuschak, T. & Eisenhaber, F. (2001) Ubiquitylation in plants: a post-genomic look at a post-translational modification. *Trends Plant Sci.*, **6**, 463–470.

Banerjee, N. & Zaitlin, M. (1992) Import of tobacco mosaic-virus coat protein into intact chloroplasts *in vitro*. *Mol. Plant Microbe Interact.*, **5**, 466–471.

Banerjee, N., Wang, J.Y. & Zaitlin, M. (1995) A single nucleotide change in the coat protein gene of tobacco mosaic virus is involved in the induction of severe chlorosis. *Virology*, **207**, 234–239.

Baulcombe, D.C. (1996) Mechanisms of pathogen-derived resistance to viruses in transgenic plants. *Plant Cell*, **8**, 1833–1844.

Baulcombe, D.C. (1999) Fast forward genetics based on virus-induced gene silencing. *Curr. Opin. Plant Biol.*, **2**, 109–113.

Baulcombe, D.C. (2001) Diced defence: RNA silencing. *Nature*, **409**, 295–296.

Beachy, R.N. (1999) Coat-protein-mediated resistance to tobacco mosaic virus: discovery mechanisms and exploitation. *Phil. Trans. R. Soc. Lond.*, **B 354**, 659–664.

Beachy, R.N. & Heinlein, M. (2000) Role of P30 in replication and spread of TMV. *Traffic*, **1**, 540–544.

Beachy, R.N. & Zaitlin, M. (1977) Characterization and in vitro translation of the RNAs from less-than-full-length, virus-related, nucleoprotein rods present in tobacco mosaic virus preparations. *Virology*, **81**, 160–169.

Beier, H., Barciszewska, M., Krupp, G., Mitnacht, R. & Gross, H.J. (1984) UAG readthrough during TMV RNA translation – isolation and sequence of two transfer RNAs^{TYR} with suppressor activity from tobacco plants. *EMBO J.*, **3**, 351–356.

Bendahmane, A., Kanyuka, K. & Baulcombe, D.C. (1999) The *Rx* gene from potato controls separate virus resistance and cell death responses. *Plant Cell*, **11**, 781–791.

Bendahmane, M., Szécsi, J., Chen, I., Berg, R.H. & Beachy, R.N. (2002) Characterization of mutant tobacco mosaic virus coat protein that interferes with virus cell-to-cell movement. *Proc. Natl. Acad. Sci. USA*, **99**, 3645–3650.

Berna, A., Gafny, R., Wolf, S., Lucas, W.J., Holt, C.A. & Beachy, R.N. (1991) The TMV movement protein – role of the c-terminal 73 amino acids in subcellular localization and function. *Virology*, **182**, 682–689.

Bernstein, E., Caudy, A.A., Hammond, S.M. & Hannon, G.J. (2001) Role for a bidentate ribonuclease in the initiation step of RNA interference. *Nature*, **409**, 363–366.

Berzal-Herranz, A., De La Cruz, A., Tenllado, F., Diaz-Ruiz, J.R., Lopez, L., Sanz, A.I., Vaquero, C., Serra, M.T. & Garcia-Luque, I. (1995) The *Capsicum* L3 gene-mediated resistance against the tobamoviruses is elicited by the coat protein. *Virology*, **209**, 498–505.

Blackman, L.M. & Overall, R.L. (2001) Structure and function of plasmodesmata. *Aust. J. Plant Physiol.*, **28**, 709–727.

Boyko, V., van der Laak, J., Ferralli, J., Suslova, E., Kwon, M.O. & Heinlein, M. (2000a) Cellular targets of functional and dysfunctional mutants of tobacco mosaic virus movement protein fused to green fluorescent protein. *J. Virol.*, **74**, 11339–11346.

Boyko, V., Ferralli, J., Ashby, J., Schellenbaum, P. & Heinlein, M. (2000b) Function of microtubules in intercellular transport of plant virus RNA. *Nat. Cell Biol.*, **2**, 826–832.

Boyko, V., Ashby, J.A., Suslova, E., Ferralli, J., Sterthaus, O., Deom, C.M. & Heinlein, M. (2002) Intramolecular complementing mutations in tobacco mosaic virus movement protein confirm a role for microtubule association in viral RNA transport. *J. Virol.*, **76**, 3974–3980.

Bucher, G.L., Tarina, C., Heinlein, M., Di Serio, F., Meins, F. Jr. & Iglesias, V.A. (2001) Local expression of enzymatically active class I beta-1,3-glucanase enhances symptoms of TMV infection in tobacco. *Plant J.*, **28**, 361–369.

Buck, K.W. (1996) Comparison of the replication of positive-stranded RNA viruses of plants and animals. *Adv. Virus Res.*, **47**, 159–251.

Buck, K.W. (1999) Replication of tobacco mosaic virus RNA. *Phil. Trans. R. Soc. Lond.*, **B 354**, 613–627.

Butler, P.J.G. (1999) Self-assembly of tobacco mosaic virus: the role of an intermediate aggregate in generating both specificity and speed. *Phil. Trans. R. Soc. Lond.*, **B 354**, 537–550.

Canto, T. & Palukaitis, P. (2001) A cucumber mosaic virus (CMV) RNA 1 transgene mediates suppression of the homologous viral RNA 1 constitutively and prevents CMV entry into the phloem. *J. Virol.*, **75**, 9114–9120.

Canto, T. & Palukaitis, P. (2002) Novel *N* gene-associated, temperature-independent resistance to the movement of *Tobacco mosaic virus* vectors neutralized by a *Cucumber mosaic virus* RNA1 transgene. *J. Virol.*, **76**, 12908–12916.

Canto, T., Prior, D.A.M., Hellwald, K.H., Oparka, K.J. & Palukaitis, P. (1997) Characterization of cucumber mosaic virus. 4. Movement protein and coat protein are both essential for cell-to-cell movement of cucumber mosaic virus. *Virology*, **237**, 237–248.

Carr, J.P. & Klessig, D.F. (1989) The pathogenesis-related proteins of plants. In: *Genetic Engineering: Principles and Methods*, Vol. 11 (ed. J.K. Setlow), pp. 65–100. Plenum, NY.

Carr, J.P. & Zaitlin, M. (1991) Resistance in transgenic tobacco plants expressing a non-structural gene sequence of tobacco mosaic virus is a consequence of markedly reduced virus replication. *Mol. Plant Microbe Interact.*, **4**, 579–585.

Carr, J.P. & Zaitlin, M. (1993) Replicase-mediated resistance. *Semin. Virol.*, **4**, 339–347.

Carr, J.P., Marsh, L.E., Lomonossoff, G.P., Sekiya, M.E. & Zaitlin, M. (1992) Resistance to tobacco mosaic virus induced by the 54-kDa gene sequence requires expression of the 54-kDa protein. *Mol. Plant Microbe Interact.*, **5**, 397–404.

Carr, J.P., Gal-On, A., Palukaitis, P. & Zaitlin, M. (1994) Replicase-mediated resistance to cucumber mosaic virus in transgenic plants involves suppression of both virus replication in the inoculated leaves and long distance movement. *Virology*, **199**, 439–447.

Chen, M.H., Sheng, J.S., Hind, G., Handa, A.K. & Citovsky, V. (2000) Interaction between the tobacco mosaic virus movement protein and host cell pectin methylesterases is required for viral cell-to-cell movement. *EMBO J.*, **19**, 913–920.

Cheng, N.-H., Su, C.-L., Carter, S.A. & Nelson, R.S. (2000) Vascular invasion routes and systemic accumulation patterns of tobacco mosaic virus in *Nicotiana benthamiana*. *Plant J.*, **23**, 349–362.

Chivasa, S. & Carr, J.P. (1998) Cyanide restores *N* gene-mediated resistance to tobacco mosaic virus in transgenic tobacco expressing salicylic acid hydroxylase. *Plant Cell*, **10**, 1489–1498.

Chivasa, S., Murphy, A.M., Naylor, M. & Carr, J.P. (1997) Salicylic acid interferes with tobacco mosaic virus replication *via* a novel, salicylhydroxamic acid-sensitive mechanism. *Plant Cell*, **9**, 547–557.

Citovsky, V. (1999) Tobacco mosaic virus: a pioneer of cell-to-cell movement. *Phil. Trans. R. Soc. Lond.*, **B 354**, 637–643.

Citovsky, V., Knorr, D., Shuster, G. & Zambryski, P. (1990) The P30 movement protein of tobacco mosaic virus is a single-strand nucleic acid binding protein. *Cell*, **60**, 637–647.

Citovsky, V., Wong, M.L., Shaw, A., Prasad, B.V.V. & Zambryski, P. (1992) Visualization and characterization of tobacco mosaic virus movement protein binding to single-stranded nucleic acids. *Plant Cell*, **4**, 397–411.

Cooper, B., Lapidot, M., Heick, J.A., Dodds, J.A. & Beachy, R.N. (1995) A defective movement protein of TMV in transgenic plants confers resistance to multiple viruses whereas the functional analog increases susceptibility. *Virology*, **206**, 307–313.

Culver, J.N. (2002) Tobacco mosaic virus assembly and disassembly: determinants in pathogenicity and resistance. *Annu. Rev. of Phytopathol.*, **40**, 287–308.

Culver, J.N. & Dawson, W.O. (1989) Tobacco mosaic virus coat protein – an elicitor of the hypersensitive reaction but not required for the development of mosaic symptoms in *Nicotiana sylvestris*. *Virology*, **173**, 755–758.

Culver, J.N., Stubbs, G. & Dawson, W.O. (1994) Structure–function relationship between tobacco mosaic virus coat protein and hypersensitivity in *Nicotiana sylvestris*. *J. Mol. Biol.*, **242**, 130–138.

Dardick, C.D. & Culver, J.N. (1997) Tobamovirus coat proteins: elicitors of the hypersensitive response in *Solanum melongena* (eggplant). *Mol. Plant Microbe Interact.*, **10**, 776–778.

Dardick, C.D., Taraporewala, Z.F., Lu, B. & Culver, J.N. (1999) Comparison of tobamovirus coat protein structural features that affect elicitor activity in pepper, eggplant and tobacco. *Mol. Plant Microbe Interact.*, **12**, 247–251.

Dardick, C.D., Golem, S. & Culver, J.N. (2000) Susceptibility and symptom development in *Arabidopsis thaliana* to *Tobacco mosaic virus* is influenced by virus cell-to-cell movement. *Mol. Plant Microbe Interact.*, **13**, 1139–1144.

Dasgupta, R., Garcia, B.H. & Goodman, R.M. (2001) Systemic spread of an RNA insect virus in plants expressing plant viral movement protein genes. *Proc. Natl. Acad. Sci. USA*, **98**, 4910–4915.

Dawson, W.O. (1999) Tobacco mosaic virus virulence and avirulence. *Phil. Trans. R. Soc. Lond. B.*, **354**, 645–651.

Dawson, W.O., Bubrick, P. & Grantham, G.L. (1988) Modifications of the tobacco mosaic virus coat protein gene affecting replication, movement, and symptomatology. *Phytopathology*, **78**, 783–789.

de la Cruz, A., Lopez, L., Tenllado, F., Diaz Ruiz, J.R., Sanz, A.I., Vaquero, C., Serra, M.T. & Garcia-Luque, I. (1997) The coat protein is required for the elicitation of the Capsicum L^2 gene-mediated resistance against the tobamoviruses. *Mol. Plant Microbe Interact.*, **10**, 107–113.

Deom, C.M., Oliver, M.J. & Beachy, R.N. (1987) The 30-kilodalton gene product of tobacco mosaic virus potentiates virus movement. *Science*, **237**, 389–394.

Deom, C.M., Quan, S. & He, X.Z. (1997) Replicase proteins as determinants of phloem-dependent long-distance movement of tobamoviruses in tobacco. *Protoplasma*, **199**, 1–8.

Derrick, P.M., Carter, S.A. & Nelson, R.S. (1997) Mutation of the tobacco mosaic tobamovirus 126- and 183-kDa proteins: effects on phloem-dependent virus accumulation and synthesis of viral proteins. *Mol. Plant Microbe Interact.*, **10**, 589–596.

Dinesh-Kumar, S.P. & Baker, B. (2000) Alternatively spliced *N* resistance gene transcripts: their possible role in tobacco mosaic virus resistance. *Proc. Natl. Acad. Sci. USA*, **97**, 1908–1913.

Ding, B., Haudenshield, J.S., Hull, R.J., Wolf, S., Beachy, R.N. & Lucas, W.J. (1992) Secondary plasmodesmata are specific sites of localization of the tobacco mosaic virus movement protein in transgenic tobacco plants. *Plant Cell*, **4**, 915–928.

Ding, X.S., Shintaku, M.H., Arnold, S.A. & Nelson, R.S. (1995) Accumulation of mild and severe strains of tobacco mosaic virus in minor veins of tobacco. *Mol. Plant Microbe Interact.*, **8**, 32–40.

Ding, X.S., Shintaku, M.H., Carter, S.A. & Nelson, R.S. (1996) Invasion of minor veins of tobacco leaves inoculated with tobacco mosaic virus mutants defective in phloem-dependent movement. *Proc. Natl. Acad. Sci. USA*, **93**, 11155–11160.

Di Serio, F., Schöb, H., Iglesias, A., Tarina, C., Bouldoires, E. & Meins, J.F. (2001) Sense- and anti-sense-mediated gene silencing in tobacco is inhibited by the same viral suppressors and is associated with accumulation of small RNAs. *Proc. Natl. Acad. Sci. USA*, **98**, 6506–6510.

Dorokhov, Y.L., Ivanov, P.A., Novikov, V.K., Agranovsky, A.A., Morozov, S.Y., Efimov, V.A., Casper, R. & Atabekov, J.G. (1994) Complete nucleotide-sequence and genome organization of a tobamovirus infecting *Cruciferae* plants. *FEBS Lett.*, **350**, 5–8.

Dorokhov, Y.L., Makinen, K., Frolova, O.Y., Merits, A., Saarinen, J., Kalkkinen, N., Atabekov, J.G. & Saarma, M. (1999) A novel function for a ubiquitous plant enzyme pectin methylesterase: the host-cell receptor for the tobacco mosaic virus movement protein. *FEBS Lett.*, **461**, 223–228.

Dos Reis Figueira, A., Golem, S., Goregaoker, S.P. & Culver, J.N. (2002) A nuclear localization signal and a membrane association domain contribute to the cellular localization of the *Tobacco mosaic virus* 126-kDa replicase protein. *Virology*, **301**, 81–89.

Douglas, T. & Young, M. (1998) Host-guest encapsulation of materials by assembled virus protein cages. *Nature*, **393**, 152–155.

Dujardin, E., Peet, C., Stubbs, G., Culver, J.N. & Mann, S. (2003) Organization of metallic nanoparticles using *Tobacco mosaic virus* templates. *Nano Letters*, **3**, 413–417.

Dunigan, D.D. & Zaitlin, M. (1990) Capping of tobacco mosaic virus RNA. *J. Biol. Chem.*, **265**, 7779–7786.

Dunigan, D.D., Golemboski, D.B. & Zaitlin, M. (1987) Analysis of the *N*-gene of *Nicotiana*. *CIBA Found. Symp.*, **133**, 120–135.

Dunigan, D.D., Dietzgen, R.G., Schoelz, J.E. & Zaitlin, M. (1988) Tobacco mosaic virus particles contain ubiquitinated coat protein subunits. *Virology*, **165**, 310–312.

Erickson, F.L., Dinesh-Kumar, S.P., Holzberg, S., Ustach, C.V., Dutton, M., Handley, V., Corr, C. & Baker, B. (1999a) Interactions between tobacco mosaic virus and the tobacco *N* gene. *Phil. Trans. R. Soc. Lond.*, **B 354**, 653–658.

Erickson, F.L., Holzberg, S., Calderon-Urrea, A., Handley, V., Axtell, M., Corr, C. & Baker, B. (1999b) The helicase domain of the TMV replicase proteins induces the *N*-mediated defence response in tobacco. *Plant J.*, **18**, 67–75.

Esau, K. & Cronshaw, J. (1967) Relation of tobacco mosaic virus with host cells. *J. Cell Biol.*, **33**, 665–678.

Fedorkin, O.N., Denisenko, O.N., Sitkov, A.S., Zelenina, D.A., Lukashova, L.I., Morozov, S.Y. & Atabekov, J.G. (1995) The tomato mosaic virus small gene product forms a stable complex with translation elongation factor EF-1alpha. *Doklady Akademii Nauk*, **343**, 703–704.

Fenczik, C.A., Padgett, H.S., Holt, C.A., Casper, S.J. & Beachy, R.N. (1995) Mutational analysis of the movement protein of odontoglossum ringspot virus to identify a host-range determinant. *Mol. Plant Microbe Interact.*, **8**, 666–673.

Fraser, R.S.S. (1983) Varying effectiveness of the N′ gene for resistance to tobacco mosaic virus in tobacco infected with virus strains differing in coat protein properties. *Physiol. Plant Pathol.*, **22**, 109–119.

Fraser, R.S.S. (ed.) (1985) Genetics of host resistance to viruses and of virulence. In: *Mechanisms of Resistance to Plant Disease*, pp. 62–79. Martinus Nijhoff/W. Junk, Dordrecht.

Fraser, R.S.S. (1998) 'Introduction to classical cross-protection' in Plant Virology Protocols. In: *Methods in Molecular Biology* (eds G.D. Foster & S.C. Taylor), **81**, Humana Press, Totowa, NJ.

Gallie, D.R. (1996) Translational control of cellular and viral mRNAs. *Plant Mol. Biol.*, **32**, 145–158.

Gallie, D.R. (2002) The 5′-leader of tobacco mosaic virus promotes translation through enhanced recruitment of eIF4F. *Nucleic Acids Res.*, **30**, 3401–3411.

Gallie, D.R., Sleat, D.E., Watts, J.W., Turner, P.C. & Wilson, T.M.A. (1987) *In vivo* uncoating and efficient expression of foreign messenger RNAs packaged in TMV-like particles. *Science*, **236**, 1122–1124.

Gal-On, A., Kaplan, I., Roossinck, M.J. & Palukaitis, P. (1994) The kinetics of infection of zucchini squash by cucumber mosaic virus indicate a function for RNA-1 in virus movement. *Virology*, **205**, 280–289.

Garcia-Arenal, F., Palukaitis, P. & Zaitlin, M. (1984) Strains and mutants of tobacco mosaic virus are both found in virus derived from single-lesion-passaged inoculum. *Virology*, **132**, 131–137.

Gera, A., Deom, C.M., Donson, J., Shaw, J.J., Lewandowski, D.J. & Dawson, W.O. (1995) Tobacco mosaic tobamovirus does not require concomitant synthesis of movement protein during vascular transport. *Mol. Plant Microbe Interact.*, **8**, 784–787.

Gibbs, A. (1999) Tobamovirus evolution. *Phil. Trans. R. Soc. Lond. B.*, **354**, 593–602.

Gillespie, T., Boevink, P., Haupt, S., Roberts, A.G., Toth, R., Valentine, T., Chapman, S. & Oparka, K.J. (2002) Functional analysis of a DNA-shuffled movement protein reveals that microtubules are dispensable for the cell-to-cell movement of *Tobacco mosaic virus*. *Plant Cell*, **14**, 1207–1222.

Gilliland, A., Singh, D.P., Hayward, J.M., Moore, C.A., Murphy, A.M., York, C.J., Slator, J. & Carr, J.P. (2003) Genetic modification of alternative respiration has differential effects on antimycin A *versus* salicylic acid-induced resistance to *Tobacco mosaic virus*. *Plant Physiol.*, **132**, 1518–1528.

Gilliland, A., Murphy, A.M., Wong, C.E., Carson, R.A.J. & Carr, J.P. (2004) Mechanisms involved in induced resistance to plant viruses. In: *Multigenic and Induced Systemic Resistance* (eds S. Tuzun & E. Bent), Kluwer Academic Publishers (in press).

Goelet, P., Lomonossoff, G.P., Butler, P.J.G., Akam, M.E., Gait, M.J. & Karn, J. (1982) Nucleotide sequence of tobacco mosaic virus RNA. *Proc. Natl. Acad. Sci. USA*, **79**, 5818–5822.

Golemboski, D.B., Lomonossoff, G.P. & Zaitlin, M. (1990) Plants transformed with a tobacco mosaic virus nonstructural gene sequence are resistant to the virus. *Proc. Natl. Acad. Sci. USA*, **87**, 6311–6315.

Gorbalenya, A.E., Koonin, E.V., Donchenko, A.P. & Blinov, V.M. (1988) A conserved NTP motif in putative helicases. *Nature*, **333**, 22–22.

Goregaoker, S.P. & Culver, J.N. (2003) Oligomerization and activity of the helicase domain of the tobacco mosaic virus 126- and 183-kilodalton replicase proteins. *J. Virol.*, **77**, 3549–3556.

Goregaoker, S.P., Lewandowski, D.J. & Culver, J.N. (2001) Identification and functional analysis of an interaction between domains of the 126/183-kDa replicase-associated proteins of *Tobacco mosaic virus*. *Virology*, **282**, 320–328.

Gossele, V., Fache, I., Meulewaeter, F., Cornelissen, M. & Metzlaff, M. (2002) SVISS – a novel transient gene silencing system for gene function discovery and validation in tobacco plants. *Plant J.*, **32**, 859–866.

Grant, S.R. (1999) Dissecting the mechanism of posttranscriptional gene silencing: divide and conquer. *Cell*, **96**, 303–306.

Grdzelishvili, V.Z., Chapman, S.N., Dawson, W.O. & Lewandowski, D.J. (2000) Mapping of the *Tobacco mosaic virus* movement protein and coat protein subgenomic RNA promoters *in vivo*. *Virology*, **275**, 177–192.

Grill, L.K. (2003) Commercial production of antibody-based therapeutics in plants. **Abstract #1145**, Proceedings of the Meeting of the Society for General Microbiology, 7–11 April 2003 (http://www.socgenmicrobiol.org.uk/meetings/past.cfm).

Hagiwara, Y., Komoda, K., Yamanaka, T., Tamai, A., Meshi, T., Funada, R., Tsuchiya, T., Naito, S. & Ishikawa, M. (2003) Subcellular localization of host and viral proteins associated with tobamovirus RNA replication. *EMBO J.*, **22**, 344–353.

Haley, A., Hunter, T., Kiberstis, P. & Zimmern, D. (1995) Multiple serine phosphorylation sites on the 30 kDa TMV cell-to-cell movement protein synthesized in tobacco protoplasts. *Plant J.*, **8**, 715–724.

Hamilton, A.J. & Baulcombe, D.C. (1999) A species of small antisense RNA in posttranscriptional gene silencing in plants. *Science*, **286**, 950–952.

Handford, M.G. (2000) Host factors and virus infection of *Arabidopsis thaliana*. PhD Thesis, University of Cambridge.

Harrison, B.D. & Wilson, T.M.A. (1999) Milestones in the research on tobacco mosaic virus. *Phil. Trans. R. Soc. Lond. B.*, **354**, 521–529.

Haseloff, J., Goelet, P., Zimmern, D., Ahlquist, P., Dasgupta, R. & Kaesberg, P. (1984) Striking similarities in amino acid sequence among nonstructural proteins encoded by RNA viruses that have dissimilar genomic organization. *Proc. Natl. Acad. Sci. USA*, **81**, 4358–4362.

Hazelwood, D. & Zaitlin, M. (1990) Ubiquitinated conjugates are found in preparations of several plant viruses. *Virology*, **177**, 352–356.

Heinlein, M. (2002) Plasmodesmata: dynamic regulation and role in macromolecular cell-to-cell signaling. *Curr. Opin. Plant Biol.*, **5**, 543–552.

Heinlein, M., Epel, B.L., Padgett, H.S. & Beachy, R.N. (1995) Interaction of tobamovirus movement proteins with the plant cytoskeleton. *Science*, **270**, 1983–1985.

Heinlein, M., Padgett, H.S., Gens, J.S., Pickard, B.G., Casper, S.J., Epel, B.L. & Beachy, R.N. (1998) Changing patterns of localization of the tobacco mosaic virus movement protein and replicase to the endoplasmic reticulum and microtubules during infection. *Plant Cell*, **10**, 1107–1120.

Hilf, M.E. & Dawson, W.O. (1993) The tobamovirus capsid protein functions as a host-specific determinant of long distance movement. *Virology*, **193**, 106–114.

Hills, G.J., Plaskitt, K.A., Young, N.D., Dunigan, D.D., Watts, J.W., Wilson, T.M.A. & Zaitlin, M. (1987) Immunogold localization of the intracellular sites of structural and nonstructural tobacco mosaic virus proteins. *Virology*, **161**, 488–496.

Hirashima, K. & Watanabe, Y. (2001) Tobamovirus replicase coding region is involved in cell-to-cell movement. *J. Virol.*, **75**, 8831–8836.

Hiriart, J.B., Aro, E.M. & Lehto, K. (2003) Dynamics of the VIGS-mediated chimeric silencing of the *Nicotiana benthamiana ChlH* gene and of the *Tobacco mosaic virus* vector. *Mol. Plant Microbe Interact.*, **16**, 99–106.

Hodgman, T.C. (1988) A new superfamily of replicative proteins. *Nature*, **333**, 22–23.

Hughes, R.K., Perbal, M.C., Maule, A.J. & Hull, R. (1995) Evidence for proteolytic processing of tobacco mosaic virus movement protein in *Arabidopsis thaliana*. *Mol. Plant Microbe Interact.*, **8**, 658–665.

Hull, R. (2002) *Matthews' Plant Virology*. Fourth edition, Academic Press NY.

Hunter, T.R., Hunt, T., Knowland, J. & Zimmern, D. (1976) Messenger RNA for the coat protein of tobacco mosaic virus. *Nature*, **260**, 759–764.

Ishikawa, M., Meshi, T., Motoyoshi, F., Takamatsu, N. & Okada, Y. (1986) *In vitro* mutagenesis of the putative replicase genes of tobacco mosaic virus. *Nucleic Acids Res.*, **14**, 8291–8305.

Ishikawa, M., Meshi, T., Ohno, T. & Okada, Y. (1991a) Specific cessation of minus-strand RNA accumulation at an early stage of tobacco mosaic virus infection. *J. Virol.*, **65**, 861–868.

Ishikawa, M., Obata, F., Kumagai, T. & Ohno, T. (1991b) Isolation of mutants of *Arabidopsis thaliana* in which accumulation of tobacco mosaic virus coat protein is reduced to low levels. *Mol. Gen. Genet.*, **230**, 33–38.

Ishikawa, M., Naito, S. & Ohno, T. (1993) Effects of the tom1 mutation of *Arabidopsis thaliana* on the multiplication of tobacco mosaic virus RNA in protoplasts. *J. Virol.*, **67**, 5328–5338.

Kahn, T.W., Lapidot, M., Heinlein, M., Reichel, C., Cooper, B., Gafny, R. & Beachy, R.N. (1998) Domains of the TMV movement protein involved in subcellular localization. *Plant J.*, **15**, 15–25.

Karger, E.M., Frolova, O.Y., Fedorova, N.V., Baratova, L.A., Ovchinnikova, T.V., Susi, P., Makinen, K., Ronnstrand, L., Dorokhov, Y.L. & Atabekov, J.G. (2003) Dysfunctionality of a tobacco mosaic virus movement protein mutant mimicking threonine 104 phosphorylation. *J. Gen. Virol.*, **84**, 727–732.

Karpova, O.V., Ivanov, K.I., Rodionova, N.P., Dorokhov, Y.L. & Atabekov, J.G. (1997) Nontranslatability and dissimilar behavior in plants and protoplasts of viral RNA and movement protein complexes formed *in vitro*. *Virology*, **230**, 11–21.

Karpova, O.V., Rodionova, N.P., Ivanov, K.I., Kozlovsky, S.V., Dorokhov, Y.L. & Atabekov, J.G. (1999) Phosphorylation of tobacco mosaic virus movement protein abolishes its translation repressing ability. *Virology*, **261**, 20–24.

Kawakami, S., Padgett, H.S., Hosokawa, D., Okada, Y., Beachy, R.N. & Watanabe, Y. (1999) Phosphorylation and/or presence of serine 37 in the movement protein of tomato mosaic tobamovirus is essential for intracellular localization and stability *in vivo*. *J. Virol.*, **73**, 6831–6840.

Kawakami, S., Hori, K., Hosokawa, D., Okada, Y. & Watanabe, Y. (2003) Defective tobamovirus movement protein lacking wild-type phosphorylation sites can be complemented by substitutions found in revertants. *J. Virol.*, **77**, 1452–1461.

Keith, J. & Fraenkel-Conrat, H. (1975) Tobacco mosaic virus carries 5′-terminal triphosphorylated guanosine blocked by 5′-linked 7-methylguanosine. *FEBS Lett.*, **57**, 31–33.

Klug, A. (1999) The tobacco mosaic virus particle: structure and assembly. *Phil. Trans. R. Soc. Lond. B.*, **354**, 531–535.

Knorr, D.A. & Dawson, W.O. (1988) A point mutation in the tobacco mosaic virus capsid protein gene induces hypersensitivity in *Nicotiana sylvestris*. *Proc. Natl. Acad. Sci. USA*, **85**, 170–174.

Koonin, E.V. & Dolja, V.V. (1993) Evolution and taxonomy of positive-strand RNA viruses: implications of comparative analysis of amino acid sequences. *Crit. Rev. Biochem. Mol. Biol.*, **28**, 375–430.

Kozak, M. (2002) Pushing the limits of the scanning mechanism for initiation of translation. *Gene*, **299**, 1–34.

Lapidot, M., Gafny, R., Ding, B., Wolf, S., Lucas, W.J. & Beachy, R.N. (1993) A dysfunctional movement protein of tobacco mosaic virus that partially modifies the plasmodesmata and limits virus spread in transgenic plants. *Plant J.*, **4**, 959–970.

Lartey, R.T., Ghoshroy, S. & Citovsky, V. (1998) Identification of an *Arabidopsis thaliana* mutation (*vsm1*) that restricts systemic movement of tobamoviruses. *Mol. Plant Microbe Interact.*, **11**, 706–709.

Leathers, V., Tanguay, R., Kobayashi, M. & Gallie, D.R. (1993) A phylogenetically conserved sequence within viral 3′ untranslated RNA pseudoknots regulates translation. *Mol. Cell. Biol.*, **13**, 5331–5347.

Lehto, K., Tikkanen, M., Hiriart, J.-B., Paakkarinen, V. & Aro, E.-M. (2003) Depletion of the photosystem II core complex in mature tobacco leaves infected by the *flavum* strain of *Tobacco mosic virus. Mol. Plant Microbe Interact.*, **36**, 1135–1144.

Leisner, S.M. & Turgeon, R. (1993) Movement of virus and photoassimilate in the phloem – a comparative analysis. *Bioessays*, **15**, 741–748.

Leonard, D.A. & Zaitlin, M. (1982) A temperature-sensitive strain of tobacco mosaic virus defective in cell-to-cell movement generates an altered viral-coded protein. *Virology*, **117**, 416–424.

Lipardi, C., Wei, Q. & Paterson, B.M. (2001) RNAi as random degradative PCR: siRNA primers convert mRNA into dsRNAs that are degraded to generate new siRNAs. *Cell*, **107**, 297–307.

Liu, Y.L., Schiff, M., Marathe, R. & Dinesh-Kumar, S.P. (2002) Tobacco *Rar1, EDS1* and *NPR1/NIM1* like genes are required for *N*-mediated resistance to tobacco mosaic virus. *Plant J.*, **30**, 415–429.

Lucas, W.J., Olesinski, A., Hull, R.J., Haudenshield, J.S., Deom, C.M., Beachy, R.N. & Wolf, S. (1993) Influence of the tobacco mosaic virus 30 kDa movement protein on carbon metabolism and photosynthate partitioning in transgenic tobacco plants. *Planta*, **190**, 88–96.

Lucas, W.J., Bouchepillon, S., Jackson, D.P., Nguyen, L., Baker, L., Ding, B. & Hake, S. (1995) Selective trafficking of KNOTTED1 homeodomain protein and its messenger RNA through plasmodesmata. *Science*, **270**, 1980–1983.

Malamy, J., Carr, J.P., Klessig, D.F. & Raskin, I. (1990) Salicylic acid: a likely endogenous signal in the resistance response of tobacco to viral infection. *Science*, **250**, 1002–1004.

Marano, M.R. & Baulcombe, D. (1998) Pathogen-derived resistance targeted against the negative-strand RNA of tobacco mosaic virus: RNA strand-specific gene silencing? *Plant J.*, **13**, 537–546.

Marathe, R., Anandalakshmi, R., Liu, Y. & Dinesh-Kumar, S.P. (2002) The tobacco mosaic virus resistance gene, *N. Mol. Plant Pathol.*, **3**, 167–172.

Más, P. & Beachy, R.N. (1998) Distribution of TMV movement protein in single living protoplasts immobilized in agarose. *Plant J.*, **15**, 835–842.

Más, P. & Beachy, R.N. (1999) Replication of tobacco mosaic virus on endoplasmic reticulum and role of the cytoskeleton and virus movement protein in intracellular distribution of viral RNA. *J. Cell Biol.*, **147**, 945–958.

Matsushita, Y., Miyakawa, O., Deguchi, M., Nishiguchi, M. & Nyunoya, H. (2002) Cloning of a tobacco cDNA for a putative transcriptional coactivator MBF1 that interacts with the tomato mosaic virus movement protein. *J. Exp. Bot.*, **53**, 1531–1532.

Matsushita, Y., Ohshima, M., Yoshioka, K., Nishiguchi, M. & Nyunoya, H. (2003) The catalytic subunit of protein kinase CK2 phosphorylates in vitro the movement protein of *Tomato mosaic virus. J. Gen. Virol.*, **84**, 497–505.

McCormick, A.A., Kumagai, M.H., Hanley, K., Turpen, T.H., Hakim, I., Grill, L.K., Tuse, D., Levy, S. & Levy, R. (1999) Rapid production of specific vaccines for lymphoma by expression of the tumor-derived single-chain Fv epitopes in tobacco plants. *Proc. Natl. Acad. Sci. USA*, **96**, 703–708.

McLean, B.G., Zupan, J. & Zambryski, P.C. (1995) Tobacco mosaic virus movement protein associates with the cytoskeleton in tobacco cells. *Plant Cell*, **7**, 2101–2114.

Melcher, U. (2003) Turnip vein-clearing virus, from pathogen to host expression profile. *Molec. Plant Pathol.*, **4**, 133–140.

Meshi, T., Motoyoshi, F., Adachi, A., Watanabe, Y., Takamatsu, N. & Okada, Y. (1988) Two concomitant base substitutions in the putative replicase genes of tobacco mosaic virus confer the ability to overcome the effects of a tomato resistance gene, TM-1. *EMBO J.*, **7**, 1575–1581.

Meshi, T., Motoyoshi, F., Maeda, T., Yoshiwoka, S., Watanabe, H. & Okada, Y. (1989) Mutations in the tobacco mosaic virus 30-kD protein gene overcome TM-2 resistance in tomato. *Plant Cell*, **1**, 515–522.

Métraux, J.-P., Signer, H., Ryals, J., Ward, E., Wyssbenz, M., Gaudin, J., Raschdorf, K., Schmid, E., Blum, W. & Inverardi, B. (1990) Increase in salicylic-acid at the onset of systemic acquired-resistance in cucumber. *Science*, **250**, 1004–1006.

Micheli, F. (2001) Pectin methylesterases: cell wall enzymes with important roles in plant physiology. *Trends Plant Sci.*, **6**, 414–419.

Morozov, S.Y., Denisenko, O.N., Zelenina, D.A., Fedorkin, O.N., Solovyev, A.G., Maiss, E., Casper, R. & Atabekov, J.G. (1993) A novel open reading frame in tobacco mosaic virus genome coding for a putative small, positively charged protein. *Biochimie*, **75**, 659–665.

Morrain, P., Béclin, C., Elmayan, T., Feuerbach, F., Godon, C., Morel, J.-B., Jouette, D., Lacombe, A.-M., Nikic, S., Picault, N., Remoue, K., Sanial, M., Vo, T.-A. & Vaucheret, H. (2000) Arabidopsis *SGS2* and *SGS3* genes are required for posttranscriptional gene silencing and natural virus resistance. *Cell*, **101**, 533–542.

Murphy, A.M. & Carr, J.P. (2002) Salicylic acid has cell-specific effects on *Tobacco mosaic virus* replication and cell-to-cell movement. *Plant Physiol.*, **128**, 552–563.

Murphy, A.M., Chivasa, S., Singh, D.P. & Carr, J.P. (1999) Salicylic acid-induced resistance to viruses and other pathogens: a parting of the ways? *Trends Plant Sci.*, **4**, 155–160.

Murphy, A.M., Gilliland, A., Wong, C.E., West, J., Singh, D.P. & Carr, J.P. (2001) Induced resistance to viruses. *Eur. J. Plant Pathol.*, **107**, 121–128.

Naylor, M., Murphy, A.M., Berry, J.O. & Carr, J.P. (1998) Salicylic acid can induce resistance to plant virus movement. *Mol. Plant Microbe Interact.*, **11**, 860–868.

Nelson, R.S. & van Bel, A.J.E. (1997) The mystery of virus trafficking into, through and out of the vascular tissue. *Prog. Bot.*, **59**, 476–533.

Nelson, R.S., Powell-Abel, P. & Beachy, R.N. (1987) Lesions and virus accumulation in inoculated transgenic tobacco plants expressing the coat protein gene of tobacco mosaic virus. *Virology*, **158**, 126–132.

Ohno, T., Takamatsu, N., Meshi, T., Okada, Y., Nishiguchi, M. & Kiho, Y. (1983) Single amino-acid substitution in 30 K protein of TMV defective in virus transport function. *Virology*, **131**, 255–258.

Ohshima, K., Taniyama, T., Yamanaka, T., Ishikawa, M. & Naito, S. (1998) Isolation of a mutant of *Arabidopsis thaliana* carrying two simultaneous mutations affecting tobacco mosaic virus multiplication within a single cell. *Virology*, **243**, 472–481.

Okada, Y. (1999) Historical overview of research on the tobacco mosaic virus genome: genome organization, infectivity and gene manipulation. *Phil. Trans. R. Soc. Lond. B.*, **354**, 569–582.

Olesinski, A.A., Almon, E., Navot, N., Perl, A., Galun, E., Lucas, W.J. & Wolf, S. (1996) Tissue-specific expression of the tobacco mosaic virus movement protein in transgenic potato plants alters plasmodesmal function and carbohydrate partitioning. *Plant Physiol.*, **111**, 541–550.

Oparka, K.J., Boevink, P. & Santa Cruz, S. (1996) Studying the movement of plant viruses using green fluorescent protein. *Trends Plant Sci.*, **1**, 412–418.

Oparka, K.J., Prior, D.A.M., Santa Cruz, S., Padgett, H.S. & Beachy, R.N. (1997) Gating of epidermal plasmodesmata is restricted to the leading edge of expanding infection sites of tobacco mosaic virus (TMV). *Plant J.*, **12**, 781–789.

Oparka, K.J., Roberts, A.G., Boevink, P., Santa Cruz, S., Roberts, L., Pradel, K.S., Imlau, A., Kotlizky, G., Sauer, N. & Epel, B. (1999) Simple, but not branched, plasmodesmata allow the nonspecific trafficking of proteins in developing tobacco leaves. *Cell*, **97**, 743–754.

Osman, T.A.M. & Buck, K.W. (1996) Complete replication in vitro of tobacco mosaic virus RNA by a template-dependent, membrane-bound RNA polymerase. *J. Virol.*, **70**, 6227–6234.

Osman, T.A.M. & Buck, K.W. (1997) The tobacco mosaic virus RNA polymerase complex contains a plant protein related to the RNA-binding subunit of yeast eIF-3. *J. Virol.*, **71**, 6075–6082.

Padgett, H.S., Watanabe, Y. & Beachy, R.N. (1997) Identification of the TMV replicase sequence that activates the *N* gene-mediated hypersensitive response. *Mol. Plant Microbe Interact.*, **10**, 709–715.

Palukaitis, P. & Zaitlin, M. (1997) Replicase-mediated resistance to plant virus disease. *Adv. Virus Res.*, **48**, 349–377.

Palukaitis, P., Garcia-Arenal, F., Sulzinski, M. & Zaitlin, M. (1983) Replication of tobacco mosaic virus. VII. Further characterization of single- and double-stranded virus-related RNAs from TMV-infected plants. *Virology*, **131**, 533–545.

Peart, J.R., Cook, G., Feys, B.J., Parker, J.E. & Baulcombe, D.C. (2002) An *EDS1* orthologue is required for *N*-mediated resistance against tobacco mosaic virus. *Plant J.*, **29**, 569–579.

Pelham, H.R.B. (1978) Leaky UAG termination codon in tobacco mosaic virus RNA. *Nature*, **272**, 469–471.

Pogue, G.P., Lindbo, J.A., Garger, S.J. & Fitzmaurice, W.P. (2002) Making an ally from an enemy: plant virology and the new agriculture. *Annu. Rev. Phytopathol.*, **40**, 45–74.

Powell-Abel, P., Nelson, R.S., De, B., Hoffmann, N., Rogers, S.G., Fraley, R.T. & Beachy, R.N. (1986) Delay of disease development in transgenic plants that express the tobacco mosaic virus coat protein gene. *Science*, **232**, 738–743.

Reichel, C. & Beachy, R.N. (1998) Tobacco mosaic virus infection induces severe morphological changes of the endoplasmic reticulum. *Proc. Natl. Acad. Sci. USA*, **95**, 11169–11174.

Reichel, C. & Beachy, R.N. (2000) Degradation of tobacco mosaic virus movement protein by the 26S proteasome. *J. Virol.*, **74**, 3330–3337.

Reichel, C., Más, P. & Beachy, R.N. (1999) The role of the ER and cytoskeleton in plant viral trafficking. *Trends Plant Sci.*, **4**, 458–462.

Reinero, A. & Beachy, R.N. (1986) Association of TMV coat protein with chloroplast membranes in virus-infected leaves. *Plant Mol. Biol.*, **6**, 291–301.

Reinero, A. & Beachy, R.N. (1989) Reduced photosystem II activity and accumulation of viral coat protein in chloroplasts of leaves infected with tobacco mosaic virus. *Plant Physiol.*, **89**, 111–116.

Roberts, A.G. & Oparka, K.J. (2003) Plasmodesmata and the control of symplastic transport. *Plant Cell Environ.*, **26**, 103–124.

Roberts, I.M., Boevink, P., Roberts, A.G., Sauer, N., Reichel, C. & Oparka, K.J. (2001) Dynamic changes in the frequency and architecture of plasmodesmata during the sink–source transition in tobacco leaves. *Protoplasma*, **218**, 31–44.

Ross, A.F. (1961a) Localized acquired resistance to plant virus infection in hypersensitive hosts. *Virology*, **14**, 329–339.

Ross, A.F. (1961b) Systemic acquired resistance induced by localized virus infections in plants. *Virology*, **14**, 340–358.

Ryabov, E.V., Robinson, D.J. & Taliansky, M. (2001) Umbravirus-encoded proteins both stabilize heterologous viral RNA and mediate its systemic movement in some plant species. *Virology*, **288**, 391–400.

Saito, T., Hosokawa, D., Meshi, T. & Okada, Y. (1987) Immunocytochemical localization of the 130K and 180K proteins (putative replicase components) of tobacco mosaic virus. *Virology*, **160**, 477–481.

Schneider, W.L. & Roossinck, M.J. (2000) Evolutionarily related sindbis-like plant viruses maintain different levels of population diversity in a common host. *J. Virol.*, **74**, 3130–3134.

Schoelz, J.E. & Zaitlin, M. (1989) Tobacco mosaic virus RNA enters chloroplasts *in vivo*. *Proc. Natl. Acad. Sci. USA*, **86**, 4496–4500.

Scholthof, K.-B.G., Shaw, J.G. & Zaitlin, M. (eds) (1999) *One Hundred Years of Contributions to Virology*, APS Press, St. Paul MN.

Shalitin, D. & Wolf, S. (2000) Interaction between phloem proteins and viral movement proteins. *Aust. J. Plant Physiol.*, **27**, 801–806.

Shaw, J.G. (1999) Tobacco mosaic virus and the study of early events in virus infections. *Proc. Roy. Soc. Lond. B.*, **354**, 603–611.

Siegel, A., Zaitlin, M. & Sehgal, O.P. (1962) The isolation of defective tobacco mosaic virus strains. *Proc. Natl. Acad. Sci. USA*, **48**, 1845–1851.

Simon-Buela, L. & Garcia-Arenal, F. (1999) Virus particles of cucumber green mottle mosaic tobamovirus move systemically in the phloem of infected cucumber plants. *Mol. Plant Microbe Interact.*, **12**, 112–118.

Skuzeski, J.M., Nichols, L.M., Gesteland, R.F. & Atkins, J.F. (1991) The signal for a leaky UAG stop codon in several plant-viruses includes the two downstream codons. *J. Mol. Biol.*, **218**, 365–373.

Sleat, D.E., Turner, P.C., Finch, J.T., Butler, P.J.G. & Wilson, T.M.A. (1986) Packaging of recombinant RNA molecules into pseudovirus particles directed by the origin of assembly sequence from tobacco mosaic virus RNA. *Virology*, **155**, 299–308.

Spitsin, S., Steplewski, K., Fleysh, N., Belanger, H., Mikheeva, T., Shivprasad, S., Dawson, W., Koprowski, H. & Yusibov, V. (1999) Expression of alfalfa mosaic virus coat protein in tobacco mosaic virus (TMV) deficient in the production of its native coat protein supports long-distance movement of a chimeric TMV. *Proc. Natl. Acad. Sci. USA*, **96**, 2549–2553.

Sulzinski, M.A., Gabard, K.A., Palukaitis, P. & Zaitlin, M. (1985) Replication of tobacco mosaic virus. VIII. Characterization of a third subgenomic TMV RNA. *Virology*, **145**, 132–140.

Susi, P., Pehu, E. and Lehto, K. (1999) Replication in the phloem is not necessary for efficient vascular transport of tobacco mosaic tobamovirus. *FEBS Lett.*, **447**, 121–123.

Taliansky, M., Roberts, I.M., Kalinina, N., Ryabov, E.V., Raj, S.K., Robinson, D.J. & Oparka, K.J. (2003) An umbraviral protein, involved in long-distance RNA movement, binds viral RNA and forms unique, protective ribonucleoprotein complexes. *J. Virol.*, **77**, 3031–3040.

Taraporewala, Z.F. & Culver, J.N. (1996) Identification of an elicitor active site within the three-dimensional structure of the tobacco mosaic tobamovirus coat protein. *Plant Cell*, **8**, 169–178.

Taraporewala, Z.F. & Culver, J.N. (1997) Structural and functional conservation of the tobamovirus coat protein elicitor active site. *Mol. Plant Microbe Interact.*, **10**, 597–604.

Taylor, D.N. & Carr, J.P. (2000) The GCD10 subunit of yeast eIF-3 binds to the methyltransferase-like domain of the 126- and 183-kDa replicase proteins of tobacco mosaic virus in the yeast two-hybrid system. *J. Gen. Virol.*, **81**, 1557–1561.

Tenllado, F., Garcia-Luque, I., Serra, M.T. & Diaz Ruiz, J.R. (1995) *Nicotiana benthamiana* plants transformed with the 54-kDa region of the pepper mild mottle tobamovirus replicase gene exhibit two types of resistance responses against viral-infection. *Virology*, **211**, 170–183.

Tenllado, F., Garcia-Luque, I., Serra, M.T. & Diaz-Ruiz, J.R. (1996) Resistance to pepper mild mottle tobamovirus conferred by the 54-kDa gene sequence in transgenic plants does not require expression of the wild-type 54-kDa protein. *Virology*, **219**, 330–335.

Toth, R.L., Pogue, G.P. & Chapman, S. (2002) Improvement of the movement and host range properties of a plant virus vector through DNA shuffling. *Plant J.*, **30**, 593–600.

Traynor, P., Young, B.M. & Ahlquist, P. (1991) Deletion analysis of brome mosaic virus 2a protein – effects on RNA replication and systemic spread. *J. Virol.*, **65**, 2807–2815.

Tsujimoto, Y., Numaga, T., Ohshima, K., Yano, M., Ohsawa, R., Goto, D.B., Niato, S. & Ishikawa, M. (2003) Arabidopsis *Tobamovirus multiplication (TOM) 2* locus encodes a transmembrane protein that interacts with TOM1. *EMBO J.*, **22**, 335–343.

Turpen, T.H. (1999) Tobacco mosaic virus and the virescence of biotechnology. *Phil. Trans. R. Soc. Lond. B.*, **354**, 665–673.

Ueki, S. & Citovsky, V. (2002) The systemic movement of a tobamovirus is inhibited by a cadmium-ion-induced glycine-rich protein. *Nature Cell Biology*, **4**, 478–485.

Valentine, T.A., Roberts, I.M. & Oparka, K.J. (2002) Inhibition of tobacco mosaic virus replication in lateral roots is dependent on an activated meristem-derived signal. *Protoplasma*, **219**, 184–196.

van Bel, A.J.E. (2003) The phloem, a miracle of ingenuity. *Plant, Cell Environ.*, **26**, 125–149.

van den Boogaart, T., Lomonossoff, G.P. & Davies, J.W. (1998) Can we explain RNA-mediated virus resistance by homology-dependent gene silencing? *Mol. Plant Microbe Interact.*, **11**, 717–723.

Voinnet, O. (2001) RNA silencing as a plant immune system against viruses. *Trends Genet.*, **17**, 449–459.

Voinnet, O., Pinto, Y.M. & Baulcombe, D.C. (1999) Suppression of gene silencing: a general strategy used by diverse DNA and RNA viruses of plants. *Proc. Natl. Acad. Sci. USA*, **96**, 14147–14152.

Waigmann, E. & Zambryski, P. (1995) Tobacco mosaic virus movement protein-mediated protein transport between trichome cells. *Plant Cell*, **7**, 2069–2079.

Waigmann, E., Chen, M.H., Bachmaier, R., Ghoshroy, S. & Citovsky, V. (2000) Regulation of plasmodesmal transport by phosphorylation of tobacco mosaic virus cell-to-cell movement protein. *EMBO J.*, **19**, 4875–4884.

Watanabe, T., Honda, A., Iwata, I., Ueda, S., Hibi, T. & Ishihama, A. (1999) Isolation from tobacco mosaic virus-infected tobacco of a solubilized template-specific RNA-dependent RNA polymerase containing a 126K/183K protein heterodimer. *J. Virol.*, **73**, 2633–2640.

Watanabe, Y., Ogawa, T. & Okada, Y. (1992) *In vivo* phosphorylation of the 30-kda protein of tobacco mosaic virus. *FEBS Lett.*, **313**, 181–184.

Waterhouse, P.M., Smith, N.A. & Wang, M.-B. (1999) Virus resistance and gene silencing: killing the messenger. *Trends Plant Sci.*, **4**, 452–457.

Waterhouse, P.M., Wang, M.-B. & Lough, T. (2001) Gene silencing as an adaptive defence against viruses. *Nature*, **411**, 834–842.

Weber, H. & Pfitzner, A.J.P. (1998) Tm-2^2 resistance in tomato requires recognition of the carboxy terminus of the movement protein of tomato mosaic virus. *Mol. Plant Microbe Interact.*, **11**, 498–503.

Weber, H., Schultze, S. & Pfitzner, A.J.P. (1993) Two amino acid substitutions in the tomato mosaic virus 30-kilodalton movement protein confer the ability to overcome the Tm-2^2 resistance gene in the tomato. *J. Virol.*, **67**, 6432–6438.

Weissman, A.M. (2001) Themes and variations on ubiquitylation. *Nature Rev. Mol. Cell Biol.*, **2**, 169–178.

Weststeijn, E.A. (1981) Lesion growth and virus localization in leaves of *Nicotiana tabacum* cv. Xanthi nc. After inoculation with tobacco mosaic virus and incubation alternately at 22°C and 32°C. *Physiol. Plant Pathol.*, **18**, 357–368.

Whitham, S., Dinesh-Kumar, S.P., Choi, D., Hehl, R., Corr, C. & Baker, B. (1994) The product of the tobacco mosaic virus resistance gene *N*: similarity to toll and the interleukin-1 receptor. *Cell*, **78**, 1101–1115.

Whitham, S., McCormick, S. & Baker, B. (1996) The *N* gene of tobacco confers resistance to tobacco mosaic virus in transgenic tomato. *Proc. Natl. Acad. Sci. USA*, **93**, 8776–8781.

Wilson, T.M.A. (1984) Cotranslational disassembly of tobacco mosaic virus *in vitro*. *Virology*, **137**, 255–265.

Wilson, T.M.A., Perham, R.N. & Butler, P.J.G. (1978) Intermediates in the disassembly of tobacco mosaic virus at alkaline pH. Infectivity, self-assembly and translational activities. *Virology*, **89**, 475–483.

Wolf, S., Deom, C.M., Beachy, R.N. & Lucas, W.J. (1989) Movement protein of tobacco mosaic virus modifies plasmodesmatal size exclusion limit. *Science*, **246**, 377–379.

Wong, C.E., Carson, R.A.J. & Carr, J.P. (2002) Chemically-induced virus resistance in *Arabidopsis thaliana* is independent of pathogenesis-related protein expression and the *NPR1* gene. *Mol. Plant Microbe Interac.*, **15**, 75–81.

Wright, K.M., Duncan, G.H., Pradel, K.S., Carr, F., Wood, S., Oparka, K.J. & Santa Cruz, S. (2000) Analysis of the *N* gene hypersensitive response induced by a fluorescently tagged tobacco mosaic virus. *Plant Physiol.*, **123**, 1375–1385.

Wu, X. & Shaw, J.G. (1997) Evidence that a replicase protein is involved in the disassembly of tobacco mosaic virus particles *in vivo*. *Virology*, **239**, 426–434.

Wu, X., Xu, Z. & Shaw, J.G. (1994) Uncoating of tobacco mosaic virus RNA in protoplasts. *Virology*, **200**, 256–262.

Wu, X., Weigel, D. & Wigge, P.A. (2002) Signaling in plants by intercellular RNA and protein movement. *Genes & Dev.*, **16**, 151–158.

Xiao, S.Y., Charoenwattana, P., Holcombe, L. & Turner, J.G. (2003) The *Arabidopsis* genes RPW8.1 and RPW8.2 confer induced resistance to powdery mildew diseases in tobacco. *Mol. Plant Microbe Interact.*, **16**, 289–294.

Xie, Z.X., Fan, B.F., Chen, C.H. & Chen, Z.X. (2001) An important role of an inducible RNA-dependent RNA polymerase in plant antiviral defense. *Proc. Natl. Acad. Sci.*, **98**, 6516–6521.

Yamanaka, T., Ohta, T., Takahashi, M., Meshi, T., Schmidt, R., Dean, C., Naito, S. & Ishikawa, M. (2000) *TOM1*, an *Arabidopsis* gene required for efficient multiplication of a tobamovirus, encodes a putative transmembrane protein. *Proc. Natl. Acad. Sci. USA*, **97**, 10107–10112.

Yamanaka, T., Imai, T., Satoh, R., Kawashima, A., Takahashi, M., Tomita, K., Kubota, K., Meshi, T., Naito, S. & Ishikawa, M. (2002) Complete inhibition of tobamovirus multiplication by simultaneous mutations in two homologous host genes. *J. Virol.*, **76**, 2491–2497.

Zaitlin, M. (1999) Elucidation of the genome organization of tobacco mosaic virus. *Phil. Trans. R. Soc. Lond. B.*, **354**, 587–591.

Zaitlin, M. (2000) *Tobacco mosaic virus. AAB descriptions of plant viruses*, **No. 370** (eds M.J. Adams, J.F. Antoniw, H. Barker, A.T. Jones, A.F. Murant & D. Robinson). AAB descriptions of plant viruses on CD-ROM and On-line. Wellesbourne, Warwick, U.K. Association of Applied Biologists (available online http://www3.res.bbsrc.ac.uk/webdpv/web/fdpv.asp?dpvnum = 370).

Zeenko, V.V., Ryabova, L.A., Spirin, A.S., Rothnie, H.M., Hess, D., Browning, K.S. & Hohn, T. (2002) Eukaryotic elongation factor 1A interacts with the upstream pseudoknot domain in the 3' untranslated region of tobacco mosaic virus RNA. *J. Virol.*, **76**, 5678–5691.

Zimmern, D. (1975) The 5' end group of tobacco mosaic virus is $m^7G^{5'}ppp^{5'}Gp$. *Nucl. Acids Res.*, **2**, 1189–1201.

Zimmern, D. (1977) Nucleotide sequence at the origin for assembly on tobacco mosaic virus RNA. *Cell*, **11**, 463–482.

Zimmern, D. & Wilson, T.M.A. (1976) Location of the origin for viral reassembly on tobacco mosaic virus RNA and its relation to stable fragment. *FEBS Lett.*, **71**, 294–298.

3 Infection with potyviruses

Minna-Liisa Rajamäki, Tuula Mäki-Valkama,
Kristiina Mäkinen and Jari P.T. Valkonen

3.1 Infection cycle (general summary)

Potyviruses have a monopartite, single-stranded, positive-sense RNA genome of
around 9500–10 000 nucleotides. It is encapsidated by a single species of virus-
encoded coat protein (CP) to a filamentous virion. The mature virus is termed the
virion (Caspar *et al.*, 1962) or *virus particle* (Matthews, 1991); the term *virion* is
used here. Potyviruses belong to the genus *Potyvirus* (family Potyviridae) and
represent around 30% of the known plant viruses (Van Regenmortel *et al.*, 2000).
The type member of the genus is *Potato virus Y* (PVY) (De Bokx & Huttinga, 1981;
Glais *et al.*, 2002a).

The main steps of the potyviral infection cycle are illustrated in Fig. 3.1. (1) Entry
of potyviral virions into plant cells, and the subsequent local infection, takes place
most typically with the help of viruliferous vector aphids that probe on plant cells
prior to actual feeding, releasing virions from the stylet. (2) Following disassembly
of virions in cytoplasm, (3) the viral genome is translated into a polyprotein that is
subsequently processed by three viral proteinases to result in up to ten mature proteins.
Most, if not all, of them are multifunctional. These proteins (or polyprotein processing
intermediates), viral RNA and putative host factors assemble to a putative replication
complex to (4) carry out negative-strand RNA synthesis. The (−) strands of the viral
RNA, in turn, (5) are used as templates for synthesis of new copies of the viral genome.
This process is known as viral replication and takes place probably in the endoplasmic reticulum (ER). Subsequently, (6) new virus genomes are intracellularly transported to plasmodesmata, the structures connecting cells separated from each other
by the cell wall. (7) Cell-to-cell movement of viral genomes through plasmodesmata
is facilitated by many potyviral proteins. To be transported over long distances in
the plant, potyviruses must invade the phloem and be loaded into sieve elements (SE).
(8) Long-distance transport distributes viruses within the plant according to the
transport route of photoassimilates that is governed by the *source–sink relationship*
between the maturing and growing tissues. In sink tissues, (9) the virus is unloaded
from SE to phloem cells, followed by replication and cell-to-cell movement. In the
systemically infected sink tissues, the virus is distributed more uniformly than in
the originally infected leaves, (10) making the systemically infected tissues good
sources for acquisition of the virus by its vector aphids.

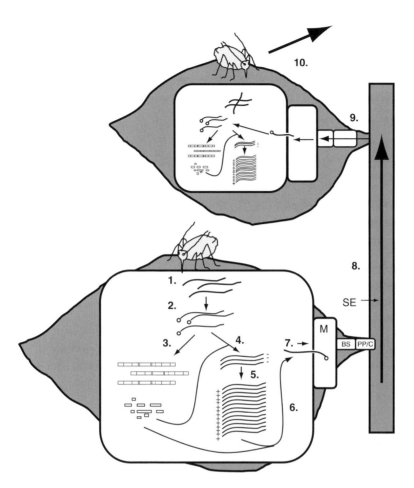

Figure 3.1 The main events in the potyvirus infection cycle: (1) Inoculation; entry into the cell in the stylet of an aphid vector, (2) dissassembly of virions, (3) translation and processing of the polyprotein, (4) synthesis of complementary minus-strand RNA, (5) synthesis of new genomic RNA strands, (6) intracellular transport of progeny viruses to plasmodesmata, (7) cell-to-cell movement and invasion of vascular tissue, (8) entry into and long-distance transport in sieve elements (SE), (9) unloading from SE and initiation of systemic infection, and (10) acquisition and transmission by aphid vectors. For more detailed description, see text. The following types of cells are indicated: M, mesophyll cell; BS, bundle sheath cell; PP, phloem parenchyma cell; C, companion cell.

The earlier studies on potyviruses were thoroughly reviewed by Shukla *et al.* (1994). Therefore, direct citations from older literature are largely excluded from this work. Furthermore, this chapter aims to review current knowledge of potyviral infection biology, which is based largely on more recent studies using infectious cDNA clones of potyviral genomes. We apologise that comprehensive citation of even the more recent potyvirus literature will not be possible within the limited space.

3.2 Architecture of virions

Flexuous, filamentous virions of potyviruses are 680–900 nm long and 11–15 nm in diameter (Fig. 3.2). Physico-chemical properties of virions have been reviewed in *Descriptions of Plant Viruses* published by the Commonwealth Mycological Institute and Association of Applied Biologists, UK. A noteworthy feature of potyvirus virions is their tendency to aggregate during virus purification, which requires special attention in the design of the purification procedure (e.g. for PVY, see De Bokx & Huttinga, 1981).

The virion is composed of about 2000 copies of a single type of CP that encapsidates a single viral RNA molecule (ssRNA). A viral genome-linked protein (VPg) is covalently bound to the 5′-end of viral RNA (Murphy *et al.*, 1991), providing polarity to the virion (Puustinen *et al.*, 2002). The potyviral CP has a core domain that is trypsin-insensitive, whereas both the N- and the C-termini of CP may be exposed on the virion surface (Shukla *et al.*, 1988). However, this was not found in a recent study in which the CP secondary structure was modelled based on tritium bombardment of the virions, augmented with theoretical predictions of protein topology. The analyses revealed eight putative α-helices and seven β-sheets, and a surface-exposed, partially unstructured N-terminus. According to the model, the C-terminal region forming a putative RNA-binding domain was not exposed. Furthermore, a central region was found to comprise a bundle of four putative α-helices that possibly play a role in the inter-subunit interactions (Baratova *et al.*, 2001).

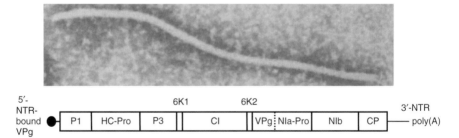

Figure 3.2 Micrograph of a virion, a schematic genetic map of a potyvirus, and (next page) the main functions of potyviral proteins and the genomic non-translated regions (NTR). The functions attributable to a specific protein are mentioned, rather than functions shown for a combination of several proteins. Please note that potyviral proteins are multifunctional and also the polyprotein processing intermediates may have functions. *All proteins and NTRs are essential for virus propagation* (Shukla *et al.*, 1994; Kekarainen *et al.*, 2002). Furthermore, *all main proteins except P3 bind RNA* (Merits *et al.*, 1998). These functions are not indicated per protein, neither are their roles as avirulence determinants in specific potyvirus–host combinations. The proteinase cleavage sites in the polyprotein are indicated by vertical lines, whereas the suboptimal cleavage site between the VPg and NIa-Pro domains is indicated by a dotted line. P1 and HC-Pro process themselves from the polyprotein, whereas other cleavage sites are processed by NIa-Pro. Horizontal lines at both ends indicate the NTRs. 3′-NTR terminates with a poly(A) tail, whereas VPg is covalently attached to the 5′-end of the viral genome.

	Name and function*	Reference
5′-NTR	**5′-Non-translated region**	
	Enhancement of translation	Carrington & Freed (1990)
P1	**P1 protein**	
	Proteinase	Verchot *et al.* (1991)
	Accessory factor for suppression of RNA silencing	Pruss *et al.* (1997)
HC-Pro	**Helper component proteinase**	
	Proteinase	Carrington *et al.* (1989)
	Cell-to-cell movement	Rojas *et al.* (1997)
	Vascular movement	Cronin *et al.* (1995)
	Suppression of RNA silencing	Anandalakshmi *et al.* (1998); Brigneti *et al.* (1998); Kasschau & Carrington (1998)
	Aphid transmission	Govier & Kassanis (1974)
	Interaction with a calmodulin-like protein	Anandalakshmi *et al.* (2000)
	Interaction with ring-finger protein and HIP2	Guo *et al.* (2003)
P3	**P3 protein**	
6K1	**6-kDa protein 1**	
CI	**Cylindrical inclusion**	
	Helicase	Laín *et al.* (1990)
	Cell-to-cell movement	Carrington *et al.* (1998)
6K2	**6-kDa protein 2**	
	Vascular movement	Rajamäki & Valkonen (1999)
VPg	**Viral genome-linked protein (VPg domain of NIa)**	
	Cell-to-cell movement	Nicolas *et al.* (1997)
	Vascular movement	Schaad *et al.* (1997)
	Interaction with eIF4E and eIF(iso)4E	Wittmann *et al.* (1997)
	Interaction with PVIP	Dunoyer *et al.* (2004)
NIa-Pro	**Nuclear inclusion protein a, proteinase domain**	
	Proteinase	Carrington & Dougherty (1987)
NIb	**Nuclear inclusion protein b**	
	RNA-dependent RNA polymerase	Hong & Hunt (1996)
	Interaction with poly(A) binding protein (PABP)	Wang *et al.* (2000)
CP	**Coat protein**	
	Encapsidation of viral RNA	McDonald & Bancroft (1977)
	Cell-to-cell movement	Dolja *et al.* (1994)
	Vascular movement	Dolja *et al.* (1995)
	Aphid transmission	Atreya *et al.* (1990)
3′-NTR	**3′-Non-translated region**	
	Symptom induction	Rodríguez-Cerezo *et al.* (1991)

*Virus propagation, symptom induction and RNA-binding functions are not indicated. See the text.

The helix pitch of the virion is 3.4–3.5 nm with 7–9 CP subunits per ring. To date, there is no information about the mutual organisation of the CP subunits in the virion or their contacts with viral RNA. However, the estimated number of nucleotides associated with a CP unit is six (structural models reviewed in Shukla *et al.*,

1994). High-resolution data based on diffraction techniques are hard to obtain due to the flexuous nature of the potyviral virion.

Phosphorylation of the potyviral CP is found *in vitro* and *in vivo* (Ivanov *et al.*, 2001, 2003). Studies using anti-phosphoserine and anti-phosphothreonine antibodies suggest that the CP units assembled to particles are phosphorylated (Fernandez-Fernandez *et al.*, 2002). Detection of serine/threonine-linked N-acetylglucosamine indicates another type of posttranslational modification of the CP (Fernández-Fernández *et al.*, 2002). These events may play a role in regulation of the putative functions of potyviral CP in the virion.

3.3   Entry of virions into the cell, disassembly and assembly

Potyviral virions typically enter plant cells via wounds on cell walls made by the feeding vector insects (aphids) (see Section 3.8). Disassembly (uncoating) of virions and release of the genomic RNA in cytoplasm are critical events for initiation of infection, but little is known about them in potyviruses. They precede translation of the viral proteins, or translation starts before the virion is entirely uncoated (see Section 3.4).

Recombinant CP expressed in bacteria readily self-polymerises to form virus-like particles (VLP) in the absence of viral RNA. A few amino acids within the CP core region are essential for VLP formation (Jagadish *et al.*, 1993). Mixing viral RNA with CP in a solution under low ionic conditions and pH 6.0–9.0 results in the formation of VLPs that, however, are non-infectious (McDonald & Bancroft, 1977). Assembly of the true, infectious virions has been observed only in infected cells. The origin-of-assembly may be the 5′-terminal region of the newly synthesised viral RNA molecule (Wu & Shaw, 1998). Virion formation is probably necessary to protect viral RNA during transmission by the vector.

3.4   Translation and polyprotein processing

3.4.1   Translation initiation

The potyviral ssRNA genome contains a 5′-terminal, non-translated region (5′-NTR) followed by a single open reading frame (ORF) (Riechmann *et al.*, 1991). It acts as an mRNA and is translated directly following release from the virion. Eukaryotic mRNA involves a $^{7Me}G^{5'}ppp^{5'}Np$ cap structure at the 5′-end, whereas the poly(A) tail at the 3′ end is associated with a poly(A) binding protein (PABP), which may interact with the cap structure (Dever, 1999). In contrast, potyviral RNA lacks the 5′-cap and is translated cap-independently. The 5′-leader sequence (i.e. 5′-NTR) directs efficient translation in co-operation with the 3′-terminal poly(A) tail (Gallie *et al.*, 1995). The 5′-leader sequence contains internal ribosome entry site (IRES) activity (Basso *et al.*, 1994) and cap-independent translation regulatory elements (CIREs) (Niepel & Gallie, 1999), of which IRES directs cap-independent translation through a mechanism that involves the eukaryotic translation initiation factor 4G (eIF4G) (Gallie, 2001).

VPg is covalently linked to the viral RNA 5′-end, but is not required for translation of potyviral polyprotein *in vitro*. It is therefore puzzling that VPg interacts with eIF4E and/or eIF(iso)4E (Wittmann *et al.*, 1997; Schaad *et al.*, 2000) and that the interaction is required for infectivity (Leonard *et al.*, 2000; Lellis *et al.*, 2002). Also, the VPg–RNA linkage is required for viral infectivity (Murphy *et al.*, 1996). Therefore, it seems that VPg may be involved in translation. Perhaps the covalent linkage of VPg with the viral RNA 5′-end, its possible association with the 3′-end of viral RNA (see Section 3.5), interaction with eIF4E4/eIF(iso)4E4 and its self-interaction (Oruetxebarria *et al.*, 2001) contribute to assembly of the translation complex in a manner that is not yet fully elucidated.

3.4.2   Polyprotein processing

The single ORF of potyviruses encodes a polyprotein of about 3000–3350 amino acids. The polyprotein is processed to ten mature proteins by three virus-encoded proteinases (reviewed in Riechmann *et al.*, 1992). For names and functions of the mature proteins, see Fig. 3.2. Two of the viral proteinases are encoded by the 5′-proximal part of the virus genome. The chymotrypsin-like serine proteinase domain of the first protein (P1) and the papain-like cysteine proteinase domain at the C-terminus of the helper component proteinase (HC-Pro, the second protein) cleave the polyprotein *in cis* at their respective C-termini. The amino acid residues at the proteolytic cleavage site for P1 show some variability among potyviruses, whereas HC-Pro cleaves in between a Gly-Gly dipeptide flanked by amino acids that are conserved in potyviruses (Shukla *et al.*, 1994).

The third proteinase (NIa-Pro) comprises the C-proximal part of the nuclear inclusion protein a (NIa). NIa-Pro is a picornavirus 3C-like proteinase that carries out the proteolytic processing events at seven sites in polyprotein, namely the junctions P3/6K1, 6K1/CI, CI/6K2, 6K2/VPg, VPg/NIa-Pro, NIa-Pro/NIb and NIb/CP (Fig. 3.2). In most cases, NIa-Pro cleaves between Gln and either Ser, Gly or Ala (Shukla *et al.*, 1994). It mediates both *cis*- and *trans*-cleavages, but the *cis*-cleavages may be preferred (Merits *et al.*, 2002). Processing is regulated by a conserved heptapeptide sequence flanking the cleavage site. While most proteolytic sites are quickly processed, the P3/6K1, CI/6K2 and VPg/NIa-Pro junctions may be processed at a slower rate (Schaad *et al.*, 1996; Merits *et al.*, 2002). Efficient genome amplification requires slow processing of the suboptimal cleavage site Glu/Gly between the VPg and NIa-Pro domains of NIa (Schaad *et al.*, 1996). Timing of polyprotein cleavages during the infection cycle may provide a mechanism by which the activities of viral proteins can be regulated in a desirable manner.

The potyviral proteins are produced in equimolar amounts due to the polyprotein expression strategy. As about 2000 CP units are needed to encapsidate a single viral RNA genome, other mature viral proteins are present in large excess at the late stage of infection. They form various types of cytoplasmic or nuclear inclusions that can be detected using light microscopy (Shukla *et al.*, 1994). Morphology of the cytoplasmic inclusions formed by cylindrical inclusion protein (CI) may be characteristic of a potyvirus species and, hence, can be observed for diagnostic purposes by electron microscopy (Edwardson, 1992). Nuclear inclusion protein b (NIb) and the VPg domain of NIa contain a nuclear localisation signal. Consequently, NIb and VPg (NIa) are found predominantly in the nucleus at later stages of infection (Schaad *et al.*, 1996; Li *et al.*, 1997; Rajamäki & Valkonen, 2003). P3 can also be translocated to the nucleus (Langenberg & Zhang, 1997) but for none of these proteins the functional significance of nuclear localisation is known.

3.5 RNA synthesis and viral genome replication

Translation, polyprotein processing and RNA synthesis (replication) may be difficult to study as separate functions in potyviruses since they occur simultaneously in infected cells. For example, it is possible that the same (+)strand of potyviral RNA is used simultaneously as an mRNA for translation of the polyprotein and as a template for synthesis of the (−)strand RNA. Furthermore, potyviral proteins are multifunctional (Fig. 3.2) and the viral protein, or RNA region, under study may be associated with several of the aforementioned activites. All potyviral proteins and the NTRs are essential for virus propagation (Kekarainen *et al.*, 2002). Also, potyviral proteins except P3 bind RNA *in vitro* (Merits *et al.*, 1998, and references therein). For the two small membrane-associated peptides (6K1 and 6K2), there is less information available in this respect. Thus, multiplication of potyviruses may be referred to as virus propagation if the specific step of multiplication process cannot be singled out.

Replication refers to viral RNA synthesis that with potyviruses takes place in the cytoplasm, and is most likely associated with ER (Martin *et al.*, 1995). The putative potyviral replication complex may be anchored to a replication site on ER via the 6K2 protein (Restrepo-Hartwig & Carrington, 1994). Interaction of NIb, the potyviral RNA-dependent RNA polymerase (RdRp), with the RNA secondary structures (Haldeman-Cahill *et al.*, 1998) and PABP (Wang *et al.*, 2000) at the viral RNA 3′-end may be important for positioning of the replication complex. Recognition of the viral RNA 3′-end and initiation of (−)strand RNA synthesis are not strictly virus-specific but can be carried out *in trans* by NIb of heterologous potyviruses (Teycheney *et al.*, 2000). This may be because the 3′-RNA secondary structures are conserved among subgroups of potyviruses (Spetz *et al.*, 2003). Besides the secondary structures formed by 3′-NTR and the 3′-part of CP-encoding RNA, the 3′ part of the CP-encoding RNA contains a *cis*-acting RNA element that is predicted to form stem loop structures and is important for replication (Mahajan *et al.*, 1996; Haldeman-Cahill *et al.*, 1998).

The (−)strand copy of the viral genome is used as a template for synthesis of many new genomic (+)strands (Wang & Maule, 1995). VPg may act as a primer for RNA synthesis since it is uridylylated *in vitro* by NIb (Puustinen & Mäkinen, unpublished). In poliovirus (related to potyviruses), the uridylylated genome-linked protein primes specifically the (+)strand RNA synthesis (Paul *et al.*, 1998; Murray & Barton, 2003).

The putative potyviral replication complex has not yet been elucidated but is believed to contain at least NIb (RdRp) (Martin *et al.*, 1995), the viral RNA helicase (CI) (Laín *et al.*, 1990), and also NIa that interacts with NIb either via the VPg (Hong *et al.*, 1995) or NIa-Pro domain (Li *et al.*, 1997; Guo *et al.*, 2001). NIa–NIb interaction stimulates polymerase activity (Fellers *et al.*, 1998) and is needed for infectivity (Daròs *et al.*, 1999). NIb (Li & Carrington, 1995) and P1 (Verchot & Carrington, 1995) can carry out viral-genome amplification functions *in trans*, indicating involvement of a replication complex formed via protein–protein interactions. There are also other interactions between potyviral proteins (Rodríguez-Cerezo *et al.*, 1993; Merits *et al.*, 1999; Guo *et al.*, 2001; López *et al.*, 2001; Roudet-Tavert *et al.*, 2002; Yambao *et al.*, 2003), as well as self-interactions (see 3.6.2), but their importance to viral replication remains to be studied. In contrast to P1 and NIb, the proteins NIa (or VPg) (Schaad *et al.*, 1996) and HC-Pro (Kasschau & Carrington, 1995) must act *in cis*; replication functions mediated by these proteins are not complemented by the corresponding functional protein supplied *in trans*. These examples provide evidence for a rather complicated process of genome amplification in potyviruses.

Potyviral replication is not dependent on *de novo* synthesis of host proteins in infected cells. In contrast, host gene expression is largely shut off during the most active stage of virus replication (Wang & Maule, 1995). However, synthesis of polyubiquitin and heathshock protein-like (HSP) proteins is induced, which may be for enhancing virus replication (Aranda *et al.*, 1996). On the other hand, plant viruses of another family (Closteroviridae) encode HSP proteins to facilitate viral movement (Peremyslov *et al.*, 1999).

Another host-driven process that needs to be suppressed for a successful replication process is RNA silencing. It is a cytoplasmic mechanism by which double-stranded RNA or double-stranded structures of ssRNA are recognised and subjected to quick, sequence-specific degradation. Subsequently, also other, single-stranded homologous RNA molecules are targeted to degradation, and the silencing is spread to other parts of the plant by a signal that is not yet fully characterised (reviewed in Baulcombe, 2002). Potyviruses suppress RNA silencing with HC-Pro (Anandalakshmi *et al.*, 1998; Brigneti *et al.*, 1998; Kasschau & Carrington, 1998) to protect the replicative, double-stranded forms and unencapsidated, genomic strands of viral RNA (Moissiard & Voinnet, 2004). Consequently, once infected, susceptible plants cannot recover (Yelina *et al.*, 2002). HC-Pro was previously known as a viral protein needed for *maintainance of viral amplification*, which is now explained with the RNA silencing suppression activities of this protein (Kasschau & Carrington, 2001). Also, potyviruses are long known to have synergistic interactions with other viruses in mixed infections, causing increased accumulation of the synergised virus and often increased symptom severity. HC-Pro is responsible for the synergism (Shi *et al.*, 1997) that can now be explained by the RNA silencing suppression activity of HC-Pro, which is stimulated by P1 (Pruss *et al.*, 1997).

The putative host proteins involved in the replication complex are not known, but PABP (Wang *et al.*, 2000) and eIF4E/eIF4(iso)4E (Léonard *et al.*, 2000; Duprat *et al.*, 2002) seem to be likely candidates. A plant calmodulin-like protein (Anandalakshmi *et al.*, 2000), a RING finger protein (HIP1) and a protein (HIP2) with unknown functions (Guo *et al.*, 2003) interact with HC-Pro but the functional significance of these interactions has not yet been resolved.

3.6   Virus movement

Potyvirus progeny are transported by intracellular macromolecule trafficking mechanisms from the putative membrane-bound sites of replication to plasmodesmata. Plasmodesmata are specialised structures that connect neighbouring plant cells, allowing cytoplasm of different cells to form the so-called symplast (Haywood *et al.*, 2002). By cell-to-cell movement through plasmodesmata, virus is transported through mesophyll, reaches and invades vascular tissue in a vein, is loaded to SE for long-distance transport and is then distributed to other parts of the plant (Fig. 3.1).

Cell-to-cell movement of viruses is an active process that usually refers to movement within epidermal and mesophyll tissues in leaves. However, viral *vascular movement* within the vascular tissue also involves cell-to-cell movement between the various types of cells constituting the vasculature (van Bel, 2003). The same potyviral protein can be involved in non-vascular and vascular phases of the movement process. For these reasons, cell-to-cell and vascular movement processes cannot be dealt with as entirely independent mechanisms. However, there are differences in the regions of viral proteins involved in these two phases of viral movement. Furthermore, plasmodesmata between vascular cells differ from those between mesophyll cells (Haywood *et al.*, 2002), and there is ample evidence for different host factors controlling the two phases of viral movement. Therefore, *vascular movement* refers to viral invasion of vascular cells prior to and after long-distance transport via SE. It includes virus loading to, transport in and unloading from SE (Fig. 3.1).

Potyviruses encode no dedicated movement protein (MP) but use several viral proteins for movement. These movement-associated proteins are HC-Pro, CI, 6K2, VPg and CP (Fig. 3.2), referred to as MP for simplicity. All of them are implicated in several steps of the viral movement process.

3.6.1   Intracellular movement

The details of intracellular transport are largely unknown in potyviruses. Studies on the membrane-associated potyviral 6K2 protein (Restrepo-Hartwig & Carrington, 1994; Rajamäki & Valkonen, 1999; Spetz & Valkonen, 2004) suggest that intracellular trafficking of potyviruses to the openings of plasmodesmata may resemble tobamoviruses (TMV) (see Chapter 2). CI and/or VPg might participate in the process since they are part of the putative replication complex that may associate with membranes via a CI-6K2 polyprotein (Merits *et al.*, 2002) or a 6K2-VPg polyprotein (Schaad *et al.*, 1996). Futhermore, CI forms special structures close to the plasmodesmal aperture (Rodríguez-Cerezo *et al.*, 1997; Roberts *et al.*, 1998; 2003) (see below).

3.6.2   Cell-to-cell movement

The average plasmodesmal dimensions are too narrow to allow free trafficking of virions or viral nucleoprotein complexes between mesophyll cells. Therefore, viruses encode MPs to target virus to plasmodesmata and transiently increase plasmodesmal permeability (Wolf *et al.*, 1989). CP and HC-Pro increase the plasmodesmal exclusion limit, facilitating transport of viral RNA between mesophyll cells. They, similar to the 30K MP of TMV (Crawford & Zambryski, 2001), may also facilitate their own transport through plasmodesmata (Rojas *et al.*, 1997).

Potyviral MPs bind RNA (Merits *et al.*, 1998). However, no data exclusively show whether the transportable form of potyviruses is a virion or another type of ribonucleoprotein complex (reviewed in Carrington *et al.*, 1996, 1998). Mutations in the conserved core region of CP prevent virion assembly and cell-to-cell movement which has been taken as an evidence that potyviruses are transported as virions (Dolja *et al.*, 1994). Alternatively, these mutations interfere with other CP-mediated movement functions. CP and viral RNA localise in the conical structures formed by CI at plasmodesmal openings in the cells at the infection frontier (Rodríguez-Cerezo *et al.*, 1997; Roberts *et al.*, 1998). The functional role of CI in cell-to-cell movement is evident since amino acid substitutions at the CI N-terminus can restrict the virus to single mesophyll cells without compromising viral replication (Carrington *et al.*, 1998). Similarly, VPg is required for viral cell-to-cell movement in mesophyll, the central region of the protein playing an important role (Nicolas *et al.*, 1997). For understanding the mode of function of CI and VPg in viral movement, data on their putative *in trans* activities would be useful. Taken together, data suggest a cell-to-cell transport model of potyviruses in which VPg is linked to the 5'-end of viral RNA, serving as a targeting signal, whereas HC-Pro and/or CP may gate the plasmodesmata open. CI might provide structural support for viral movement through its organised structures, positioning the newly synthesised, CP-associated viral genomes in the central channels of CI structures, through which they are subsequently inserted into plasmodesmata (Carrington *et al.*, 1998).

Formation of virions and conical structures in infected cells shows that CP and CI self-interact, as was also shown using recombinant CP and CI proteins in experimental systems (Guo *et al.*, 2001; López *et al.*, 2001). Likewise, VPg (Oruetxebarria *et al.*, 2001) and HC-Pro (Guo *et al.*, 1999; Plisson *et al.*, 2003) show self-interaction and mediate potyviral movement functions, but in their case the functional connection, if any, of these two functions is unclear. Phosphorylation of potyviral VPg and CP seems to be important for their functional regulation, as described for the TMV 30K MP (Karpova *et al.*, 1999; Waigmann *et al.*, 2000). The host protein kinase CK2 phosphorylates a widely conserved threonine residue of potyviral CP. Mutations that affect the consensus target sequence of CK2 in CP and which consequently prevent phosphorylation restrict the virus to single mesophyll or epidermal cells in inoculated leaves (Ivanov *et al.*, 2003). Similarly, substitution of specific amino acids in VPg debilitate its ability to support viral accumulation and/or vascular movement (Rajamäki & Valkonen, 2002) and have a major impact on phosphorylation of VPg (Puustinen *et al.*, 2002).

3.6.3   Vascular movement

Vascular movement of viruses is less studied compared to cell-to-cell movement, due to the complex structure of the plant vasculature (reviewed in Haywood *et al.*, 2002). Most studies have been carried out with (non-vector transmissible) tobamo- and potexviruses (reviewed in Van Bel, 2003). Vascular movement of insect-transmissible viruses, such as potyviruses, takes place in phloem sap via SE, following the route partitioning photoassimilates from the mature source tissues to young, developing sink tissues (Van Bel, 2003). Virus transport in SE is believed to occur passively, although few studies have critically addressed this question.

Similarly, few detailed studies on phloem loading and unloading of potyviruses are available. Unloading from minor veins (devoid of rib tissue) is observed in some hosts, whereas in others such as *Nicotiana* species, unloading from major veins (containing rib tissue) may be preferred (Ding *et al.*, 1998; Rajamäki & Valkonen, 2003 and unpublished data).

There are specific host and viral factors required for vascular movement of potyviruses. A cystein-rich protein (PVIP) found in different plants families interacts with VPg of many potyviruses and may represent an essential host factor for viral movement (Dunoyer *et al.*, 2004). In many potyvirus–host combinations, efficient virus propagation and movement are observed in mesophyll tissue, but the virus fails to establish systemic infection. Genetic analyses on the hosts show that restricted virus movement may be attributable to a single or a few recessive host factors (e.g. Schaad & Carrington, 1996; Hämäläinen *et al.*, 2000). While these results may indicate mutations that have occured in host factors making them incompatible for viral movement functions, they could as well be due to inducible resistance mechanisms. In most cases, the defective step of vascular movement has not been determined, but blockage of potyvirus loading to SE is found in some instances (Rajamäki & Valkonen, 2002). In general, the number of plasmodesmata varies between different cell types, as does the permeability of plasmodesmata which also varies depending on the developmental stage of phloem (Crawford & Zambryski, 2001). Plasmodesmata between SE and companion cells may allow diffusion of rather large molecules, whereas the cellular interface between the SE–companion cell complex and phloem parenchyma seems to constitute the strictest and most

selective barrier through which virus must move (reviewed in Ruiz-Medrano
et al., 2001).

It is not known in which form (virion, or other type of ribonucleoprotein complex)
potyviruses are transported in SE. However, CP, HC-Pro and VPg are crucial for
vascular transport. The N- and C-termini of CP mediate vascular movement functions
(Dolja *et al.*, 1994). The C-terminus may be more important (Dolja *et al.*, 1995;
Arazi *et al.*, 2001) and its defective functions cannot be complemented *in trans*
(Dolja *et al.*, 1995). In contrast, the vascular movement functions of HC-Pro can be
provided *in trans* (Cronin *et al.*, 1995).

A central domain in HC-Pro is involved in vascular movement and RNA silencing suppression, which suggests that viral vascular movement requires suppression of
RNA silencing in phloem (Kasschau & Carrington, 2001), possibly in companion
cells that may be hyperactive in RNA silencing (Marathe *et al.*, 2000; Savenkov &
Valkonen, 2002).

The central region of VPg determines virus strain and host-specific interactions
required for vascular movement (reviewed in Rajamäki & Valkonen, 2002). It is
noteworthy that mutations within this central region of VPg affect virus accumulation in inoculated cells (Borgstrom & Johansen, 2001), cell-to-cell movement in
inoculated leaves (Nicolas *et al.*, 1997), or vascular movement (Schaad *et al.*, 1997;
Rajamäki & Valkonen, 1999, 2002), depending on the virus–host combination.
Mutations at the VPg C-terminus also affect viral accumulation and/or movement
(Rajamäki & Valkonen, 2002) and interfere with phosphorylation of VPg
(Puustinen *et al.*, 2002). VPg accumulates in companion cells of the upper
non-inoculated leaves in the apparent absence of other virus proteins and RNA
(Rajamäki & Valkonen, 2003). Thus, data suggest that VPg is transported in
vasculature in advance of the infection frontier, perhaps to suppress host responses
that might otherwise prevent virus transport through companion cells and
establishment of systemic infection. Mutations in CI (Carrington *et al.*, 1998) and
6K2 (Rajamäki & Valkonen, 1999; Spetz & Valkonen, 2004) interfere with
systemic infection but the reasons are not yet understood.

3.7   Induction of symptoms

Potyvirus-infected plants show various types of symptoms, from symptomless
infection to vein chlorosis, mild to severe mosaic symptoms and leaf malformation.
It is difficult to describe symptoms that could be considered generally characteristic of
potyviruses and helpful for viral identification. However, potyviruses rarely cause
bright yellowing and calico mosaic symptoms or upward leaf-rolling. Plants with
seemingly symptomless leaves may suffer from growth reduction, and plants with
seemingly unaffected growth of canopy may produce lower tuber or tuberous root
yields than healthy plants. Sometimes potyviruses cause necrotic symptoms in
their hosts, which may be an expression of host defence that fails to restrict virus
movement but triggers cell death in infected tissues.

The mechanisms of disease induction by potyviruses have remained largely unknown. Potyviruses cause a dramatic shut-off of host gene expression in cells where they replicate (Wang & Maule, 1995), which seriously disturbes cellular physiology and probably causes many of the varied symptoms. Interference with host gene expression is particularly apparent in potyvirus-infected petunia flowers, in which virus infection (i.e. virus-encoded silencing suppressor) prevents natural silencing of the chalcone synthetase gene in the white sectors of the flower. Subsequently, the white, virus-infected sectors acquire spots and stripes of pigmentation (Teycheney & Tepfer, 2001). Various types of developmental defects associated with potyvirus infection of plants may also be caused by RNA silencing suppression. The P1/HC-Pro of *Turnip mosaic virus* (TuMV) causes developmental abnormalities similar to those observed with TuMV infection in *Arabidopsis thaliana* due to inhibition of micro-RNA guided cleavage of mRNAs encoding for various transcription factors that, in turn, regulate genes required for the development of vegetative and reproductive organs (Kasschau *et al.*, 2003).

Many regions of the potyviral genome (Fig. 3.2) have been implicated as determinants of symptom induction in specific virus–host combinations. They include HC-Pro (vein clearing symptoms in tobacco plants; Atreya *et al.*, 1992), the region encoding the C-proximal part of P3 (yellow mosaic symptoms in *Brassica juncea*; Jenner *et al.*, 2003), the C-proximal part of P3 and 6K1 (chlorotic mottle symptoms in leaves of *Nicotiana benthamiana*; Sáenz *et al.*, 2000), P3 and CI-6K2-VPg encompassing regions (wilting response in tabasco pepper; Chu *et al.*, 1997), and a genomic segment encoding NIa and NIb (vein clearing symptoms induced in pea plants; Johansen *et al.*, 1996). The 5′-NTR and 3′-NTR, respectively, induce chlorotic mosaic symptoms in *N. clevelandii* leaves (Simón-Buela *et al.*, 1997) or vein mottling and blotch symptoms in tobacco leaves (Rodríguez-Cerezo *et al.*, 1991). In many studies, it has remained unknown whether mutations in the viral proteins or NTRs affected symptom expression independent of their effects on virus accumulation and/or movement. However, the vein chlorosis symptoms induced in *N. benthamiana* following systemic infection with *Potato virus A* were abolished by a 6xHis insertion introduced to the 6K2 protein, which was not associated with changes in virus accumulation in the systemically infected leaves or the rate of long distance movement of the virus (Spetz & Valkonen, 2004).

The N-proximal part of CP is the determinant of veinal chlorosis symptoms in potyvirus-infected cucumber leaves (Ullah & Grumet, 2002). Chlorosis and mosaic symptoms are indicative of interference with the proper function of chloroplasts in virus-infected plants, as studied with TMV (Zaitlin & Jagendorf, 1960). Potyviral RNA, HC-Pro and CP are found in chloroplasts of infected leaves (Gunasinghe & Berger, 1991). Targeting of CP to chloroplasts in transgenic plants was associated with severe white chlorosis symptoms (Naderi & Berger, 1997). The aforementioned data and discovery of a host protein in chloroplast that interacts with potyviral CP (McClintock *et al.*, 1998) suggest that translocation of CP to chloroplasts may induce chlorosis symptoms.

3.8   Transmission

3.8.1   Transmission by aphids

Aphids transmit potyviruses in a stylet-borne, non-persistent manner carrying the virions in the distal part (tip) of the stylet (Wang *et al.*, 1996). Acquisition of virus takes only few seconds to minutes as aphids probe on leaf epidermis and ingest cell contents. Likewise, inoculation occurs quickly during salivation (Collar *et al.*, 1997; Martín *et al.*, 1997). Potyviral virions attached to the aphid stylet remain infectious only for some minutes to hours and, therefore, transmission occurs over relatively short distances. However, owing to the quick acquisition and inoculation by aphids, potyviruses can be spread very efficiently by their vectors in the field. Transmission is nearly impossible to prohibit using control measures directed against the aphids, but spraying the leaves with mineral oil may reduce transmission efficiency (Pirone & Harris, 1977; Shukla *et al.*, 1994).

HC-Pro and CP play a crucial role in aphid transmission of potyviruses. Transmission occurs only when HC-Pro is provided prior to or simultaneously with the virions. This is why HC-Pro was originally designated as a *helper component* (Govier & Kassanis, 1974). HC-Pro contains two motifs important for aphid transmission, namely Lys-Ile-Thr-Cys (KITC; Atreya & Pirone, 1993) and Pro-Thr-Cys (PTK; Huet *et al.*, 1994) at N- and C-proximal regions respectively. In CP, a motif of three amino acids at the variable N-terminus is required for aphid transmission. This motif may show some sequence variability but is usually found as Asp-Ala-Gly (DAG) (Atreya *et al.*, 1990; reviewed in López-Moya *et al.*, 1999). Amino acid context of the DAG motif affects transmission efficiency (López-Moya *et al.*, 1999). The co-ordinated functions of HC-Pro and CP in virus transmission are hypothesised to depend on HC-Pro acting as a bridge between virion and aphid stylet. The model posits that the PTK and DAG motifs determine a mutual interaction between HC-Pro and the virion respectively (Blanc *et al.*, 1997; Peng *et al.*, 1998), whereas the KITC motif of HC-Pro may mediate interaction with a putative receptor in aphid stylet (Blanc *et al.*, 1998).

3.8.2   Seed transmission

Some 20 potyviruses are seed transmitted besides also being aphid transmitted. A few of these viruses are pollen transmitted as well, including *Bean common mosaic virus*, *Lettuce mosaic virus* and *Pea seed-borne mosaic virus* (PSbMV). Transmission of potyviruses via seed requires infection of the embryo, which occurs by direct invasion via ovule, or by indirect invasion mediated by infected gametes (reviewed in Johansen *et al.*, 1994). Futhermore, the virus must maintain infectivity during seed dormancy. Probably potyviruses do commonly infect reproductive parts of the plant, including seed coat, but, nevertheless, they are not seed transmitted because they fail to infect the embryo (Roberts *et al.*, 2003; Rajamäki & Valkonen, 2004).

Seed transmission of PSbMV is best studied among potyviruses. PSbMV invades pea embryos directly via the suspensor at an early developmental stage and multiplies in the embryonic tissues (Wang & Maule, 1992). For entrance to the suspensor, PSbMV accumulates passively, without replication, in the endosperm at the base of the suspensor (Roberts *et al.*, 2003). Suspensor is a transient structure acting as a channel to provide nutritional support for the growing embryo. Therefore, there is only a narrow *physiological window* for the virus to reach the embryo before the suspensor degenerates; plants need to be infected at a young developmental stage (Wang & Maule, 1994). For example, PSbMV must infect pea plants prior to flowering to be seed transmitted (Hampton & Mink, 1975). Comparison of the highly and poorly seed-transmissible isolates of PSbMV revealed that 5′-NTR, HC-Pro and CP were the major determinants of seed transmission (Johansen *et al.*, 1996). In general, efficiency of viral seed transmission depends on the genetic factors of both the host and the virus, and also on environmental factors (Roberts *et al.*, 2003). Therefore, as with aphid-transmissibility, seed-transmissibility may occasionally be found with virus isolates that are typically non-transmissibile.

3.8.3   Mechanical transmission

Potyviruses are usually readily transmitted by mechanical inoculation, e.g. rubbing sap extracted from infected leaves on the leaves of healthy plants dusted with an abrasive (Shukla *et al.*, 1994). Transmissibility may differ depending on the source or recipient host, owing to interfering host substances (Shukla *et al.*, 1994; Valkonen *et al.*, 1996). Mechanical inoculation can also be used with viral RNA extracted from virions, infectious $^{7Me}G^{5′}ppp^{5′}Np$-capped *in vitro* RNA transcripts derived from full-length cDNAs of potyviruses, or with potyviral cDNA clones (plasmids) containing the viral genome under the control of a plant-specific promoter (e.g., *Cauliflower mosaic virus* 35S promoter). However, inoculation of viral transcripts and cDNAs can be significantly enhanced using biolistic inoculation methods, for which various types of devices are available (Gal-On *et al.*, 1997), also commercially.

Despite their transmissibility by mechanical inoculation under experimental conditions, contact-based transmission between plants probably plays a negligible role in potyvirus spread in the field. In contrast to potyviruses, TMV and potexviruses depend on mechanical (contact-based) transmission and have very stable virions persisting and retaining infectivity on contaminated surfaces for days or months (Matthews, 1991).

3.9   Variability and evolution

Plant viruses are genetically heterogeneous. A virus-infected plant contains a population of nearly identical viral sequences, termed quasispecies (Eigen, 1996; Smith *et al.*, 1997). The quasispecies nature of RNA viruses, such as potyviruses, is due to their RdRp that lacks proof-reading activity, which makes viral RNA synthesis prone to errors. Mutations may be base substitutions, insertions, deletions or inversions.

Recombination is a process in which genetic material is exchanged between the RNA strands of viruses during replication. Recombination is prevalent in the genus *Potyvirus* (Cervera *et al.*, 1993; Revers *et al.*, 1996; Ohshima *et al.*, 2002). For example, in the type member PVY, two variants related to the tobacco vein necrosis strain (PVY^N) have emerged during the last two decades. Both variants (PVY^{NTN} and PVY^{N}W) differ from PVY^N in virulence on potato and contain a mosaic structure of sequences related to PVY^N and the ordinary strains (PVY^O; Fig. 3.3). Thus, they

Figure 3.3 (A) Heavy growth reduction in potato cv. Bintje grown from a seed tuber infected with *Potato virus Y* (left) compared to virus-free seed (right) (Courtesy Jari Valkonen). (B) Highly reduced tuber yield (left) due to virus infection, as compared to the yield of a virus-free plant (right) (Courtesy Seif El-Amin).

appear to be single or multiple recombinants between PVY^O and PVY^N (Glais *et al.*, 2002b). Paalme *et al.* (2004) showed experimentally, using two phenotypically identical and sequence-wise nearly identical genomes of potyvirus isolates, that their recombination can result in a new virus strain, as defined by the symptoms distinctly different from the parental virus isolates.

Variability and high mutation rates of viruses are advantageous in adaptation to a changing environment (Drake & Holland, 1999). The success of new variants generated via mutation or recombination depends on genetic drift and selection. Genetic drift is a random process in which random effects influence the probability of each variant to produce its next progeny. During selection, variants that fit best to a given environment (host and vector) are favoured over those that are less compatible (García-Arenal *et al.*, 2001). Thus, variability helps viruses to expand their host range, for example, by overcoming host resistance (Roossinck, 1997). In potyviruses, selection for higher fitness in the host under experimental conditions may occur with cost of losing aphid transmissibility. This is observed following repeated cycles of virus propagation in test plants by mechanical inoculation (Shukla *et al.*, 1994). In vegetatively propagated plants such as potato, it may become less prudent for the virus to maintain aphid transmissibility, especially if it compromises virulence in the host. Isolates of PVY strain group C (PVY^C) may represent an example, since different from other strain groups of PVY, PVY^C seems to contain a proportionally large number of isolates that are non-aphid-transmissible (Blanco-Urgoiti *et al.*, 1998). Depending on whether mutations occur in the aphid-transmissibility motifs of CP or HC-Pro, the consequently non-transmissible viral variant may show increased (Andrejeva *et al.*, 1999) or decreased (Legavre *et al.*, 1996) accumulation in leaves respectively. Thus, fitness (level of virus accumulation) in the host is not necessarily positively correlated with aphid transmissibility.

Potyviruses belong to the picorna-like genetic lineage of viruses and are all phylogenetically related (Berger *et al.*, 1997; Van Regenmortel *et al.*, 2000). They are the most common among plant viruses and have the widest distribution over the plant kingdom. As a viral genus, their evolutionary success is extraordinary, suggesting that they continue to adapt to new hosts and environments with high efficiency. Hence, the robust genome expression strategy based on a single, mRNA-like genomic component encoding a polyprotein, and a seemingly wasteful production of viral proteins in great excess compared to what is necessary do not appear to be disadvantageous but perhaps are the strength of potyviruses. Translation, replication, movement and transmission mechanisms of potyviruses are still far from being fully elucidated at the molecular level, and all important crops continue to suffer from yield losses due to potyviruses. Therefore, potyviruses remain as an important subject of both basic and applied studies in virology.

Acknowledgments

The authors acknowledge financial support from Academy of Finland for their studies on potyviruses.

References

Anandalakshmi, R., Pruss, G.J., Ge, X., Marathe, R., Mallory, A.C., Smith, T.H. & Vance, V.B. (1998) A viral suppressor of gene silencing in plants. *Proc. Natl. Acad. Sci. USA*, **95**, 13079–13084.

Anandalakshmi, R., Marathe, R., Ge, X., Herr, J. M., Jr, Mau, C., Mallory, A., Pruss, G., Bowman, L. & Vance, V. B. (2000) A calmodulin-related protein that suppresses posttranscriptional gene silencing in plants. *Science*, **290**, 142–144.

Andrejeva, J., Puurand, ü., Merits, A., Rabenstein, F., Järvekülg, L. & Valkonen, J.P.T. (1999) Potyvirus HC-Pro and CP proteins have coordinated functions in virus–host interactions and the same CP motif affects virus transmission and accumulation. *J. Gen. Virol.*, **80**, 1133–1139.

Aranda, M.A., Escaler, M., Wang, D. & Maule, A.J. (1996) Induction of Hsp70 and polyubiquitin expression associated with plant virus replication. *Proc. Natl. Acad. Sci. USA*, **93**, 15289–15293.

Arazi, T., Shiboleth, Y.M. & Gal-On, A. (2001) A nonviral peptide can replace the entire N-terminus of Zucchini yellow mosaic potyvirus coat protein and permits viral systemic infection. *J. Virol.*, **75**, 6329–6336.

Atreya, C. D. & Pirone, T. P. (1993) Mutational analysis of the helper component-proteinase gene of a potyvirus: effects of amino acid substitutions, deletions, and gene replacement on virulence and aphid transmissibility. *Proc. Natl. Acad. Sci. USA*, **90**, 11919–11923.

Atreya, C. D., Raccah, B. & Pirone, T. P. (1990) A point mutation in the coat protein abolishes aphid transmissibility of a potyvirus. *Virology*, **178**, 161–165.

Atreya, C.D., Atreya, P.L., Thornbury, D.W. & Pirone, T.P. (1992) Site-directed mutations in the potyvirus HC-PRO gene affect helper component activity, virus accumulation, and symptom expression in infected tobacco plants. *Virology*, **191**, 106–111.

Baratova, L.A., Efimov, A.V., Dobrov, E.N., Fedorova, N.V., Hunt, R., Badun, G.A., Ksenofontov, A.L., Torrance, L. & Järvekülg, L. (2001) *In situ* spatial organization of potato virus A coat protein subunits as assessed by tritium bombardment. *J. Virol.*, **75**, 9696–9702.

Basso, J., Dallaire, P., Charest, P.J., Devantier, Y. & Laliberté, J.-F. (1994) Evidence for an internal ribosome entry site within the 5′ non-translated region of turnip mosaic potyvirus RNA. *J. Gen. Virol.*, **75**, 3157–3165.

Baulcombe, D. (2002) Viral suppression of systemic silencing. *Trends Microbiol.*, **10**, 306–308.

Berger, P.H., Wyatt, S.D., Shiel, P.J., Silbernagel, M.J., Druffel, K. & Mink, G.I. (1997) Phylogenetic analysis of the potyviridae with emphasis on legume-infecting potyviruses. *Arch. Virol.*, **142**, 1979–1999.

Blanc, S., Ammar, E. D., García-Lampasona, S., Dolja, V. V., Llave, C., Baker, J. & Pirone, T. P. (1998) Mutations in the potyvirus helper component protein: effects on interactions with virions and aphid stylets. *J. Gen. Virol.*, **79**, 3119–3122.

Blanc, S., López-Moya, J. J., Wang, R. Y., García-Lampasona, S., Thornbury, D. W. & Pirone, T. P. (1997) A specific interaction between coat protein and helper component correlates with aphid transmission of a potyvirus. *Virology*, **231**, 141–147.

Blanco-Urgoiti, B., Sánchez, F., Pérez de San Román, C., Dopazo, J. & Ponz, F. (1998) Potato virus Y group C isolates are a homogenous pathotype but two different genetic strains. *J. Gen. Virol.*, **79**, 2037–2042.

Borgstrom, B. & Johansen, I.E. (2001) Mutations in *Pea seedborne mosaic virus* genome-linked protein VPg alter pathotype-specific virulence in *Pisum sativum*. *Mol. Plant Microbe Interact.*, **14**, 707–714.

Brigneti, G., Voinnet, O., Li, W.-X., Ji, L.-H., Ding, S.-W. & Baulcombe, D.C. (1998) Viral pathogenicity determinants are suppressors of transgene silencing in *Nicotiana benthamiana*. *EMBO J.*, **17**, 6739–6746.

Carrington, J. C. & Dougherty, W. G. (1987) Small nuclear inclusion protein encoded by plant potyvirus genome is a protease. *J. Virol.*, **61**, 2540–2548.

Carrington, J.C. & Freed, D.D. (1990) Cap-independent enhancement of translation by a plant potyvirus 5′ nontranslated region. *J. Virol.*, **64**, 1590–1597.

Carrington, J. C., Cary, S. M., Parks, T. D. & Dougherty, W. G. (1989) A second proteinase encoded by a plant potyvirus genome. *EMBO J.*, **8**, 365–370.

Carrington, J.C., Kasschau, K.D., Mahajan, S.K. & Schaad, M.C. (1996) Cell-to-cell and long-distance transport of viruses in plants. *Plant Cell*, **8**, 1669–1681.

Carrington, J.C., Jensen, P.E. & Schaad, M.C. (1998) Genetic evidence for an essential role for potyvirus CI protein in cell-to-cell movement. *Plant J.*, **14**, 393–400.

Caspar, D.L.D., Dulbecco, R., Klug, A., Lwoff, A., Stoker, M.G., Tournier, P. & Wildy, P. (1962) Proposals. *Cold Spring Har. Symp. Quant. Biol.*, **27**, 49.

Cervera, M.T., Riechmann, J.L., Martin, M.T. & Garcia, J.A. (1993) 3-terminal sequence of the plum pox virus PS and õ6 isolates: evidence for RNA recombination within the potyvirus group. *J. Gen. Virol.*, **74**, 329–334.

Chu, M., Lopez-Moya, J.J., Llave-Correas, C. & Pirone, T.P. (1997) Two separate regions in the genome of the tobacco etch virus contain determinants of the wilting response of tabasco pepper. *Mol. Plant Microbe Interact.*, **10**, 472–480.

Collar, J.L., Avilla, C. & Fereres, A. (1997) New correlations between aphid stylet paths and nonpersistent virus transmission. *Environ. Entomol.*, **26**, 537–544.

Crawford, K.M. & Zambryski, P.C. (2001) Non-targeted and targeted protein movement through plasmodesmata in leaves in different developmental and physiological states. *Plant Physiol.*, **125**, 1802–1812.

Cronin, S., Verchot, J., Haldeman-Cahill, R., Schaad, M.C. & Carrington, J.C. (1995) Long-distance movement factor: a transport function of the potyvirus helper component proteinase. *Plant Cell*, **7**, 549–559.

Daròs, J.A., Schaad, M.C. & Carrington, J.C. (1999) Functional analysis of the interaction between VPg-proteinase (NIa) and RNA polymerase (NIb) of tobacco etc potyvirus, using conditional and suppressor mutants. *J. Virol.*, **73**, 8732–8740.

De Bokx, J.A. & Huttinga, H. (1981) Potato virus Y. *Commonwealth Mycological Institute/Association of Applied Biologists Descriptions of Plant Viruses*, No. 242, Kew, Surrey, UK.

Dever, T.E. (1999) Translation initiation: adept at adapting. *Trends Biochem. Sci.*, **24**, 398–403.

Ding, X.S., Carter, S.A., Deom, C.M. & Nelson, R.S. (1998) Tobamovirus and potyvirus accumulation in minor veins of inoculated leaves from representatives of the Solanaceae and Fabaceae. *Plant Physiol.*, **116**, 125–136.

Dolja, V.V., Haldeman, R., Robertson, N.L., Dougherty, W.G. & Carrington, J.C. (1994) Distinct functions of capsid protein in assembly and movement of tobacco etch potyvirus in plants. *EMBO J.*, **13**, 1482–1491.

Dolja, V.V., Haldeman-Cahill, R., Montgomery, A.E., Vandenbosch, K.A. & Carrington, J.C. (1995) Capsid protein determinants involved in cell-to-cell and long-distance movement of tobacco etch potyvirus. *Virology*, **206**, 1007–1016.

Drake, J.W. & Holland, J.J. (1999) Mutation rates among RNA viruses. *Proc. Natl. Acad. Sci. USA*, **99**, 13910–13913.

Dunoyer, P., Thomas, C., Harrison, S., Revers, F. & Maule, A. (2004) A cysteine-rich plant protein potentiates *Potyvirus* movement through an interaction with the virus genome-linked protein VPg. *J. Virol.*, **78**, 2301–2309.

Duprat, A., Caranta, C., Revers, F., Menand, B., Browning, K.S. & Robaglia, C. (2002) The *Arabidopsis* eukaryotic initiation factor (iso)4E is dispensable for plant growth but required for susceptibility to potyviruses. *Plant Journal*, **32**, 927–934.

Edwardson, J.R. (1992) Inclusion bodies. *Archives of Virology, Supplement S5*, pp. 25–30.

Eigen, M. (1996) On the nature of virus quasispecies. *Trends Microbiol.*, **4**, 216–217.

Fellers, J., Wan, J., Hong, Y., Collins, G.B. & Hunt, A.G. (1998) *In vitro* interactions between a potyvirus-encoded, genome-linked protein and RNA-dependent RNA polymerase. *J. Gen. Virol.*, **79**, 2043–2049.

Fernández-Fernández, M.R., Camafeita, E., Bonay, P., Méndez, E., Albar, J.P. & García, J.A. (2002) The capsid protein of a plant single-stranded RNA virus is modified by O-linked N-acetylglucosamine. *J. Biol. Chem.*, **277**, 135–140.

Gallie, D.R. (2001) Cap-independent translation conferred by the 5′ leader of tobacco etch virus is eukaryotic initiation factor 4G dependent. *J. Virol.*, **75**, 12141–12152.

Gallie, D.R., Tanguay, R.L. & Leathers, V. (1995) The tobacco etch viral 5′ leader and poly(A) tail are functionally synergistic regulators of translation. *Gene*, **165**, 233–238.

Gal-On, A., Meiri, E., Elman, C., Gray, D.J. & Gaba, V. (1997) Simple hand-held devices for the efficient infection of plants with viral-encoding constructs by particle bombardment. *J. Virol. Methods*, **64**, 103–110.

García-Arenal, F., Fraile, A. & Malpica, J. M. (2001) Variability and genetic structure of plant virus populations. *Annu. Rev. Phytopathol.*, **39**, 157–186.

Glais, L., Kerlan, C. & Robaglia, C. (2002a) Variability and evolution of *Potato virus Y*, the type species of the *Potyvirus* genus. In: *Plant Viruses as Molecular Pathogens* (eds J.A. Khan & J. Dijkstra), pp. 225–253, The Haworth Press, Binghamton, NY, USA.

Glais, L., Tribodet, M. & Kerlan, C. (2002b) Genomic variability in *Potato potyvirus Y* (PVY): evidence that PVY^{N}W and PVY^{NTN} variants are single to multiple recombinants between PVY^O and PVY^N isolates. *Arch. Virol.*, **147**, 363–378.

Govier, D.A. & Kassanis, B. (1974) Evidence that a component other than the virus particle is needed for aphid transmission of potato virus Y. *Virology*, **57**, 285–286.

Gunasinghe, U.B. & Berger, P.H. (1991) Association of potato virus Y gene products with chloroplasts in tobacco. *Mol. Plant Microbe Interact.*, **4**, 452–457.

Guo, D., Merits, A. & Saarma, M. (1999) Self-association and mapping of interaction domains of helper component-proteinase of potato A potyvirus. *J. Gen. Virol.*, **80**, 1127–1131.

Guo, D., Rajamäki, M.-L., Saarma, M. & Valkonen, J.P.T. (2001) Towards a protein interaction map of potyviruses: protein interaction matrixes of two potyviruses based on the yeast two-hybrid system. *J. Gen. Virol.*, **82**, 935–939.

Guo, D., Spetz, C., Saarma, M. & Valkonen, J.P.T. (2003) Two potato proteins, including a novel RING-finger protein, interact with the potyviral multifunctional protein HC-Pro. *Mol. Plant Microbe Interact.*, **16**, 405–410.

Haldeman-Cahill, R., Daros, J.-A. & Carrington, J.C. (1998) Secondary structures in the capsid protein coding sequence and 3′ non-translated region involved in the amplification of the tobacco etch virus genome. *J. Gen. Virol.*, **72**, 4072–4079.

Hämäläinen, J.H., Kekarainen, T., Gebhardt, C., Watanabe, K.N. & Valkonen, J.P.T. (2000) Recessive and dominant genes interfere with the vascular transport of *Potato virus A* in diploid potatoes. *Mol. Plant Microbe Interact.*, **13**, 402–412.

Hampton, R.O. & Mink, G.I. (1975) Pea seed-borne mosaic virus. *Commonwealth Mycological Institute/ Association of Applied Biologists Descriptions of Plant Viruses, No..* Kew, Surrey, UK.

Haywood, V., Kragler, F. & Lucas, W.J. (2002) Plasmodesmata: pathways for protein and ribonucleoprotein signaling. *Plant Cell*, Suppl., S303–S325.

Hong, Y. & Hunt, A.G. (1996) RNA polymerase activity catalyzed by a potyvirus-encoded RNA-dependent RNA polymerase. *Virology*, **226**, 146–151.

Hong, Y., Levay, K., Murphy, J.F., Klein, P.G., Shaw, J.G. & Hunt, A.G. (1995) A potyvirus polymerase interacts with the viral coat protein and VPg inyeast cells. *Virology*, **214**, 159–166.

Huet, H., Gal-On, A., Meir, E., Lecoq, H. & Raccah, B. (1994) Mutations in the helper-component (HC) gene of zucchini yellow mosaic virus (ZYMV) affect aphid transmissibility. *J. Gen. Virol.*, **72**, 1407–1414.

Ivanov, K.I., Puustinen, P., Merits, A., Saarma, M. & Mäkinen, K. (2001) Phosphorylation down-regulates the RNA binding function of the coat protein of Potato virus A. *J. Biol. Chem.*, **276**, 13530–13540.

Ivanov, K.I., Puustinen, P., Rönnstrand, L., Valmu, L., Vihinen, H., Gabrenaite, R., Kalkkinen, N. & Mäkinen, K. (2003) Phosphorylation of the potyvirus capsid protein by plant protein kinase CK2 and its relevance for virus infection. *The Plant Cell*, **15**, 2124–2139.

Jagadish, M.N., Huang, D. & Ward, C.W. (1993) Site-directed mutagenesis of a potyvirus coat protein and its assembly in *Escherichia coli*. *J. Gen. Virol.*, **74**, 893–896.

Johansen, E., Edwards, M. C. & Hampton, R. O. (1994) Seed transmission of viruses: current perspectives. *Ann. Rev. Phytopathol.*, **32**, 363–386.

Johansen, I. E., Dougherty, W. G., Keller, K. E., Wang, D. & Hampton, R. O. (1996) Multiple viral determinants affect seed transmission of pea seedborne mosaic virus in *Pisum sativum*. *J. Gen. Virol.*, **77**, 3149–3154.

Karpova, O.V., Rodinova, N.P., Ivanov, K.I., Kozolovsky, S.V., Dorokhov, Yu, L. & Atabekov, J.G. (1999) Phosphorylation of tobacco mosaic virus movement proteins abolishes its translation repressing ability. *Virology*, **216**, 20–24.

Kasschau, K.D. & Carrington, J.C. (1995) Requirement for HC-Pro processing during genome amplification of tobacco etch potyvirus. *Virology*, **209**, 268–273.

Kasschau, K.D. & Carrington, J.C. (1998) A counterdefensive strategy of plant viruses: suppression of posttranscriptional gene silencing. *Cell*, **95**, 461–470.

Kasschau, K.D. & Carrington, J.C. (2001) Long-distance movement and replication maintenance functions correlate with silencing suppression activity of potyviral HC-Pro. *Virology*, **285**, 71–81.

Kasschau, K.D., Xie, Z., Allen, E., Llave, C., Chapman, E.J., Krizan, K.A. & Carrington, J.C. (2003) P1/HC-PRo, a viral suppressor of RNA silencing, interferes with *Arabidopsis* development and miRNA function *Dev. Cell*, **4**, 205–217.

Kekarainen, T., Savilahti, H. & Valkonen, J.P.T. (2002) Functional genomics on *Potato virus A*: a virus genome-wide map of sites essential for virus propagation. *Genome Res.*, **12**, 584–594.

Laín, S., Riechmann, J.L. & García, J.A. (1990) RNA helicase: a novel activity associated with a protein encoded by a positive strand RNA virus. *Nucleic Acids Res.*, **18**, 7003–7006.

Langenberg, W.G. & Zhang, L. (1997) Immunocytology shows the presence of tobacco etch virus P3 protein in nuclear inclusions. *J. Struct. Biol.*, **118**, 243–247.

Legavre, T., Maia, I.G., Casse-Delbart, F., Bernardi, F. & Robaglia, C. (1996) Switches in the mode of transmission select for or against a poorly aphid-transmissible strain of potato virus Y with reduced helper component and virus accumulation. *J. Gen. Virol.*, **77**, 1343–1347.

Lellis, A.D., Kasschau, K.D., Whitham, S.E. & Carrington, J.C. (2002) Loss-of-susceptibility mutants of *Arabidopsis thaliana* reveal an essential role for eIF(iso)4E during potyvirus infection. *Curr. Biol.*, **12**, 1046–1051.

Léonard, S., Plante, D., Wittmann, S., Daigneault, N., Fortin, M.G. & Laliberté, J.F. (2000) Complex formation between potyvirus VPg and translation eukaryotic initiation factor 4E correlates with virus infectivity. *J. Virol.*, **74**, 7730–7737.

Li, X.H. & Carrington, J.C. (1995) Complementation of tobacco etch potyvirus mutants by active RNA polymerase expressed in transgenic cells. *Proc. Natl. Acad. Sci. USA*, **92**, 457–461.

Li, X.H., Valdez, P., Olvera, R. & Carrington, J.C. (1997) Functions of the tobacco etch virus RNA polymerase (NIb): subcellular transport and protein–protein interaction with VPg/proteinase (NIa). *J. Virol.*, **71**, 1598–1607.

López, L., Urzainqui, A., Domínguez, E. & García, J.A. (2001) Identification of an N-terminal domain of the plum pox potyvirus CI RNA helicase involved in self-interaction in a yeast two-hybrid system. *J. Gen. Virol.*, **82**, 677–686.

López-Moya, J. J., Wang, R. Y. & Pirone, T. P. (1999) Context of the coat protein DAG motif affects potyvirus transmissibility by aphids. *J. Gen. Virol.*, **80**, 3281–3288.

Mahajan, S., Dolja, V.V. & Carrington, J.C. (1996) Roles of the sequence encoding tobacco etch virus capsid protein in genome amplification: requirements for the translation process and a *cis*-active element. *J. Virol.*, **70**, 4370–4379.

Marathe, R., Anandalakshmi, R., Smith, T.H., Pruss, G.J. & Vance, V.B. (2000) RNA viruses as inducers, suppressors and targets of post-transcriptional gene silencing. *Plant Mol. Biol.*, **43**, 295–306.

Martin, M.T., Cervera, M.T. & Garcia, J.A. (1995) Properties of the active plum pox potyvirus RNA polymerase complex in defined glycerol gradient fractions. *Virus Res.*, **37**, 127–137.

Martín, B., Collar, J.L., Tjallingii, W.F. & Fereres, A. (1997) Intracellular ingestion and salivation by aphids may cause the acquisition and inoculation of non-persistently transmitted plant viruses. *J. Gen. Virol.*, **78**, 2701–2705.

Matthews, R.E.F. (1991) *Plant Virology*, Third edn, Academic Press, San Diego, CA, USA. McClintock, K., Lamarre, A., Parsons, V., Laliberte, J.F. & Fortin, M.G. (1998) Identification of a 37-kDa plant protein that interacts with the turnip mosaic potyvirus capsid protein using anti-ideotypic antibodies. *Plant Mol. Biol.*, **37**, 197–204.

McDonald, J.G. & Bancroft, J.B. (1977) Assembly studies of potato virus Y and its coat protein. *J. Gen. Virol.*, **35**, 251–263.

Merits, A., Guo, D. & Saarma, M. (1998) VPg, coat protein and five non-structural proteins of potato A potyvirus bind RNA in a sequence-unspecific manner. *J. Gen. Virol.*, **79**, 3123–3127.

Merits, A., Guo, D., Järvekülg, L. & Saarma, M. (1999) Biochemical and genetic evidence for interactions between potato A potyvirus-encoded proteins P1 and P3 and proteins of the putative replication complex. *Virology*, **263**, 15–22.

Merits, A., Rajamäki, M.-L., Lindholm, P., Runeberg-Roos, P., Kekarainen, T., Puustinen, P., Mäkeläinen, K., Valkonen, J.P.T. & Saarma, M. (2002) Proteolytic processing of potyviral proteins and polyprotein processing intermediates in insect and plant cells. *J. Gen. Virol.*, **83**, 1211–1221.

Moissard, G. & Voinnet, O. (2004) Viral suppression of RNA silencing in plants. *Mol. Plant Pathol.*, **5**, 71–82.

Murphy, J.F., Rychlik, W., Rhoads, R.A., Hunt, A.G. & Shaw, J.G. (1991) A tyrosine residue in the small nuclear inclusion protein of tobacco vein mottling virus links the VPg to the viral RNA. *J. Virol.*, **65**, 511–513.

Murphy, J.F., Klein, P.G., Hunt, A.G. & Shaw, J.G. (1996) Replacement of the tyrosine residue that links a potyviral VPg to the viral RNA is lethal. *Virology*, **220**, 535–538.

Murray, K.E. & Barton, D.J. (2003) Poliovirus CRE-dependent VPg uridylylation is required for positive-strand RNA synthesis but not for negative-strand RNA synthesis. *J. Virol.*, **77**, 4739–4750.

Naderi, M. & Berger, P.H. (1997) Effects of chloroplast targeted potato virus Y coat protein on transgenic plants. *Physiol. Mol. Plant Pathol.*, **50**, 67–83.

Nicolas, O., Dunnington, S.W., Gotow, L.F., Pirone, T.P. & Hellmann, G.M. (1997) Variations in the VPg protein allow a potyvirus to overcome *va* gene resistance in tobacco. *Virology*, **237**, 452–459.

Niepel, M. & Gallie, D.R. (1999) Identification and characterization of the functional elements within the tobacco etch virus 5′ leader required for cap-independent translation. *J. Virol.*, **73**, 9080–9088.

Ohshima, K., Yamaguchi, Y., Hirota, R., Hamamoto, T., Tomimura, K., Tan, Z., Sano, T., Azuhata, F., Walsh, J. A., Fletcher, J., Chen, J., Gera, A. & Gibbs, A. (2002) Molecular evolution of *Turnip mosaic virus*: evidence of host adaptation, genetic recombination and geographical spread. *J. Gen. Virol.*, **83**, 1511–1521.

Oruetxebarria, I., Guo, D., Merits, A., Mäkinen, K., Saarma, M. & Valkonen, J.P.T. (2001) Identification of the genome-linked protein of a *Potato virus A*, with comparison to other members of genus *Potyvirus*. *Virus Res.*, **73**, 103–112.

Paalme, V., Gammelgård, E., Järvekülg, L. & Valkonen, J.P.T. (2004) *In vitro* recombinants of two nearly identical potyviral isolates express novel virulence and symptom phenotypes in plants. *J. Gen. Virol.*, **85**, 739–747.

Paul, A.V., van Boom, J.H., Filippov, D. & Wimmer, E. (1998) Protein-primed RNA synthesis by purified poliovirus RNA polymerase. *Nature*, **393**, 280–284.

Peng, Y.-H., Kadoury, D., Gal-On, A., Huet, H., Wang, Y. & Raccah, B. (1998) Mutations in the HC-Pro gene of zucchini yellow mosaic potyvirus: effects on aphid transmission and binding to purified virions. *J. Gen. Virol.*, **79**, 897–904.

Peremyslov, V.V., Hagiwara, Y. & Dolja, V.V. (1999) HSP70 homolog functions in cell-to-cell movement of a plant virus. *Proc. Natl. Acad. Sci. USA*, **96**, 14771–14776.

Pirone, T.P. & Harris, K.F. (1977) Nonpersistent transmission of plant viruses by aphids. *Annu. Rev. Phytopathol.*, **15**, 55–73.

Pruss, G., Ge, X., Shi, X.M., Carrington, J.C. & Vance, V.B. (1997) Plant viral synergism: the potyviral genome encodes a broad-range pathogenicity enhancer that transactivates replication of heterologous viruses. *Plant Cell*, **9**, 859–868.

Puustinen, P., Rajamäki, M.-L., Ivanov, K., Valkonen, J.P.T. & Mäkinen, K. (2002) Detection of potyviral genome linked protein VPg in virions and its phosphorylation with host kinases. *J. Virol.*, **76**, 12703–12711.

Rajamäki, M.-L. & Valkonen, J.P.T. (1999) The 6K2 protein and the VPg of *Potato virus A* are determinants of systemic infection in *Nicandra physaloides*. *Mol. Plant Microbe Interact.*, **12**, 1074–1081.

Rajamäki, M.-L. & Valkonen, J.P.T. (2002) Viral genome-linked protein (VPg) controls accumulation and phloem-loading of a potyvirus in inoculated potato leaves. *Mol. Plant Microbe Interact.*, **15**, 138–149.

Rajamäki, M.-L. & Valkonen, J.P.T. (2003) Localization of a potyvirus and the viral genome-linked protein in upper non-inoculated leaves at an early stage of systemic infection. *Mol. Plant Microbe Interact.*, **16**, 25–34.

Rajamäki, M.-L. & Valkonen, J.P.T. (2004) Detection of a natural point mutation in *Potato virus A* that overcomes resistance to vascular movement in *Nicandra physaloides*, and studies on seed-transmissibility of the mutant virus. *Ann. Appl. Biol.*, **144**, 77–86.

Restrepo-Hartwig, M.A. & Carrington, J.C. (1994) The tobacco etch potyvirus 6-kilodalton protein is membrane associated and involved in viral replication. *J. Virol.*, **68**, 2388–2397.

Revers, F., Le Gall, O., Candresse, T., Le Romancer, M. & Dunez, J. (1996) Frequent occurrence of recombinant potyvirus isolates. *J. Gen. Virol.*, **77**, 1953–1965.

Riechmann, J.L., Laín, S. & García, J.A. (1991) Identification of the initiation codon of plum pox potyvirus genomic RNA. *Virology*, **185**, 544–552.

Riechmann, J.L., Laín, S. & García, J.A. (1992) Highlights and prospects of potyvirus molecular biology. *J. Gen. Virol.*, **73**, 1–16.

Roberts, I.M., Wang, D., Findlay, K. & Maule, A.J. (1998) Ultrastructural and temporal observations of the potyvirus cylindrical inclusions (CIs) show that the CI protein acts transiently in aiding virus movement. *Virology*, **245**, 173–181.

Roberts, I.M., Wang, D., Thomas, C.L. & Maule, A.J. (2003) Pea seed-borne mosaic virus seed transmission exploits novel symplastic pathways to infect the pea embryo and is, in part, dependent upon chance, *Protoplasma*, **222**, 31–43.

Rodríguez-Cerezo, E., Klein, P.G. & Shaw, J.G. (1991) A determinant of disease symptoms severity is located in the 3′-terminal noncoding region of the RNA of a plant virus. *Proc. Natl. Acad. Sci. USA*, **88**, 9863–9867.

Rodríguez-Cerezo, E., Ammar, E.D., Pirone, T.P. & Shaw, J.G. (1993) Association of the nonstructural P3 viral protein with cylindrical inclusions in potyvirus-infected cells. *J. Gen. Virol.*, **74**, 1945–1949.

Rodríguez-Cerezo, E., Findlay, K., Shaw, J.G., Lomonossoff, G.P., Qiu, S.G., Linstead, P., Shanks, M. & Risco, C. (1997) The coat and cylindrical inclusion proteins of a potyvirus are associated with connections between plant cells. *Virology*, **236**, 296–306.

Rojas, M.R., Zerbini, F.M., Allison, R.F., Gilbertson, R.L. & Lucas, W.J. (1997) Capsid protein and helper component-proteinase function as potyvirus cell-to-cell movement proteins. *Virology*, **237**, 283–295.

Roossinck, M.J. (1997) Mechanisms of plant virus evolution. *Annu. Rev. Phytopathol.*, **35**, 191–209.

Ruiz-Medrano, R., Xoconostle-Cázares, B. & Lucas, W.J. (2001) The phloem as a conduit for inter-organ communication. *Curr. Opin. Plant Biol.*, **4**, 202–209.

Sáenz, P., Cervera, M.T., Dallot, S., Quiot, L., Quiot, J.-B., Riechmann, J.L. & García, J.A. (2000) Identification of a pathogenicity determinant of *Plum pox virus* in the sequence encoding the C-terminal region of protein P3 + 6K1. *J. Gen. Virol.*, **81**, 557–566.

Savenkov, E. I. & Valkonen, J. P. T. (2002) Silencing of a viral RNA silencing suppressor in transgenic plants. *J. Gen. Virol.*, **83**, 2325–2335.

Schaad, M.C. & Carrington, J.C. (1996) Suppression of long-distance movement of tobacco etch virus in a nonsusceptible host. *J. Virol.*, **70**, 2556–2561.

Schaad, M.C., Haldeman-Cahill, R., Cronin, S. & Carrington, J.C. (1996) Analysis of the VPg-proteinase (NIa) encoded by tobacco etch potyvirus: effects of mutations on subcellular transport, proteolytic processing and genome amplification. *J. Virol.*, **70**, 7039–7048.

Schaad, M.C., Lellis, A.D. & Carrington, J.C. (1997) VPg of tobacco etch potyvirus is a host genotype-specific determinant for long-distance movement. *J. Virol.*, **71**, 8624–8631.

Schaad, M.C., Anderberg, R.J. & Carrington, J.C. (2000) Strain-specific interaction of the tobacco etch virus NIa protein with the translation initiation factor eIF4E in yeast two-hybrid system. *Virology*, **273**, 300–306.

Shi, X.M., Miller, H., Verchot, J., Carrington, J.C. & Bowman Vance, V. (1997) Mutations in the region encoding the central domain of helper component-proteinase (HC-Pro) eliminate potato virus X/potyviral synergism. *Virology*, **231**, 35–42.

Shukla, D.D., Strike, P.M., Tracy, S.L., Gough, K.H. & Ward, C.M. (1988) The N and C termini of the coat proteins of potyviruses are surface-located and the N terminus contains major virus-specific epitopes. *J. Gen. Virol.*, **69**, 1497–1508.

Shukla, D.D., Ward, C.W. & Brunt, A.A. (1994) *The Potyviridae*. C.A.B. International, Wallingford, UK. 516p.

Simón-Buela, L., Guo, H.S. & García, J.A. (1997) Long sequences in the 5′ noncoding region of plum pox virus are not necessary for viral infectivity but contribute to viral competitiveness and pathogenesis. *Virology*, **233**, 157–162.

Smith, D.B., McAllister, J., Casino, C. & Simmonds, P. (1997) Virus 'quasispecies': making a mountain of a molehill? *J. Gen. Virol.*, **78**, 1511–1519.

Spetz, C. & Valkonen, J.P.T. (2004) Potyviral 6K2 protein long distance movement and symptom induction functions are independent and host-specific. *Mol. Plant–Microbe Interact.*, **17**, in press.

Spetz, C., Taboada, A.M., Darwich, S., Ramsell, J., Salazar, L.F. & Valkonen, J.P.T. (2003) Molecular resolution of a complex of potyviruses infecting solanaceous crops at the centre of origin in Peru. *J. Gen. Virol.*, **84**, 2565–2578.

Teycheney, P.Y., Aaziz, R., Dinant, S., Salánki, K., Tourneur, C., Baláz, E., Jacquemond, M. & Tepfer, M. (2000) Synthesis of (−)-strand RNA from the 3′ untranslated region of plant viral genomes expressed in transgenic plants upon infection with related viruses. *J. Gen. Virol.*, **81**, 1121–1126.

Teycheney, P.Y. & Tepfer, M. (2001) Virus-specific spatial differences in the interference with silencing of the *chs-A* gene in non-transgenic petunia. *J. Gen. Virol.*, **82**, 1239–1243.

Ullah, Z. & Grumet, R. (2002) Localization of *Zucchini yellow mosaic virus* to the veinal regions and role of the viral coat protein in veinal chlorosis conditioned by the *zym* potyvirus resistance locus in cucumber. *Physiol. Mol. Plant Pathol.*, **60**, 79–89.

Valkonen, J.P.T., Kyle, M.M. & Slack, S.A. (1996) Comparison of resistance to potyviruses within Solanaceae: infection of potatoes with tobacco etch potyvirus and peppers with potato A and Y potyviruses. *Ann. Appl. Biol.*, **129**, 25–38.

van Bel, A.J.E. (2003) The phloem, a miracle of ingenuity. *Plant, Cell and Environment*, **26**, 125–149.

Van Regenmortel, M.H.V., Fauquet, C.M., Bishop, D.H.L., Carstens, E.B., Estes, M.K., Lemon, S.M., Maniloff, J., Mayo, M.A., McGeoch, D.J., Pringle, C.R. & Wickner, R.B. (2000) *Virus Taxonomy: The Seventh Report of the International Committee on Taxonomy of Viruses*. Academic Press, San Diego, CA, USA.

Verchot, J. & Carrington, J.C. (1995) Evidence that the potyvirus P1 proteinase functions *in trans* as an accessory factor for genome amplification. *J. Virol.*, **69**, 3668–3674.

Verchot, J., Koonin, E. V. & Carrington, J. C. (1991) The 35-kDa protein from the N-terminus of the potyviral polyprotein functions as a third virus-encoded proteinase. *Virology*, **185**, 527–535.

Waigmann, E., Chen, M.H., Bachmaier, R., Ghosroy, S. & Citovsky, V. (2000) Regulation of plasmodesmal transport by phosphorylation of tobacco mosaic virus cell-to-cell movement protein. *EMBO J.*, **19**, 4875–4884.

Wang, D. & Maule, A. J. (1992) Early embryo invasion as a determinant in pea of the seed transmission of pea seed-borne mosaic virus. *J. Gen. Virol.*, **73**, 1615–1620.

Wang, D. & Maule, A. J. (1994) A model for seed transmission of a plant virus: genetic and structural analyses of pea embryo invasion by pea seed-borne mosaic virus. *Plant Cell*, **6**, 777–787.

Wang, D. & Maule, A. J. (1995) Inhibition of host gene expression associated with plant virus replication. *Science*, **267**, 229–231.

Wang, R.Y., Ammar, E.D., Thornbury, D.W., Lopez-Moya, J.J. & Pirone, T.P. (1996) Loss of potyvirus transmissibility and helper component activity correlate with non-retention of virions in aphid stylets. *J. Gen. Virol.*, **77**, 861–867.

Wang, X., Ullah, Z. & Grumet, R. (2000) Interaction between zucchini yellow mosaic potyvirus RNA-dependent RNA polymerase and host poly-(A) binding protein. *Virology*, **275**, 433–443.

Wittmann, S., Chatel, H., Fortin, M.G. & Laliberté, J.-F. (1997) Interaction of the viral protein genome linked of turnip mosaic potyvirus with the translational eukaryotic initiation factor (iso) 4E of *Arabidopsis thaliana* using the yeast two-hybrid system. *Virology*, **234**, 84–92.

Wolf, S., Deom, C.M., Beachy, R.N. & Lucas, W.J. (1989) Movement protein of tobacco mosaic virus modifies plasmodesmal size exclusion limit. *Science*, **246**, 377–379.

Wu, X. & Shaw, J. (1998) Evidence that assembly of a potyvirus begins near the 5′ termius of the viral RNA. *J. Gen. Virol.*, **79**, 1525–1529.

Yambao, Ma.L.M., Masuta, C., Nakahara, K. & Uyeda, I. (2003) The central and C-terminal domains of VPg of *Clover yellow vein virus* are important for VPg-HCPro and VPg-VPg interaction. *J. Gen. Virol.*, **84**, 2861–2869.

Yelina, N.E., Savenkov, E.I., Solovyev, A.G., Morozov, S.Y. & Valkonen, J.P.T. (2002) Long distance movement, virulence, and RNA silencing suppression controlled by a single protein in hordei- and potyviruses: complementary functions between virus families. *J. Virol.*, **76**, 12981–12991.

Zaitlin, M. & Jagendorf, A.T. (1960) Photosynthetic phosphorylation and Hill reaction activities of chloroplasts isolated from plants infected with tobacco mosaic virus. *Virology*, **12**, 477–486.

4   The *Ralstonia solanacearum*–plant interaction

Christian Boucher and Stéphane Genin

4.1   The pathogen

4.1.1   *A major plant pathogen with an unusually wide host range*

Ralstonia solanacearum, previously known as *Pseudomonas solanacearum*, was originally described by Smith (1896) as the causative agent of bacterial wilt of solanaceous plants. It has been intensively studied since then, leading to the production of an abundant literature, including several comprehensive reviews (Kelman, 1953; Buddenhagen, 1986; Hayward, 1991, 2000; Boucher *et al.*, 1992). This bacterium is responsible for some of the most devastating bacterial plant diseases in the world – including Granville wilt of tobacco, brown rot of potato and Moko disease of banana. The high economic and social impact of this organism results from its wide geographical distribution in all warm and tropical countries of the globe (Hayward, 1991). This impact also results from the very wide host range of *R. solanacearum* which comprises over 200 plant species, representing over 50 botanical families and covering both monocots and dicots extending from annual plants to trees and shrubs. Apart from the crop species already mentioned (tobacco, potato and banana), *R. solanacearum* is a major pathogen on tomato, pepper, eggplant, strawberry, bean, ginger, mulberry and other crops (Hayward, 1994; Fig. 4.1 AB).

Recently, geographical distribution of the pathogen has been extended to more temperate countries from Europe and North America as the result of the dissemination of strains belonging to race 3, biovar 2 (see Section 4.1.4). These strains differentiated in Andean plateaus and as such are adapted to cooler environmental conditions. Such strains were probably disseminated through movement of infected potato seed and edible tubers.

4.1.2   *Taxonomical status of the species and infraspecific classification*

R. solanacearum is classified as a β-proteobacterium together with two closely related plant pathogens *R. (Pseudomonas) syzigii* and the blood disease bacterium (BDB) (Taghavi *et al.*, 1996). Based on several criteria, *R. solanacearum* appears to be a complex species and because different strains may represent different threats to agriculture, depending on environmental conditions or geographical location, different systems have been proposed for classification. This was particularly

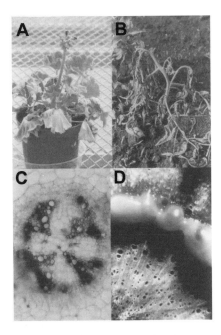

Figure 4.1 Wilt symptoms caused by *R. solanacearum* race 1 strains on geranium (A) and tomato (B). Vascular colonization of wilted tomato plants: stem section showing a bacterial exudate oozing out from xylem vessels (C) and tranversal section of a hypocotyl showing four distinct vascular bundles, each containing numerous vessels colonized with blue-stained bacteria (D). Pictures courtesy of M.T. Momol, University of Florida/IFAS (A), P. Frey, INRA (B) and J. Vasse, INRA (C and D).

needed to conduct epidemiological studies and to prevent the dissemination of new strains in geographical areas exempted of specific strains. A classification of races based on host range was first proposed by Buddenhagen *et al.* (1962). In this system, race 1 strains are pathogenic on solanaceous hosts and diploid banana, race 2 strains are only pathogenic on *Heliconia* and triploid banana and race 3 strains infect primarily potato and occasionally tomato and a few weeds such as *Solanum dulcamara*. More recently, this system has been extended with race 4, to respectively include strains specific of ginger and mulberry (Buddenhagen & Kelman, 1964; Aragaki & Quinon, 1965; He *et al.*, 1983). Independently of the race system, a biovar classification system (originally called biotype) has been developed based on the capability of the bacteria to metabolize or oxidize a set of hexose alcohols and disaccharides (Hayward, 1964). No direct relationship could be established between the race and the biovar classifications, with the exception of race 3 which matches fairly well with biovar 2 (Hayward, 2000). Such a discrepency suggests that the two classification systems could be artificial and that they do not reflect evolutionary relationships between the species. Decisive information on this point came from the development of molecular techniques which have allowed investigation of genome organization. Use of RFLP markers to analyse the genetic diversity at various randomly chosen loci in the genome allowed identification of 40 distinct genomic groups (also called Multi Locus Genotypes) (Cook *et al.*, 1989; Cook & Sequeira, 1994). This system clearly distinguishes two main subdivisions in the species which are highly correlated with the geographical origin of the strains, thus suggesting that following geographical isolation of a common ancestor strain, the species has evolved into two independent phyla. Strains from the first subdivision, named *Asiaticum*, originate mainly from Australia and Asia, whereas strains from the second subdivision, named *Americanum*, originate mostly from South and Central America. These conclusions are fully supported by the analysis of the nucleotide sequence of 16S rDNA (Li *et al.*, 1993). More recently, a third subdivision corresponding to African isolates has been identified using the related molecular criteria; however, the relative positioning of the African phylum compared to the Americanum and Asiaticum phyla is still uncertain (Poussier *et al.*, 2000a,b).

4.1.3   R. solanacearum, *a vascular pathogen that promotes xylem vessel occlusion*

As a soil-borne plant pathogen *R. solanacearum* naturally infects plants via the roots. Several studies combining light and electron microscopy have been used to unravel the infection process (Schmit, 1978; Wallis & Trutner, 1978; Vasse *et al.*, 1995, 2000). This process starts with the bacterial colonization of root exudation sites such as root extremities and natural wounds resulting from the emergence of secondary roots. The bacteria are attracted towards these sites and the polar flagella of the bacterium are probably involved in this early step of colonization (see Section 4.2.3). Infection then progresses via multiplication of bacteria on the surface of epidermal cells, followed by invasion of the inner root cortex and the formation of intercellular microcolonies. Bacteria then actively traverse the natural barrier of the endodermis and penetrate into the vascular cylinder where they multiply within the vascular parenchyma that surrounds the xylem vessels (Fig. 4.1 CD). After colonization of the xylem vessels, they multiply heavily in these vessels thus allowing their migration towards the aerial part of the plant. At this stage, the bacteria remain located within the xylem vessels and at the high cell concentration stage that they reach and depending on the *phcA* quorum sensing system described in Section 4.2.4.1, they reorient their metabolism to produce large amount of a high molecular weight exopolysaccharide (see Section 4.2.1) that mechanically plugs the vessels thereby preventing water flow in the plant and leading to the appearence of the typical wilt symptoms.

The functional analysis of pathogenicity genes indicates that several hydrolytic enzymes might be necessary to promote the intercellular progression of the bacterium within the inner cortex and during translation towards the xylem vessels. Type III-dependent effectors may act at this stage by inhibiting plant defense mechanisms and may later be involved in reorienting host metabolism so that plant cells provide the bacterium with valuable nutrients.

4.1.4   *Epidemiology and environmental survival*

It is not possible to give a general scheme of the epidemiology of bacterial wilt since the bacterium uses different strategies for dissemination and survival in the environment, depending on the bacterial strain, its host, environmental conditions, agricultural practices and other factors that are not yet clearly identified. In an extensive review that is far from being exhaustive, Hayward (1991) mentions movement of contaminated soil, contaminated water and contaminated plants, spread from plant to plant through root to root contact and pruning, aerial transmission through rain splashes or insects as factors playing an important role in bacterial dissemination. Table 4.1 summarizes selected references illustrating the different modes of

Table 4.1 Some modes of dissemination of *Ralstonia solanacearum* (reproduced from *Bacterial Wilt: the disease and its causative agent, Pseudomonas solanacearum*, with permission of CAB International)

Agency	Other information	Key references
Vegetative planting material	Latently infected potato tubers	Hayward (1991)
	Ginger rhizomes	Lum (1973)
	Banana corms from Central America to the Philippines	Rillo (1979); Buddenhagen (1986)
	Heliconia rhizomes from Hawaii to Quensland, Australia	Hyde *et al.* (1992)
True seed	Groundnut seed in China	Zhang *et al.* (1993)
	Groundnut seed in Indonesia	Machmud & Middleton (1990)
	Tomato seed from India to Nepal	Shakya (1993)
	Tomato seed, other locations	Hayward (1991)
Tomato transplants	From the southeastern USA to nothern states and Canada	Vaughan (1944); Gitaitis *et al.* (1992)
Strawberry seedlings	In Taiwan	Hsu (1991)
Insect transmission	From diseased to healthy banana inflorescences in moko disease	Buddenhagen & Elsasser (1962)
Mechanical transmission		
by clipping	On tomato in Georgia and Florida, USA	McCarter & Jaworski (1969)
by pruning knives	In moko disease of banana in Central America	Sequeira (1958)
by harvesting equipment	On *Perilla crispa* in Taiwan	Hsu (1991)
by root wounding during cultural practices	General	Kelman (1953)
Root-knot nematode	Reported on several hosts	Kelman (1953); Lucas *et al.* (1955); Johnson & Powell (1969); Napiere (1980)
Root-to-root transmission	Release of bacteria into the soil from diseased roots in the vicinity of healthy roots of neighbouring plants	Kelman & Sequeira (1965)
Aerial transmission through rain-splash dispersal of an epiphytic population	On tobacco in Japan	Hara & Ono (1985); Ono (1983)

dissemination of the bacteria. The ability of the bacterium to survive in the environment in the absence of a host plant is another key element in the epidemiology of bacterial wilt.

Recently, following accidental introduction of *R. solanacearum* to potatoes in northern European Union countries (Hayward *et al.*, 1998), intensive studies have been conducted to investigate dispersal and survival of race 3 bacteria in the environment (Elphinstone *et al.*, 1998; Janse *et al.*, 1998). We use the main conclusions from these studies to illustrate here the complexity of the epidemiological behavior of the organism to maintain and propagate itself after introduction in a new geographical context (see Fig. 4.2).

It is most likely that introduction of the bacterium in northern European countries occurred via importation of consumption potato tubers originating from endemically infected countries. Water effluents resulting from domestic or industrial processing of this material led to contamination of water streams. This initial inoculum is multiplied by infecting asymptomatically the weed *Solananum dulcamara* that is commonly found along rivers and that forms aquatic, adventitious roots, therefore prompting the release of bacteria in the water stream and infection of additional plants downstream. Use of such contaminated waters for irrigation of plots devoted to potato seed production is the most probable source of primary potato infections. Because potato can develop asymptomatic latent infection, infected potato seed thus obtained became a main source of race 3 dissemination among many western European countries.

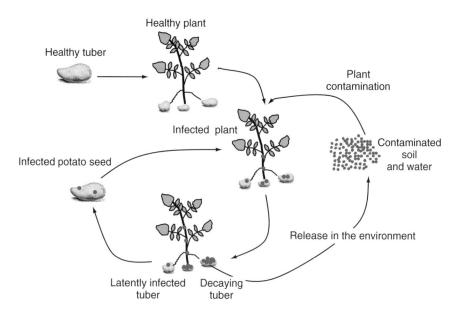

Figure 4.2 Schematic representation of bacterial wilt epidemiological cycle of a potato race 3 strain.

Because knowledge concerning maintenance of the bacterium in infected soils is of crucial importance for the design of strategies to fight the disease and for the management of contaminated areas, it is not surprising that many studies addressing this point have been conducted under different environmental conditions (for examples see [Graham *et al.*, 1979; Granada & Sequeira, 1983; Moffett *et al.*, 1983; Shekawat & Pérombélon, 1991]). Although some of the resulting conclusions may be somewhat controversial on specific points, all the authors agree that the bacterium is usually still present in the soil several months after removal of the infected host. More recent studies undertaken in Belgium, in the United Kingdom and in the Netherlands have established that the bacterium can over-winter in water from irrigation ditches and also in soil (Elphinstone, 1996; van Elsas *et al.*, 2000, 2001). Over-wintering is favored by the presence of perennial weeds such as stinging nettle that can be infected by the bacterium (Wenneker *et al.*, 1999). These observations are the basis of a strict legislation that has been developed to prevent further expansion of the disease and to tentatively eradicate the pathogen from infested areas (Council Directive 77/93/EEC).

Recently, introduction of a race 3 isolate, pathogenic on potato, has been reported in Wisconsin and South Dakota. In these cases, the bacterium was imported on geranium cuttings originating from Kenya (Williamson *et al.*, 2002).

4.2 Molecular studies of pathogenicity determinants

Over the past two decades, genetic and molecular studies unraveled important mechanisms underlying *R. solanacearum* pathogenicity. The availability of the complete genome sequence from the biovar 4 race 1 strain GMI1000 (Salanoubat *et al.*, 2002) has also established this plant pathogen as a model system to study the molecular determinants controlling bacterial pathogenicity.

4.2.1 Exopolysaccharide I

The onset of wilting in tomato plants is correlated with altered water transport probably caused by the high bacterial cell densities in the plant vascular sytem and by the important production of exopolysaccharide (EPS) (reviewed in Denny, 1995). The primary EPS made by *R. solanacearum* is an acidic, high molecular mass heteropolysaccharide (Orgambide *et al.*, 1991) known as EPS I. Genes coding for the biosynthetic pathway for EPS I are encoded by the 20 kb *eps* operon, which is subjected to a complex genetic control (see Section 4.2.4.3). EPS I is the most important virulence factor of *R. solanacearum* since *eps* mutants do not cause wilt symptoms even when introduced directly into stem wounds (Denny & Baek, 1991), although they are not completely non-pathogenic. Recent studies showing that these *eps* mutants are also severely reduced in systemic colonization of tomato plants when inoculated via unwounded roots (Saile *et al.*, 1997; Araud-Razou *et al.*, 1998) led to the suggestion that EPS I may contribute to minimizing or avoiding the recognition of bacterial surface structures by plant defense mechanisms.

4.2.2   Protein secretion systems

Like most bacterial pathogens of plants and animals, *R. solanacearum* pathogenicity relies on two main protein secretion systems. The first one is the Type II secretion system, also named as the General Secretory Pathway, a widely conserved Sec-dependent secretion pathway (Pugsley, 1993) that exports most major exoproteins, including the plant cell wall degrading enzymes. The second is the Type III secretion system, a specialized export system that delivers toxic proteins directly into the cytoplasm of plant cells (see Section 4.2.2.2). Both the secretion systems are essential for pathogenesis because mutants defective in either system are severely impaired in colonization ability and multiplication *in planta* (Boucher *et al.*, 1985; Kang *et al.*, 1994). However, mutants defective in individual exported proteins have subtle or no virulence phenotypes, suggesting that these secreted proteins are collectively important for disease and that many of them are likely to be functionally redundant.

4.2.2.1   Plant cell wall degrading enzymes and proteins secreted
through the General Secretory Pathway

To date, six exoproteins secreted through the General Secretory Pathway have been studied at the molecular level (see Table 4.2). Most of them are degradative enzymes of the pectin and cellulose components of plant cell walls and are presumed to contribute to virulence not only by releasing nutrients for the pathogen but also by aiding in initial invasion of plant roots, bacterial spread and colonization of vessels.

Cellulolytic enzymes. *R. solanacearum* produces an endoglucanase, Egl, that can hydrolyze the β-1,4 glycosidic linkage of cellulose (Roberts *et al.*, 1988). Egl mutants appear to be reduced in their ability to colonize the stems of infected plants but remain pathogenic (Roberts *et al.*, 1988). Another exoglucanase, a β-1,4-exocello-biohydrolase, that releases cellobiose from the non-reducing ends of the chains has been found during annotation of the whole genome sequence of the organism but has not been studied at the molecular level.

Pectinolytic enzymes. *R. solanacearum* produces one pectin methylesterase (Pme) and three polygalacturonases (PGs). Pme removes methyl groups from pectin, thereby facilitating subsequent breakdown of this primary cell wall component by PGs, which, in turn, degrade the pectin polymers. *R. solanacearum* has two types of PG: an endo-PG, named PglA or PehA (Schell *et al.*, 1988; Allen *et al.*, 1991), that cleaves the pectin polymer at random releasing large fragments, and two exo-PG, the exo-poly-α-D-galacturonosidase PehB (Allen *et al.*, 1991) and exopolygalacturonase PehC (Gonzalez & Allen, 2003), that release galacturonic acid dimers and monomers respectively.

Table 4.2 Inventory of *R. solanacearum* proteins known to transit through the Type II and Type III secretion systems

Protein	Size (kDa)	Predicted features	Effect on pathogenicity
Type II secretion system (General Secretory Pathway)			
Pme	44	Pectin methyl esterase	None
PglA (PehA)	53	Endo-polygalacturonase	Reduced stem colonization
PehB	74	Exo-poly-α-D-galacturonosidase	Minor
PehC	70	Exo-polygalacturonase	None
Egl	43	Endo-1,4-β-glucanase	Minor
CbhA	61	1,4-β-cellobiosidase	Unknown
Tek	28		None
Type III secretion system			
PopA	33	Harpin, glycin-rich thermoresistant protein	None
PopB	18	Functional nuclear localisation signals	None
PopC	106	22 Leucine-rich-repeats	None
PopP1	41	AvrRxv/YopJ family, putative cysteine protease	Confers avirulence towards certain *Petunia* lines
PopP2	53	AvrRxv/YopJ family, putative cysteine protease	Confers avirulence towards *Arabidopsis*
		Functional nuclear localisation signals	RRS1-R plants (ecotype Columbia)
RipA	121	AWR family	Minor
RipG	64	17.5 Leucine-rich-repeats (Gala subfamily)	None
RipT	34	AvrPphB/YopT family, putative cysteine protease	None

As mentioned above, the contribution of these enzymes to *R. solanacearum* virulence is relatively minor. Inactivation of *pme* does not affect virulence on several host plants (Tans-Kersten *et al.*, 1998). A triple mutant strain defective for PehA, PehB and PehC has recently been generated, that is completely non-pectinolytic but is only slightly reduced in virulence on tomato following soil inoculation assays (Gonzalez & Allen, 2003). Interestingly, a galacturonate transporter gene, *exuT*, lies immediately downstream of *pehC*. In order to evaluate the nutritional role *in planta* of the PGs, an *exuT* mutant was created which still produces all three isozymes of PGs but cannot take up PG degradation products (Gonzalez & Allen, 2003). This *exuT* mutant had wild-type virulence on tomato, demonstrating that metabolism of galacturonic acid does not contribute significantly to bacterial multiplication inside the plant (Gonzalez & Allen, 2003). These data rather support the view that the primary virulence functions of PGs is a maceration of plant pectic substances during the early stages of interaction to facilitate *R. solanacearum* invasion and spread.

Tek protein.   Tek is a 28-kDa protein which is the most abundant exoprotein found in *R. solanacearum* supernatants (Denny *et al.*, 1996). It is derived from a 58-kDA precursor whose very basic C-terminal portion becomes the 28-kDa Tek mature protein via multistep processing (Denny *et al.*, 1996). Because *tek* is located 3 kb downstream of the *xpsR* gene controlling EPS biosynthesis (see Section 4.2.4.3), its association with virulence is suspected but a non-polar *tek* mutant still produces wild-type amounts of EPS I and remains fully virulent (Denny *et al.*, 1996). The function of this protein remains unknown. However, the whole genome sequence analysis recently revealed the presence of a gene located 2 kb downstream of the *tek* gene and encoding a protein highly similar to Tek (75% similarity). This genetic redundancy suggests that the potential involvement of the Tek proteins in virulence should be re-examined through the analysis of a mutant strain in which both genes have been disrupted.

4.2.2.2 *The Type III secretion system (TTSS)*

hrp *genes.*   In *R. solanacearum*, the TTSS is encoded by the hypersensitive reaction and pathogenicity (*hrp*) genes. Through the screening of ~8000 *R. solanacearum* Tn5-induced mutants on host plants, Boucher *et al.* (1985) identified 12 strains altered simultaneously in their ability to cause disease on the tomato host and to induce the hypersensitive reaction on the non-host tobacco plant. In *P. syringae* pv. *phaseolicola*, the molecular characterization of mutants showing similar null phenotypes on plants led to the discovery of *hrp* genes (Lindgren *et al.*, 1986). It is now well established that *hrp* genes are present in all Gram-negative plant pathogens, except *Agrobacterium* sp., and are organized in large gene clusters (reviewed in Lindgren, 1997). Based on similarities in *hrp* gene organization and regulation, plant pathogenic bacteria have been classified into two groups, group I (*Erwinia* sp. and *P. syringae*) and group II (*R. solanacearum* and *Xanthomonas* sp.).

Assembly of the Type III secretion structure.   The *hrp* locus of strain GMI1000 consists of 26 open reading frames (ORFs) organized into seven operons and is located on the megaplasmid (Arlat *et al.*, 1992; van Gijsegem *et al.*, 1995). The systematical non-polar mutagenesis of the ORFs in the *hrp* gene cluster identified 13 structural genes required for the biogenesis of the TTSS (van Gijsegem *et al.*, 2002), which comprise a set of eight conserved and five non-conserved *hrp* genes. Conserved *hrp* genes (termed *hrc* genes for *hrp*-conserved) are common to other bacterial plant pathogens and encode components of the TTSS which are present in animal pathogenic bacteria (Gough *et al.*, 1992; Bogdanove *et al.*, 1996 and for a comprehensive review Cornelis & van Gijsegem, 2000). With the exception of HrcC, an outer membrane protein which belongs to the secretin family (Genin & Boucher, 1994), Hrc proteins share sequence similarity with components of the flagellar protein export system (van Gijsegem *et al.*, 1995). Thus, the flagellar assembly apparatus probably represents an evolutionary ancestor of the TTSS (Aizawa, 2001). Another class of *hrp* genes encode regulatory proteins (HrpB, HrpG) that are responsible for the expression of all TTSS-associated genes *in planta* or in *hrp*-inducing medium (see Section 4.2.4.4).

The TTSS is prolonged by an extracellular structure called the Hrp pilus, which has been identified in *R. solanacearum* and other plant pathogenic bacteria (Roine *et al.*, 1997; van Gijsegem *et al.*, 2000). *hrpY* encodes the major constituent of the *R. solanacearum* Hrp pilus: it has a diameter of 7 nm and is required for Type III secretion *in vitro* (van Gijsegem *et al.*, 2000). It has been shown that the Hrp pilus serves as a conduit for secreted proteins (Jin & He, 2001) and the current models view this structure as a component of the protein translocation apparatus, which crosses the plant cell wall and injects Type III effectors into the host cell (Romantschuk *et al.*, 2001; Büttner *et al.*, 2002).

Type III effector proteins.   Through the TTSS, various effector proteins are secreted and/or translocated into the plant cell but their biochemical function and their contribution to pathogenicity remains largely unknown. The completion of the *R. solanacearum* genome sequence has allowed the identification of >40 potential Type III effectors belonging to the Hrp regulon (Cunnac *et al.*, submitted). To date, eight of these proteins are known to transit through the TTSS, such as the harpin PopA (Arlat *et al.*, 1994), PopB, PopC (Guéneron *et al.*, 2000) and PopP1 (Lavie *et al.*, 2002) which were shown to be secreted *in vitro*. Recently, direct evidence for translocation into plant cells was reported for the avirulence gene product PopP2 (see p. 102) and the RipA, RipG and RipT effectors (Cunnac *et al.*, submitted), see Table 4.2.

It is presumed that these TTSS effectors are able, collectively, to suppress host defense responses and somehow to promote disease development. In *R. solanacearum*, *hrp*-dependent functions are not essential to the infection process, since it is known that several *hrp* mutants retain their ability to naturally infect and colonize tomato plants (Trigalet & Trigalet-Demery, 1990; Vasse *et al.*, 2000). Following soil inoculation, these mutants can be isolated from the stems of infected plants, although their respective populations in plants always remained very low compared to those in plants inoculated with the wild-type strain. Therefore, the growth of *hrp* mutants *in planta* is certainly limited by the low availability of nutrients and/or general plant defense responses. This suggests that, in addition to subverting the plant surveillance mechanisms, TTSS effectors could play a role in diverting certain plant metabolites from the plant to the bacteria. Recent advances in the functional characterization of harpins support this hypothesis: harpins were shown to associate with lipid bilayers and form pores *in vitro* (Lee *et al.*, 2001; H. Keller, personal communication), thus leading to the suggestion that they may function in the release of nutrients from the host cells.

Avirulence proteins: Type III effectors recognized by plant resistance genes.   The best studied Type III effectors in plant pathogens are probably the products of avirulence (*avr*) genes which were discovered 20 years ago without knowing that they encode TTSS-substrates. *avr* genes were identified by functional assays, because a cloned *avr* gene can convert a virulent strain into an avirulent one when tested on a resistant (or non-host) plant that carries the appropriate resistance gene (reviewed in Leach & White, 1996). *avr* genes are naturally present in avirulent strains, but in fact the original function of the Avr proteins must have been to promote disease in susceptible plants that lack the corresponding disease resistance genes. The first cloned *avr* determinant in *R. solanacearum* was *avrA*, which confers avirulence towards tobacco at the host species level (Carney & Denny, 1990). Recently, the products of the *popP1* and *popP2* genes, both belonging to the AvrRxv/YopJ family of Type III effectors, were shown to be avirulence factors recognized by resistant Petunia and *Arabidopsis* plants respectively (Lavie *et al.*, 2002; Deslandes *et al.*, 2003).

Pop2 is of special interest since it is involved in the first *gene-for-gene* relationship described for *R. solanacearum*. PopP2 interacts physically with *Arabidopsis* RRS1-R, a TIR-NBS-LRR resistance protein that also possesses a WRKY domain characteristic of several plant transcription factors (Deslandes *et al.*, 2003). The nuclear localization of RRS1-R was shown to be dependent on the presence of nuclear localization signals borne by the PopP2 effector (Deslandes *et al.*, 2003). These data suggest a model in which the disease-resistant protein may reach the plant nucleus by a *piggyback* mechanism following recognition of and interaction with the PopP2 effector in order to activate a battery of plant defense genes.

4.2.3 Motility and attachment to host cell surfaces

4.2.3.1 Swimming motility

R. solanacearum possesses swimming motility mediated by one to four polar flagella. The fact that motility is co-regulated with known virulence functions by the Phc cell-density sensing system was the first hint that motility may also contribute to virulence. This question was recently adressed by Tans-Kersten *et al.* (2001) who constructed two non-motile mutants by disrupting the *fliC* (encoding the subunit of the flagellar filament) and *fliM* (encoding the flagellar motor switch protein) genes. FliC mutants (and to a lesser extent FliM mutants) appear to wilt tomato plants following soil inoculations much more slowly than wild types but this difference cannot be observed after wounded petiole inoculations, suggesting that swimming motility makes its most important contribution to virulence during early stages of host plant invasion (Tans-Kersten *et al.*, 2001).

4.2.3.2 Type IV pili

R. solanacearum also produces Type IV pili (Tfp) that determine the so-called "twitching motility" (Liu *et al.*, 2001), a form of flagella-independent translocation of bacteria over solid surfaces. Tfp and twitching motility have been studied most in *P. aeruginosa*, where >30 genes are required for synthesis and function of polar and retractable Tfp (Wall & Kaiser, 1999). *R. solanacearum* Tfp is composed primarily of a single pilin protein, PilA, assembled to a flexuous polar filament (Kang *et al.*, 2002). A *pilA* mutant is reduced in virulence on tomato plants, whether unwounded roots or wounded petioles are inoculated (Kang *et al.*, 2002). This *pilA* mutant is also not naturally competent for transformation and does not exhibit polar attachment to tobacco or tomato cells (Kang *et al.*, 2002), as typically observed during co-cultivation of bacteria with plant cell suspensions (Aldon *et al.*, 2000). Taken together, these results demonstrate the multiple roles of Tfp in the biology of *R. solanacearum*. Pilus formation probably contributes to pathogenesis during both invasion of roots and inside the plant by promoting adherence to host cell surfaces, colonization of root surfaces, migration to wound sites and biofilm development. There is indeed suspicion that *R. solanacearum* is able to form biofilms on host xylem vessel walls; these specialized aggregates may protect the pathogen from host defenses and aid bacterial survival during latent infections and saprophytic life.

4.2.4   *Regulation of pathogenicity*

R. solanacearum pathogenicity functions are regulated by a sophisticated, multi-component network that is sensitive to the environment. A general picture summarizing the complex genetic relationships between >20 regulatory genes as evidenced by molecular studies during the last 15 years in several research groups is presented in Fig. 4.3. An exhaustive presentation of the components and regulatory targets of this complex network can be found in the review of Schell (2000).

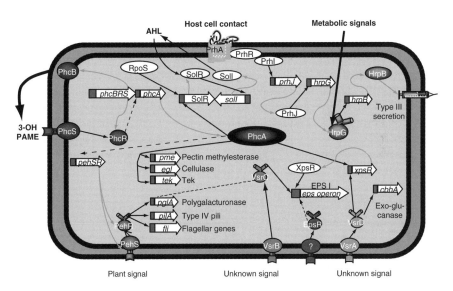

Figure 4.3 Model of the regulation network controlling virulence functions in *R. solanacearum*. This model is based on a compilation of data obtained from studies on different strains. Genes and gene promoters are represented by large white arrows and red boxes respectively. Ellipses represent regulatory proteins and their action on target promoters (direct or indirect) is symbolized by thin black arrows. Full black arrows indicate positive transcriptional control, dotted arrows indicate negative control. See details in text.

4.2.4.1   PhcA, a global regulator controlling phenotypic conversion

The master regulator of several major physiological processes (including patho-genicity) in the *R. solanacearum* life cycle is the LysR-type transcriptional regulator PhcA (Brumbley *et al.*, 1993). Spontaneous or induced mutations in *phcA* result in the pleiotropic changes called *phenotypic conversion* (PC), a phenomenon described 50 years ago by Kelman. PC-type mutants are easily recognized on agar medium containing tetrazolium chloride because their colonies are round, red and butyrous, which contrast with the irregularly round, white or pink, fluidal parent colonies (Kelman, 1954). This change in colony type is due to the loss of one or more com-ponents of EPS I (Orgambide *et al.*, 1991). Further molecular studies have demon-strated that PhcA controls directly or indirectly, in addition to EPS production, the expression of other functions contributing to virulence, such as plant cell wall degrading enzymes, swimming motility or Type IV pili (Schell, 2000; Kang *et al.*, 2002). In fact, PhcA acts as a switch which, depending on its status (activated or not), elevates production of some gene products while repressing others. Inactivation of PhcA through mutational events (Brumbley *et al.*, 1993; Poussier *et al.*, 2003) or stringent control of its activated state (see Section 4.2.4.2) therefore triggers dramatic physiological changes via global shifts in gene expression.

4.2.4.2   An atypical cell-sensing system

Tle levels of active PhcA are controlled by an endogenous signal molecule, 3-hydroxypalmitic acid methyl ester (3-OH PAME) (Flavier *et al.*, 1997a). The synthesis of this molecule is catalyzed by PhcB, whereas two other proteins, PhcS and PhcR, a histidine kinase sensor and a response regulator respectively, are involved in sensing and responding to 3-OH PAME (Clough *et al.*, 1997). Only when extracellular 3-OH PAME accumulates above 5 nM (i.e. at high cell density in a restricted space, such as the plant vascular system) is repression of PhcA relieved, resulting in appropriate gene activation (for those genes involved in EPS biosynthesis and Pme, Egl and Tek exoproteins synthesis) or repression (genes involved in motility, polygalacturonase or siderophore production) (Flavier *et al.*, 1997a). Unlike many other known bacterial autoregulators, 3-OH PAME is active both in solution and in the vapor phase.

R. solanacearum was also shown to produce a typical N-acyl-homoserine lactone-dependent autoinduction system consisting of *luxR* and *luxI* homologues, designated *solR* and *solI* respectively (Flavier *et al.*, 1997b). This second autoregu-latory system is regulated by the 3-OH PAME system via PhcA but its function is currently unknown.

4.2.4.3   A multicomponent network regulating virulence functions

Expression of the *eps* locus and genes encoding extracellular enzyme is subject to stringent control exerted by a complex PhcA-dependent regulatory network (Fig. 4.3). For example, the regulation of the *eps* promoter involves at least seven proteins which are organized in three converging signal transduction cascades (see Schell, 2000 and Fig. 4.3). The downstream regulatory component is the response regulator VsrC which, in conjunction with the highly basic XpsR protein, directly binds with the *eps* promoter to activate its transcription (Garg *et al.*, 2000). Interestingly, this network comprises several couples of two-component regulatory systems which, through their sensor components (PhcS, VsrA, VsrB), can potentially integrate various environmental signals, whose nature still remains unknown. Such an architecture thus provides a way to branch and/or combine some plant-dependent activation signals in the 3-OH PAME-dependent signaling pathway. An illustration of such regulatory interactions is the case of PhcA, which requires a signal-activated VsrD regulator to ensure full transcription activation of the *xpsR* promoter (Huang *et al.*, 1998).

Gene disruption of several of these regulatory components lead to significant decrease in virulence but this may simply reflect the fact that such mutations are highly pleiotropic. For example, the PehSR two-component regulatory system was found to control the production of pectinolytic enzymes (Allen *et al.*, 1997) but it also regulates bacterial swimming motility and production of Type IV pili which participate in important aspects of the *R. solanacearum* life cycle (Allen *et al.*, 1997; Kang *et al.*, 2002).

4.2.4.4   hrp *gene activation in response to plant cell contact*
Expression of Type III secretion (*hrp*) genes and Type III effectors is controlled by an AraC family transcriptional activator named HrpB (Genin *et al.*, 1992; Guéneron *et al.*, 2000). Transcription of these genes is regulated by at least two environmental signals: it is repressed when bacteria are grown in the presence of organic nitrogen sources and induced either *in planta* or when bacteria are grown in an *apoplast-mimicking* minimal medium. Genetic dissection of the regulatory circuits governing expression of these genes led to the identification of a regulatory cascade comprising five proteins in addition to HrpB. PrhI is an ECF sigma70 factor that, together with the plant signal receptor PrhA and the PrhR protein, forms a signal transduction module traversing three compartments (outer membrane, periplasm and cytoplasmic membrane) (Marenda *et al.*, 1998; Aldon *et al.*, 2000; Brito *et al.*, 2002). In response to a signal from the plant cell surface, it rapidly modulates transcription of the target gene *prhJ*. The PrhJ regulator, in turn, activates HrpG, an OmpR-related regulatory protein (Brito *et al.*, 1999). Whereas HrpG is required for transcriptional activation of *hrpB* in response to both plant and nutritional signals, PrhA, PrhR, PrhI and PrhJ proteins are involved only in transduction of the plant signal (Brito *et al.*, 1999; Brito *et al.*, 2002).

The exact nature of the plant signal sensed by the PrhA receptor is not known but there is evidence that it is a non-diffusible molecule that may be an intrinsic constituent of the plant cell wall (Aldon *et al.*, 2000). This is in support of the hypothesis that a contact-dependent signal contributes to triggering a polarized transfer of specific effector proteins into the plant cell.

4.2.5   *Genome-wide identification of candidate genes potentially involved in pathogenicity*

Beside a large group of Type III-effector proteins identified on the basis of genome-wide *in silico* promoter analysis (Cunnac *et al.*, submitted), other potentially important pathogenicity factors were recently identified in the genome of *R. solanacearum* strain GMI1000. For example, there is an abundance of genes encoding hemagglu-tinin-related proteins, a class of adhesins produced by diverse pathogenic bacteria and that was shown to contribute to the attachment, aggregation and virulence of *Erwinia chrysanthemi* (Rojas *et al.*, 2002). Another probable attachment factor, an atypical fucose-binding lectin, has also recently been characterized biochemically (Sudakevitz *et al.*, 2002).

Genes potentially involved in the synthesis of plant hormones (ethylene, auxin, and the cytokinin *trans*-zeatin) can be found in the *R. solanacearum* genome. It is likely that these plant signaling molecule analogues produced by the pathogen play a role during the interaction of the bacterium with its hosts. These aspects clearly require further investigations at the molecular level.

Another important issue concerns the identification of the targets of the multiple virulence genes already identified in *R. solanacearum* and to determine how the expression of these genes is coordinated during the infection process. The genomic era now opens up the possibility of investigating this question by analyzing the transcriptome of these various regulatory mutants using DNA microarrays. Such genome wide searches for regulatory gene targets will certainly contribute to identification of novel pathogenicity factors and to draw a far better global picture of the multiple determinants contributing to bacterial wilt disease.

Acknowledgment

The authors thank Julie Cullimore for editing this chapter.

References

Aizawa, S.I. (2001) Bacterial flagella and Type III secretion systems. *FEMS Microbiol. Lett.*, **202**, 157–164.

Aldon, D., Brito, B., Boucher, B. & Genin, S. (2000) A bacterial sensor of plant cell contact controls the transcriptional induction of *Ralstonia solanacearum* pathogenicity genes. *EMBO J.*, **19**, 2304–2314.

Allen, C.A., Huang, Y. & Sequeira, L. (1991) Cloning of genes affecting polygalacturonase production in *Pseudomonas solanacearum. Mol. Plant Microbe Interact.*, **4**, 157–154.

Allen, C.A., Gay, J. & Simon-Buela, L. (1997) A regulatory locus, *pehSR*, controls polygalacturonase production and other virulence functions in *Ralstonia solanacearum. Mol. Plant–Microbe Interact.*, **10**, 1054–1064.

Aragaki, M. & Quinon, V.L. (1965) Bacterial wilt of ornamental gingers (*Hedychium* spp.) caused by *Pseudomonas solanacearum. Plant Dis. Rep.*, **49**, 378–379.

Araud-Razou, I., Vasse, J., Montrozier, H., Etchebar, C. & Trigalet, A. (1998) Detection and visualization of the major acidic polysaccharide of *Ralstonia solanacearum* and its role in tomato root infection and vascular colonization. *Eur. J. Plant Pathol.*, **104**, 795–809.

Arlat, M., Gough, C.L., Zischek, C., Barberis, P.A., Trigalet, A. & Boucher, C.A. (1992) Transcrip-tional organization and expression of the large *hrp* gene cluster of *Pseudomonas solanacearum. Mol. Plant Microbe Interact.*, **5**, 187–193.

Arlat, M., van Gijsegem, F., Huet, J.C., Pernollet J.C. & Boucher, C.A. (1994) PopA1, a protein which induces a hypersensitivity-like response on specific Petunia genotypes, is secreted via the Hrp pathway of *Pseudomonas solanacearum*. *EMBO J.*, **13**, 543–553.

Bogdanove, A.J., Beer, S.V., Bonas, U., Boucher, C.A., Collmer, A., Coplin, D.L., Cornelis, G.R., Huang, H.C., Hutcheson, S.W., Panopoulos, N.J. & van Gijsegem, F. (1996) Unified nomenclature for broadly conserved *hrp* genes of phytopathogenic bacteria. *Mol. Microbiol.*, **20**, 681–683.

Boucher, C.A., Barberis, P., Trigalet, A.P. & Demery, D.A. (1985) Transposon mutagenesis of *Pseudomonas solanacearum*: isolation of Tn5-induced avirulent mutants. *J. Gen. Microbiol.*, **131**, 2449–2457.

Boucher, C.A., Gough, C.L. & Arlat, M. (1992) Molecular genetics of pathogenicity determinants of *Pseudomonas solanacearum* with special emphasis on *hrp* genes. *Annu. Rev. Phytopathol.*, **30**, 443–461.

Brito, B., Marenda, M., Barberis, P., Boucher, C. & Genin, S. (1999) *prhJ* and *hrpG*, two new components of the plant signal-dependent regulatory cascade controlled by PrhA *in Ralstonia solanacearum*. *Mol. Microbiol.*, **31**, 237–251.

Brito, B., Aldon, D., Barberis, P., Boucher, C. & Genin, S. (2002) A signal transfer system through three compartments transduces the plant cell contact-dependent signal controlling *Ralstonia solanacearum hrp* genes. *Mol. Plant Microbe Interact.*, **15**, 109–119.

Brumbley, S.M., Carney, B.F. & Denny, T.P. (1993) Phenotype conversion in *Pseudomonas solanacearum* due to spontaneous inactivation of PhcA, a putative LysR transcriptional regulator. *J. Bacteriol.*, **175**, 5477–5487.

Buddenhagen, I.W. (1986) Bacterial wilt revisited. In: *Bacterial Wilt in Asia and South Pacific* (ed. G.J. Persley). ACIAR Proceedings, **13**, 126–139. ACIAR Canberra.

Buddenhagen, I.W. & Elsasser, T.A. (1962) An insect-spread bacterial wilt epiphytotic of Bluggoe banana. *Nature*, **194**, 164–165.

Buddenhagen, I.W. & Kelman, A. (1964) Biological and physiological aspects of bacterial wilt caused by *Pseudomonas solanacearum*. *Annu. Rev. Phytopathol.*, **2**, 203–230.

Buddenhagen, I.W., Sequeira, L. & Kelman, A. (1962) Designation of races in *Pseudomonas solanacearum*. *Phytopathology*, **52**, 726.

Büttner, D., Nennstiel, D., Klusener, B. & Bonas, U. (2002) Functional analysis of HrpF, a putative Type III translocon protein from *Xanthomonas campestris* pv. *vesicatoria*. *J. Bacteriol.*, **184**, 2389–2398.

Carney, B.F. & Denny, T.P. (1990) A cloned avirulence gene from *Pseudomonas solanacearum* determines incompatibility on *Nicotiana tabacum* at the host species level. *J. Bacteriol.*, **172**, 4836–4843.

Clough, S.J., Lee, K.E., Schell, M.A. & Denny, T.P. (1997) A two-component system in *Ralstonia (Pseudomonas) solanacearum* modulates production of PhcA-regulated virulence factors in response to 3-hydroxypalmitic acid methyl ester. *J. Bacteriol.*, **179**, 3639–3648.

Cook, D. & Sequeira, L. (1994) Strain differentiation of *Pseudomonas solanacearum* by molecular genetic methods. In: *Bacterial Wilt: the Disease and its Causative Agent, Pseudomonas solanacearum* (eds A.C. Haward & G.L. Hartman), pp. 77–93. CAB International.

Cook, D., Barlow, E. & Sequeira, L. (1989) Genetic diversity of *Pseudomonas solanacearum*: detection of restriction fragment length polymorphisms with DNA probes that specify virulence and hypersensitive response. *Mol. Plant Microbe. Interact.*, **2**, 113–121.

Cornelis, G.R. & van Gijsegem, F. (2000) Assembly and function of Type III secretory systems. *Annu. Rev. Microbiol.*, **54**, 735–774.

Cunnac, S., Occhialini, A., Barberis, P., Boucher, C. & Genin, S. Functional analysis the *Ralstonia solanacearum* Hrp regulon identifies novel effector proteins translocated to plant host cells through the Type III secretion system. Submitted for publication.

Denny, T.P. (1995) Involvement of bacterial polysaccharides in plant pathogenesis. *Annu. Rev. Phytopathol.*, **33**, 173–197.

Denny, T.P. & Baek, S.R. (1991) Genetic evidence that extracellular polysaccharide is a virulence factor of *Pseudomonas solanacearum*. *Mol. Plant Microbe Interact.*, **4**, 198–206.

Denny, T.P., Ganova-Raeva, L.M., Huang, J. & Schell, M.A. (1996) Cloning and characterization of *tek*, the gene encoding the major extracellular protein of *Pseudomonas solanacearum*. *Mol. Plant Microbe Interact.*, **9**, 272–281.

Deslandes, L., Olivier, J., Peeters, N., Feng, D.X., Khounlotham, M., Boucher, C., Somssich, I., Genin, S. & Marco, Y. (2003) Physical interaction between RRS1-R, a protein conferring resistance to bacterial wilt, and PopP2, a Type III effector targeted to the plant nucleus. *Proc. Natl. Acad. Sci. USA*, **100**, 8024–8029.

Elphinstone, J.G. (1996) Survival and possibility of extinction of *Pseudomonas solanacearum* (Smith) in cool climates. *Potato Res.*, **39**, 403–410.

Elphinstone, J.G., Stanford, H.M. & Stead, D.E. (1998) Detection of *Ralstonia solanacearum* in potato tubers, *Solanum dulcamara* and irrigation water. In: *Bacterial Wilt Disease: Molecular and Biological Aspects* (eds P. Prior, C. Allen & J. Elphinstone), pp. 133–139. Springer-Verlag, Heidelberg, Germany.

Flavier, A.B., Clough, S.J., Schell, M.A. & Denny, T.P. (1997a) Identification of β-hydroxypalmitic acid methyl ester as a novel autoregulator controlling virulence in *Ralstonia solanacearum*. *Mol. Microbiol.*, **26**, 251–259.

Flavier, A.B., Schell, M.A., Ganova-Raeva, L. & Denny, T.P. (1997b) Hierarchical autoinduction in *Ralstonia solanacearum*: control of acyl-homoserine lactone production by a novel autoregulatory system responsive to 3-hydroxypalmitic acid methyl ester. *J. Bacteriol.*, **179**, 7089–7097.

Garg, R.P., Huang, J., Yindeeyoungyeon, W., Denny, T.P. & Schell, M.A. (2000) Multicomponent transcriptional regulation at the complex promoter of the exopolysaccharide I biosynthetic operon of *Ralstonia solanacearum*. *J. Bacteriol.*, **182**, 6659–6666.

Genin, S. & Boucher, C. (1994) A superfamily of proteins involved in different secretion pathways in Gram-negative bacteria: modular structure and specificity of the N-terminal domain. *Mol. Gen. Genet.*, **243**, 112–118.

Genin, S., Gough, C.L. Zischek, C. & Boucher, C.A. (1992) Evidence that the *hrpB* gene encodes a positive regulator of pathogenicity genes from *Pseudomonas solanacearum*. *Mol. Microbiol.*, **6**, 3065–3076.

Gitaitis, R., McCarter, S. & Jones, J. (1992) Disease control in tomato transplants produced in Georgia and Florida. *Plant Dis.*, **76**, 651–656.

Gonzalez, E.T. & Allen, C. (2003) Characterization of a *Ralstonia solanacearum* operon required for polygalacturonase degradation and uptake of galacturonic acid. *Mol. Plant Microbe Interact.*, **16**, 536–544.

Gough, C.L., Genin, S., Zischek, C. & Boucher, C. (1992) *hrp* genes of *Pseudomonas solanacearum* are homologous to pathogenicity determinants of animal pathogenic bacteria and are conserved among plant pathogenic bacteria. *Mol. Plant Microbe Interact.*, **5**, 384–389.

Graham, J., Jones, D.A. & Lloyd, A.B. (1979) Survival of *Pseudomonas solanacerum* race 3 in plant debris and in latently infected potato tubers. *Phytopathology*, **69**, 1100–1103.

Granada, J.A. & Sequeira, L. (1983) Survival of *Pseudomonas solanacearum* in soil, rhizosphere and plant roots. *Can. J. Microbiol.*, **29**, 433–440.

Guéneron, M., Timmers, A.C., Boucher, C. & Arlat, M. (2000) Two novel proteins, PopB, which has functional nuclear localization signals, and PopC, which has a large leucine-rich repeat domain, are secreted through the *hrp*-secretion apparatus of *Ralstonia solanacearum*. *Mol. Microbiol.*, **36**, 261–277.

Hara, H. & Ono, K. (1985) Ecological studies on the bacterial wilt of tobacco caused by *Pseudomonas solanacearum* E.F. Smith. VI. Dissemination in infected field and survival on tobacco leaf of the pathogen exuded from the upper part of the infected tobacco plants. *Bull. Okayama Tob. Exp. Station*, **44**, 87–92.

Hayward, A.C. (1964) Characteristics of *Pseudomonas solanacearum*. *J. Appl. Bacteriol.*, **27**, 265–277.

Hayward, A.C. (1991) Biology and epidemiology of bacterial wilt caused by *Pseudomonas solanacearum*. *Annu. Rev. Phytopathol.*, **29**, 65–87.

Hayward, A.C. (1994) The hosts of *Pseudomonas solanacearum*. In: *Bacterial Wilt: the Disease and its Causative Agent, Pseudomonas solanacearum* (eds A.C. Hayward & G.L. Hartman), pp. 9–24. CAB International.

Hayward, A.C. (2000) *Ralstonia solanacearum*. In: *Encyclopedia of Microbiology* (ed. J. Lederberg), Vol. **4**, pp. 32–42. Academic Press.

Hayward, A.C., Elphinstone, J.G., Caffier, D., Janse, J., Steohani, E., French, E.R. & Wright, A.J. (1998) Round table on bacterial wilt (brown rot) of potato. In: *Bacterial Wilt Disease* (eds Ph. Prior, C. Allen & J. Elphinstone), pp. 420–430. Springer, INRA editions.

He, L.Y., Sequeira, L. & Kelman, A. (1983) Characteristics of strains of *Pseudomonas solanacearum* from China. *Plant Dis.*, **67**, 1357–1361.

Hsu, S.T. (1991) Ecology and control of *Pseudomonas solanacearum* in Taiwan. *Plant Prot. Bull., Taiwan*, **33**, 72–79.

Huang, J., Yindeeyoungyeon, W., Garg, R.P., Denny, T.P. & Schell, M.A. (1998) Joint transcriptional control of *xpsR*, the unusual signal integrator of the *Ralstonia solanacearum* virulence gene regulatory network, by a response regulator and a LysR-type transcriptional activator. *J. Bacteriol.*, **180**, 2736–2743.

Hyde, K.D., McCulloch, B., Akiew, E., Peterson, R.A. & Diatloff, A. (1992) Strategies used to eradicate bacterial wilt of *Heliconia* (race 2) in Cairns, Australia, following introduction of the disease from Hawaii. *Aust. Plant Patho.*, **21**, 29–31.

Janse, J.D., Araluppen, F.A.X., Schans, J., Wenneker, M. & Westerhuis, W. (1998) Experiences with bacterial brown rot *Ralstonia solanacearu* biovar 2, race 3 in the Netherlands. In: *Bacterial Wilt Disease: Molecular and Biological Aspects* (eds P. Prior, C. Allen and J. Elphinstone), pp. 146–152. Springer-Verlag, Heidelberg, Germany.

Jin, Q. & He, S.Y. (2001) Role of the Hrp pilus in Type III protein secretion in *Pseudomonas syringae*. *Science*, **294**, 2556–2558.

Johnson, H.A. & Powell, N.T. (1969) Influence of root knot nematodes on bacterial wilt development in flue-cured tobacco. *Phytopathology*, **59**, 486–491.

Kang, Y., Huang, J., Mao, G., He, L.Y. & Schell, M.A. (1994) Dramatically reduced virulence of mutants of *Pseudomonas solanacearum* defective in export of extracellular proteins across the outer membrane. *Mol. Plant Microbe Interact.*, **7**, 370–377.

Kang, Y., Liu, H., Genin, S., Schell, M.A. & Denny, T.P. (2002) *Ralstonia solanacearum* requires type 4 pili to adhere to multiple surfaces and for natural transformation and virulence. *Mol. Microbiol.*, **46**, 427–437.

Kelman, A. (1953) The bacterial wilt caused by *Pseudomonas solanacearum*. *North Carolina Agricultural Experiment Station Technical Bulletin*, **99**, 194pp.

Kelman, A. (1954) The relationship of pathogenicity of *Pseudomonas solanacearum* to colony appearance on tetrazolium medium. *Phytopathology*, **44**, 693–695.

Kelman, A. & Sequeira, L. (1965) Root-to-root spread of *Pseudomonas solanacearum*. *Phytopathology*, **55**, 304–309.

Lavie, M., Shillington, E., Eguiluz, C., Grimsley, N. & Boucher, C. (2002) PopP1, a new member of the YopJ/AvrRxv family of Type III effector proteins, acts as a host-specificity factor and modulates aggressiveness of *Ralstonia solanacearum*. *Mol. Plant Microbe Interact.*, **15**, 1058–1068.

Leach, J.E. & White, F.F. (1996) Bacterial avirulence genes. *Annu. Rev. Phytopathol.*, **34**, 153–179.

Lee, J., Klusener, B., Tsiamis, G., Stevens, C., Neyt, C., Tampakaki, A.P., Panopoulos, N.J., Noller, J., Weiler, E.W., Cornelis, G.R., Mansfield, J.W. & Nurnberger, T. (2001) HrpZ(Psph) from the plant pathogen *Pseudomonas syringae* pv. *phaseolicola* binds to lipid bilayers and forms an ion-conducting pore *in vitro*. *Proc. Natl. Acad. Sci. USA*, **98**, 289–294.

Li, X., Dorsch, M., Del Dot, T., Sly, L.I., Strckebrandt, E. & Hayward, A.C. (1993) Phylogenetic studies of the rRNA group II pseudomonads based on 16S rRNA gene sequences. *J. Appl. Bacteriol.*, **74**, 324–329.

Lindgren, P.B. (1997) The role of *hrp* genes during plant–bacterial interactions. *Annu. Rev. Phytopathol.*, **35**, 129–152.

Lindgren, P.B., Peet, R.C. & Panopoulos, N.J. (1986) Gene cluster of *Pseudomonas syringae* pv. *phaseolicola* controls pathogenicity of bean plants and hypersensitivity of nonhost plants. *J. Bacteriol.*, **168**, 512–522.

Liu, H., Kang, Y., Genin, S., Schell, M.A. & Denny, T.P. (2001) Twitching motility of *Ralstonia solanacearum* requires a Type IV pilus system. *Microbiology*, **147**, 3215–3229.

Lucas, G.B., Sasser, J.N. & Kelman, A. (1955) The relationship of root-knot nematodes to Granville wilt resistance in tobacco. *Phytopathology*, **45**, 537–540.

Lum, K.Y. (1973) Cross inoculation studies of *Pseudomonas solanacearum* from ginger. *MARDI Res. Bull.*, **1**, 15–21.

Machmud, M. & Middleton, K.J. (1990) Seed infection and transmission of *Pseudomonas solanacearum* on groundnut. In: *Bacterial Wilt of Groundnut* (eds K.J. Middleton & A.C. Hayward), **31**, 57. ACIAR Proceedings ACIAR, Canberra.

Marenda, M., Brito, B., Callard, D., Genin, S., Barberis, P., Boucher, C. & Arlat, M. (1998) PrhA controls a novel regulatory pathway required for the specific induction of *Ralstonia solanacearum* *hrp* genes in the presence of plant cells. *Mol. Microbiol.*, **27**, 437–453.

McCarter, S.M. & Jaworski, C.A. (1969) Field studies on spread of *Pseudomonas solanacearum* and tobacco mosaic virus in tomato plants by clipping. *Plant Dis. Rep.*, **53**, 942–945.

Moffett, M.L., Giles, J.E. & Wood, B.A. (1983) Survival of *Pseudomonas solanacearum* biovars 2 and 3 in soil: effect of moisture and soil type. *Soil Biol. Biochem.*, **15**, 587–591.

Napiere, C.M. (1980) Varying inoculum levels of bacteria-nematodes and the severity of tomato bacterial wilt. *Ann. Tropical Res.*, **2**, 129–134.

Ono, K. (1983) Ecological studies on the bacterial wilt of tobacco caused by *Pseudomonas solanacearum* E.F. Smith. III. Distribution and spread of the pathogen in infected tobacco field under rainfall. *Bull. Okayama Tob. Exp. Station*, **42**, 149–153.

Orgambide, G., Montrozier, H., Servin, P., Roussel, J., Trigalet-Demery, D. & Trigalet, A. (1991) High heterogeneity of the exopolysaccharides of *Pseudomonas solanacearum* strain GMI 1000 and the complete structure of the major polysaccharide. *J. Biol. Chem.*, **266**, 8312–8321.

Poussier, S., Prior, P., Luisetti, J., Hayward, C. & Fegan, M. (2000a) Partial sequencing of the *hrpB* and endoglucanase genes confirms and expends the known diversity within the *Ralstonia solanacearum* species complex. *System. Appl. Microbiol.*, **23**, 479–486.

Poussier, S., Trigalet-Demery, D., Vandewalle, P., Goffinet, B., Luisetti, J. & Trigalet, A. (2000b) Genetic diversity of *Ralstonia solanacearum* assessed by PCR-RFLP of the *hrp* gene region, AFLP and 16S rRNA sequence analysis, and identification of an African subdivision. *Microbiology*, **146**, 1679–1692.

Poussier, S., Thoquet, P., Trigalet-Demery, D., Barthet, S., Meyer, D., Arlat, M. & Trigalet, A.P. (2003) Host plant-dependent phenotypic reversion of *Ralstonia solanacearum* from non-pathogenic to pathogenic forms via alterations in the *phcA* gene. *Mol. Microbiol.*, **49**, 991–1003.

Pugsley, A.P. (1993) The complete general secretory pathway in Gram-negative bacteria. *Microbiol. Rev.*, **57**, 50–108.

Rillo, A.R. (1979) Bacterial wilt of banana in the Philippines. *FAO Plant Protection Bulletin*, **27**, 105–108.

Roberts, D.P., Denny, T.P. & Schell, M.A. (1988) Cloning of the *egl* gene of *Pseudomonas solanacearum* and analysis of its role in phytopathogenicity. *J. Bacteriol.*, **170**, 1445–1451.

Roine, E., Wei, W., Yuan, J., Nurmiaho-Lassila, E.L., Kalkkinen, N., Romantschuk, M. & He, S.Y. (1997) Hrp pilus: an *hrp*-dependent bacterial surface appendage produced by *Pseudomonas syringae* pv. *tomato* DC3000. *Proc. Natl. Acad. Sci. USA*, **94**, 3459–3464.

Rojas, C.M., Ham, J.H., Deng, W.L., Doyle, J.J. & Collmer, A. (2002) HecA, a member of a class of adhesins produced by diverse pathogenic bacteria, contributes to the attachment, aggregation, epidermal cell killing, and virulence phenotypes of *Erwinia chrysanthemi* EC16 on *Nicotiana clevelandii* seedlings. *Proc. Natl. Acad. Sci. USA*, **99**, 13142–13147.

Romantschuk, M., Roine, E. & Taira, S. (2001) Hrp pilus – reaching through the plant cell wall. *Eur. J. Plant Pathol.*, **107**, 153–160.

Saile, E., McGarvey, J.A., Schell, M. & Denny, T.P. (1997) Role of extracellular polysaccharide and endoglucanase in root invasion and colonization of tomato plants by *Ralstonia solanacearum*. *Phytopathology*, **87**, 1264–1271.

Salanoubat, M., Genin, S., Artiguenave, F., Gouzy, J., Mangenot, S., Arlat, M., Billault, A., Brottier, P., Camus, J.C., Cattolico, L., Chandler, M., Choisne, N., Claudel-Renard, C., Cunnac, S., Demange, N., Gaspin, C., Lavie, M., Moisan, A., Robert, C., Saurin, W., Schiex, T., Siguier, P., Thebault, P., Whalen, M., Wincker, P., Levy, M., Weissenbach, J. & Boucher, C.A. (2002) Genome sequence of the plant pathogen *Ralstonia solanacearum*. *Nature*, **415**, 497–502.

Schell, M.A. (2000) Regulation of virulence and pathogenicity genes *in Ralstonia solanacearum* by a complex network. *Annu. Rev. Phytopathol.*, **38**, 263–292.

Schell, M.A., Roberts, D.P. & Denny T.P. (1988) Analysis of the *Pseudomonas solanacearum* polygalacturonase encoded by *pglA* and its involvement in phytopathogenicity. *J. Bacteriol.*, **170**, 4501–4508.

Schmit, J. (1978) Microscopic study of early stages of infection by *Pseudomonas solanacearum* E.F.S. on *in vitro* grown seedlings. In: *Proc. 4th International Conf. Plant Pathol. Bacteriol.*, pp. 841–856. INRA editor.

Sequeira, L. (1958) Bacterial wilt of bananas: dissemination of the pathogen and control of the disease. *Phytopathology*, **48**, 64–69.

Shakya, D.D. (1993) Occurrence of *Pseudomonas solanacearum* in tomato seeds imported into Nepal. In: *Bacterial Wilt. ACIAR Proceedings* (eds G.L. Hartman & A.C. Hayward), **45**, 371–372. ACIAR, Canberra.

Shekawat, G.S. & Pérombélon, M.C.M. (1991) Factors affecting survival in soil and virulence of *Pseudomonas solanacearum*. *J. Phytopathol.*, **98**, 258–267.

Smith, E.F. (1896) A bacterial disease of tomato, pepper, eggplant and Irish potato (*Bacillus solanacearum* nov. sp.). *United States Department of Agriculture, Division of Vegetable Physiology and Pathology Bulletin*, **12**, 1–28.

Sudakevitz, D., Imberty, A. & Gilboa-Garber, N. (2002). Production, properties and specificity of a new bacterial L-Fucose- and D-Arabinose-binding lectin of the plant pathogen *Ralstonia solanacearum*, and its comparison to related plant and microbial lectins. *J. Biochem.*, **132**, 353–358.

Taghavi, M., Hayward, C., Sly, L.I. & Fegan, M. (1996) Analysis of the phylogenetic relationships of strains of *Burkholderia solanacearum*, *Pseudomonas syzygii*, and the blood disease bacterium of banana based on 16S rRNA gene sequences. *Int. J. Syst. Bacteriol.*, **46**, 10–15.

Tans-Kersten, J., Guan, Y. & Allen, C. (1998) *Ralstonia solanacearum* pectinmethylesterase is required for growth on methylated pectin but not for bacterial wilt virulence. *Appl. Environ. Microbiol.*, **64**, 4918–4923.

Tans-Kersten, J., Huang, H. & Allen, C. (2001) *Ralstonia solanacearum* needs motility for invasive virulence on tomato. *J. Bacteriol.*, **183**, 3597–3605.

Trigalet, A. & Demery, D. (1986) Invasiveness in tomato plants of Tn5-induced mutants of *Pseudomonas solanacearum*. *Physiol. Mol. Plant Pathol.*, **28**, 423–430.

Trigalet, A. & Trigalet-Demery, D. (1990) Use of avirulent mutants of *Pseudomonas solanacearum* for the biological-control of bacterial wilt of tomato plants. *Physiol. Molec. Plant Pathol.*, **36**, 27–38.

van Elsas, J.D., Kastelein, P., van Bekkum, P., van der Wolf, J.M., de Vries, P.M. & van Overbeek, L.S. (2000) Survival of *Ralstonia solancearum* biovar 2, the causative agent of bacterial rot, in field and microcosm soils in temperate countries. *Phytopathology*, **90**, 1358–1366.

van Elsas, J.D., Kastelein, P., de Vries, P.M. & van Overbeek, L.S. (2001) Effects of ecological factors on the survival and physiology of *Ralstonia solanacearum* bv.2 in irrigation water. *Can. J. Microbiol.*, **47**, 842–854.

van Gijsegem, F., Gough, C., Zischek, C., Niqueux, E., Arlat, M., Genin, S., Barberis, P., German, S., Castello, P. & Boucher, C. (1995) The *hrp* gene locus of *Pseudomonas solanacearum*, which controls the production of a Type III secretion system, encodes eight proteins related to components of the bacterial flagellar biogenesis complex. *Mol. Microbiol.*, **15**, 1095–1114.

van Gijsegem, F., Vasse, J., Camus, J.C., Marenda, M. & Boucher, C. (2000) *Ralstonia solanacearum* produces *hrp*-dependent pili that are required for PopA secretion but not for attachment of bacteria to plant cells. *Mol. Microbiol.*, **36**, 249–260.

van Gijsegem, F., Vasse, J., De Rycke, R., Castello, P. & Boucher, C. (2002) Genetic dissection of *Ralstonia solanacearum hrp* gene cluster reveals that the HrpV and HrpX proteins are required for Hrp pilus assembly. *Mol. Microbiol.*, **44**, 935–946.

Vasse, J., Frey, P. & Trigalet, A. (1995) Microscopic studies of intercellular infection and protoxyleme invasion of tomato roots by *Pseudomonas solanacearum. Mol. Plant Microbe Interact.*, **8**, 241–251.

Vasse, J., Genin, S., Frey, P., Boucher, C. & Brito, B. (2000) The *hrpB* and *hrpG* regulatory genes of *Ralstonia solanacearum* are required for different stages of the tomato root infection process. *Mol. Plant Microbe Interact.*, **13**, 259–267.

Vaughan, E.K. (1944) Bacterial wilt of tomato caused by *Phytomonas solanacearum. Phytopathology*, **34**, 443–458.

Wall, D. & Kaiser, D. (1999) Type IV pili and cell motility. *Mol. Microbiol.*, **32**, 1–10.

Wallis, F.M. & Trutner, S.J. (1978) Histopathology of tomato plants infected with *Pseudomonas solanacearum*, with emphasis on ultrastructure. *Physiol. Plant Pathol.*, **28**, 293–402.

Wenneker, M., Verdel, M.S.W., Groeneveld, R.M.W., Kmpenaar, C., Van Beuningen, A.R. & Janse, J.D. (1999). *Ralstonia (Pseudomonas) solanacearum* race 3 (biovar2) in surface water and natural hosts: first report on stinging nettle (*Urtica dioica*). *Eur. J. Plant Pathol.*, **105**, 307–315.

Williamson, L., Nakaho, K., Hudelson, B. & Allen, C. (2002) *Ralstonia solanacearum* race 3, biovar 2 strains isolated from geranium are pathogenic on potato. *Plant Dis.*, **86**, 987–991.

Zhang, Y.X., Hua, J.Y. & He, L.Y. (1993) Effect of infected groundnut seeds on transmission of *Pseudomonas solanacearum. ACIAR Bacterial Wilt Newsletter*, **9**, 9–10.

5   The *Pseudomonas syringae*–bean system

Susan S. Hirano and Christen D. Upper

5.1   Introduction

What is there to describe about the life cycle of bacteria? They divide. One becomes two, and that is all that there is to it. Bacteria are essentially clonal organisms that replicate by fission, passing identical (or because of infrequent mutations, nearly identical) genomes forward in time to all of their progeny. They also die. And that is sufficient to describe the *life cycle* of the pathogen. Or is it?

Of course, the preceding paragraph is totally insufficient to explain the life style of *Pseudomonas syringae* pv. *syringae* (Pss); i.e. the many interactions of Pss with its host plant and the environment that allow populations of these bacteria to live successfully in association with plants and to cause bacterial brown spot disease. Individual bacteria, many nearly identical, make up these populations, and the interactions among bacteria, plant, and environment are ultimately controlled by genes within host and pathogen. In recent years, great effort has been expended, with more than a little success, to shed some light within the black box of plant–microbe interactions. Unraveling the genomes of pathogen and host and understanding the molecular basis of these interactions have begun to unlock the black box. Understanding these interactions at the molecular level, however, is but one facet of understanding the forces that shape the overall system.

The Disease Triangle is a time-honored concept in plant pathology. The disease triangle is intended to illustrate the fact that there are three components required for disease to occur: susceptible host, virulent pathogen, and conducive environmental conditions. Interactions among the three components occur at different scales in time and space. Numerous factors influence and regulate these interactions. The role the environment plays in these interactions remains in some of the darkest, most obscure corners of the box, which have completely escaped illumination. If there is one lesson we wish to impart to our readers, it is the recognition that the environment plays as critical a role in disease development as do the genetics of the pathogen and host. Aside from its importance in disease epidemiology, the environment plays a significant role in shaping the genomes of organisms in evolutionary time. Is it not the variations in environments that lead to variations in genotypes and phenotypes? Darwin thought so.

We first describe what we know of the Pss–bean system from a population perspective, and place within this framework what we know about the roles and effects of pathogenicity-associated genes in interactions of Pss with bean. We present our views based on what we observe. Some readers may take issue with what we write. This is encouraged as our goal in describing the Pss–bean system is not just to provide information or facts, but also to use many of the observations we have made during decades of studying this system to raise questions and provoke thought.

5.2   The system

Pss is a Gram-negative, rod-shaped, aerobic bacterium with polar flagella. These phenotypes are not what make Pss unique among bacterial species. There is nothing outstanding about its nutritional requirements. The bacterium utilizes many sugars and other substrates and is easily cultured on general media in the laboratory. Pss colonizes aerial parts of plants as do numerous other bacterial species (Hirano & Upper, 2000). Pss is a pathogen of plants but so are a few other bacterial species. What sets Pss apart from all other bacterial species and apart from its closest relatives – other pathovars within the species *P. syringae* (PS) – is the nature of the disease it causes on plants.

On *Phaseolus vulgaris* Pss causes bacterial brown spot disease. Although we have often used the name snap bean, common in the US Midwest, to describe this crop, others may prefer *Phaseolus* bean, green bean, french bean, bush bean, or common bean. In this chapter, we will refer to these plants as beans. Foliar symptoms of brown spot disease are relatively small, brownish necrotic lesions frequently surrounded by a small chlorotic halo (Fig. 5.1). Necrotic tissue may fall out causing a shot-hole appearance to leaves. Lesions on pods, the economically important phase of the disease, are commonly small, reddish-brown to dark necrotic spots. When disease develops on young pods, the pods may become distorted at the site of lesion formation.

In the field, symptoms of bacterial brown spot disease do not always conform to the classical or text book descriptions. Even for experienced eyes, it is quite often necessary to isolate bacteria from unusual spots to decide if they are really brown spot lesions. (This variation in lesion morphology is the case for many diseases.) The variation in symptoms that one encounters in the field is surely telling us something about the variations in interactions among pathogen, host, and the environment. If not all such interactions result in identical outcomes, is it reasonable to expect that interactions leading to lesions should follow quantitatively identical paths?

5.3   Population sizes of Pss on populations of leaf habitats

The bean cultivars we have used to study this system are determinate plants that produce approximately ten true leaves and a concerted flush of flowers and pods. During the life span of the crop (~55 days from seedling emergence to pod

Figure 5.1 Field symptoms of bacterial brown spot on *Phaseolus* bean caused by *P. syringae* pv. *syringae*. Reprinted with permission from Hirano *et al*. (1995).

harvest), immense populations of leaf (and pod) habitats are produced for colonization by populations of Pss. We begin our discussion about the life style of Pss based on two general observations:

First, population sizes of Pss vary with respect to time (Fig. 5.2). The more informative aspects of this variability are the relative magnitude and timing of the changes between and within growing seasons. Comparison of seasonal profiles of pathogen population sizes reveals that some growing seasons are more conducive than others for colonization of bean leaves by Pss. Examination of within-season profiles tells us that most day-to-day changes are negligible or small. These relatively inactive periods are occasionally punctuated by brief periods of large to very large changes, both increases and decreases occur. What causes these temporal variabilities in pathogen population sizes? Do these changes relate to disease in some way?

Second, population sizes of Pss vary extensively within a given plant canopy (Fig. 5.3). The data shown in Fig. 5.2 are average pathogen population sizes.

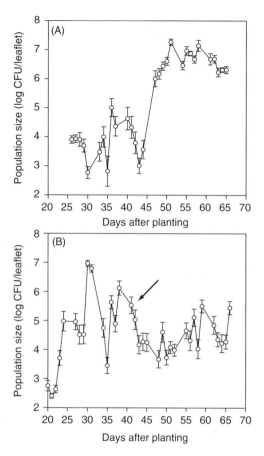

Figure 5.2 Temporal variability in mean population sizes of naturally occurring *P. syringae* on *Phaseolus* bean (cultivar Cascade). The data in Figs 5.2A and 5.2B are from two plantings of bean established 27 days apart during a single growing season. At each sampling time, 30 leaflets were randomly collected from the top of the canopy. The leaflets were processed individually by dilution plating of leaf homogenates. The limit of sensitivity of the plating assay was 2.279 log CFU per leaflet. The data are the mean and standard error for each set of 30 leaflets. Reprinted with permission from Hirano *et al.* (1994). (Arrow: see Fig. 5.3 for example of distribution of Ps population sizes on individual samples within sets of 30 leaflets.)

The values represent the overall population trend across an entire plant canopy, i.e. the *metapopulation*. It is, however, only a summary of what is happening on each of the many millions of leaflets within that plant canopy. Dynamics of the metapopulation provide only half of the information needed to describe dynamics of bacterial population as a whole. The remainder of the information is found in the form and extent of variability in bacterial populations on individual leaflets. Pss population sizes across populations of leaves are very often well approximated by a lognormal

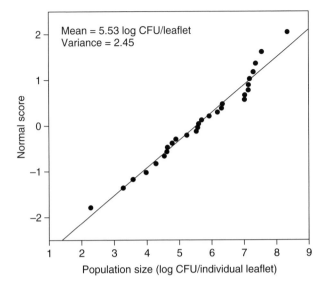

Figure 5.3 Variability in population sizes of *P. syringae* on individual bean leaflets. Bacterial population sizes are for each of the 30 leaflets collected at 41 days after planting for the planting shown in Fig. 5.2B. The nearly linear relationship between the $\log_{10}$-transformed bacterial population sizes and normal scores demonstrates the lognormal distribution of population sizes.

distribution (Hirano *et al.*, 1982). (That is, the logarithm of population sizes is usually normally distributed.) Thus, we can easily reduce description of this vast variability in Pss populations on individual leaflets to describe the metapopulation with just two parameters, the mean and variance of the lognormal. What are the causes of such variability in pathogen population sizes and how do they relate to the likelihood of brown spot disease development?

5.4 Population processes: searching for causes of variability in pathogen population sizes

The numbers of bacteria on leaves are determined by the balance among four population processes: immigration and multiplication increase population sizes, emigration and death diminish them. This simple concept applies to any number of spatial scales; numbers of bacteria in an entire field, bacteria in/on an entire leaf or at any (micro-) site within or upon a leaf. Thus, the change in population size on a given leaf between times t_1 and t_2 can be expressed as:

$$P_1 + \sum_{t_1}^{t_2} (\text{immigration, emigration, multiplication, death}) = P_2 \qquad (5.1)$$

where P_1 and P_2 represent Pss population sizes at t_1 and t_2 respectively.

Although conceptually simple, development of bacterial populations on leaves is highly complex because changes in population sizes with time are highly variable. Rates change in response to a multitude of factors related to the pathogen, host, and environment (statement 5.2).

$$\text{Pathogen} \leftrightarrow \text{Host} \leftrightarrow \text{Environment}$$
$$\downarrow \qquad \downarrow \qquad \downarrow$$

Rates of change of immigration, emigration, growth, and death   (5.2)

Coupling statements 5.1 and 5.2 above, we envision a matrix of pathogen, host, and environmental factors and interactions thereof that have varying quantitative effects on rates of immigration, emigration, growth, and death, which in turn, are reflected in variability in Pss population sizes among and within leaves. The system is clearly multifaceted, interactive, and interconnected. When a piece of the system is removed from these multifaceted interactions for laboratory study, it is possible to disconnect the very interactions that give rise to the phenomenon intended for study. Parts of the system may appear to function quite differently in the absence of these interactions than in their presence. This suggests the need for confirming conclusions drawn from studies of parts of the system by reintroducing the parts into the context of the system as a whole.

The data in Fig. 5.2 illustrate that under some conditions the sum of immigration and multiplication exceeded the sum of emigration and death and pathogen population sizes increased. Under other conditions, emigration plus death exceeded immigration and multiplication and pathogen population sizes decreased. When changes in population sizes were not detected, all of the processes were probably operating but their effects were counterbalanced to produce no net change.

5.4.1   Immigration and multiplication

Let us consider those cases when significant build-up in pathogen populations occurred (e.g. Fig. 5.2A: ~270-fold increase from 44 to 47 days after planting (DAP); Fig. 5.2B: ~300-fold increase from 29 to 30 DAP). How could large numbers of Pss accumulate on leaves? We considered two possibilities: large numbers of Pss could result from the accumulation of large numbers of immigrants. Alternatively, large pathogen population sizes could result from multiplication of relatively small founding populations.

We used a number of different experimental approaches to determine which of the two possibilities is more likely to explain pathogen population build-up (Upper & Hirano, 2002; Upper *et al.*, 2003, unpublished). In one approach, deposition rates of naturally occurring Pss in a bean field were monitored continuously for 11 days in one experiment and 13 days in another (Upper *et al.*, unpublished). Deposition rates were estimated by exposing petri dishes filled with a semi-selective medium for Pss. On average, the number of Pss arriving on an area similar to that of a bean leaflet during rain-free periods was about 11–12 CFU per day. These values should be viewed as gross averages as there was much variability in the numbers of immigrants detected among the sampling sites within and between days. The numbers of immigrants were slightly larger during rain periods than during rain-free periods. However, we do not know what proportion of these immigrants actually adhered to the leaves and what proportion remained in the rainwater as it fell to the ground.

The results from these and other experiments (Lindemann & Upper, 1985) place limits on expected immigration rates of Pss in the bean canopy. Can these relatively small rates account for the 10–100-fold increases in phyllosphere population sizes of the bacterium that occur within a day or over a few consecutive days? If initial population sizes are on the order of 1–10 cells per leaflet, then 10–100-fold increases in 24 hours could be due to immigration. Even if initial population sizes were as large as 100 cells per leaflet, 10-fold increases might conceivably be due to immigration on rare occasions. We could not totally rule out such a possibility. In the case of initial population sizes of say, 10^4 cells, however, a 100-fold increase would require addition of 10^6 cells to the population on the average leaf. Such an addition is approximately 10^5 times greater than the median immigration rate, and several hundred fold greater than the extreme rate found during 24 days of extensive sampling. On this basis, we can rule out immigration as the cause of nearly all of the very large increases we have observed. If immigration rates are too small to contribute sufficient numbers to account for the rapid and large blooms in phyllosphere population sizes of Pss, then we are left with bacterial multiplication as the only remaining alternative.

In a second approach, we measured spread (sum of immigration and multiplication) of Pss from source to sink areas planted with beans (Upper *et al.*, 2003). For Pss to spread successfully to new leaf habitats, the bacterium must immigrate *and* then multiply. Individual plots $(30 \times 30 \, \text{m})$ consisted of three nested concentric squares with the inner 6-m square serving as the sink. Each sink was surrounded by a barrier zone, usually 6-m wide, which, in turn, was surrounded by a 6-m wide source area inoculated with a doubly marked strain of Pss at the time of planting. The nature of the barrier zone and sink was altered in various ways and their effects on spread were determined during the conduct of nine field experiments spanning six growing seasons.

Spread was surprisingly rapid. In all the experiments, the marked strain was detected at low frequencies in bean sinks at the time that samples were first taken (within a day or two following source seedling emergence). Significant differences were found in the amounts of spread that occurred based on measurements of population sizes of the marked strain in the sinks. None of the differences could be attributed to the nature (bare ground, snap beans, or soybeans) or width (6 m versus 20 m) of the barrier zone. The significant differences could all be attributed to the suitability of the habitat in the sinks coupled with weather conditions favorable for multiplication of Pss. For example, spread to a susceptible cultivar of bean was greater than that to a less susceptible cultivar and to the non-host soybean. Spread to sinks planted with noninoculated seeds was greater than that to sinks planted with seeds inoculated with a mixture of phyllosphere bacteria. At least on a scale equivalent to within-field spread, the suitability of new leaf habitats for multiplication of the bacteria, but not immigration, appeared to limit the amount of spread. We arrive once again at the conclusion that multiplication is largely responsible for the accumulation of large numbers of Pss in a bean canopy, in this case based on the leaves to which the bacteria had spread.

Although immigration plays a quantitatively much smaller role than multiplication in contributing to numbers of Pss in a bean canopy, this population process is absolutely essential. Immigrants must arrive on bean leaflets before the process of multiplication can occur there. Leaves are rather open systems exposed to the atmosphere and anything that drifts, flies, walks, or drives through it. Immigrants may arrive via rain-splash and aerosol deposition during dry, sunny, windy weather conditions. When leaves are wet (e.g. with dew), insects traversing leaf surfaces become contaminated with bacteria and passively vector bacteria as they flit and fly about. Pss is seed borne, and hence may also arrive on leaves via its initial presence on infested seeds.

Consider all of the above in the context of the life cycle of the bacteria – all they do is divide. Yet, their rapid, clonal multiplication is sufficient to allow them to spread and colonize entire canopies in a matter of days. As we will see below, conditions that limit growth have a major effect in limiting population sizes of Pss.

5.4.2   Emigration and death

Population sizes of Pss also decrease at times. Some of the large and rapid decreases can be attributed to massive wash-off during rainstorms (Lindemann & Upper, 1985). Under these conditions, emigration is clearly the dominant process. When population sizes decrease during rain-free periods, we expect that death rates exceed emigration rates. Emigration rates during dry, sunny weather are relatively small, just over an order of magnitude greater than immigration rates (Lindemann & Upper, 1985). Because such measurements have not been made continuously, and rates of bacterial removal by other processes are unknown, we have insufficient direct evidence to make any sort of definitive statement about the relative importance of emigration and death during dry conditions.

5.4.3   The four processes working together

We have, thus far, identified pathogen multiplication as the dominant process during blooms in pathogen population sizes. That is, large increases in population sizes of Pss are due to *net* growth (multiplication $\ggg$ immigration + emigration + death). This tells us how population sizes can become large, but it does not fully explain temporal changes in pathogen population sizes. If all the bacteria did was divide at a constant rate, population sizes would become large and remain large. In the real world, they are highly variable with time. The frequency of divisions determines bacterial growth rate. Under conditions not suitable for bacterial growth, the apparent growth rate may be slightly to significantly negative. That is, death + emigration $\gg$ growth.

The variability in population sizes with time results from variation in net growth rates of Pss over time. What causes the net growth rate to change? The answer lies in the way the plant, the bacterium, and the environment interact to influence rates and extents of the population processes. Bacteria respond to changes in the environment. Are alterations in multiplication rates due to signals generated from changes in the environment? We think so. Such environmental influences are undoubtedly regulated by a complex network of genes switching on and off as bacterial multiplication rates change in the phyllosphere. Before we can ask how net growth is regulated, we need to know how the bacteria respond to the environment.

5.4.4   Enter the environment

What conditions are conducive to rapid bursts of growth, what conditions attenuate or stop net bacterial growth? We measured population sizes of naturally occurring Pss daily or nearly daily, from plant emergence to pod harvest over many plantings in different growing seasons, and hence weather conditions (e.g. Fig. 5.2) (Hirano *et al.*, 1994). Comprehensive weather data were collected at the site of the experimental plots. From comparison of the dynamics of Pss population sizes to the weather records, it became clear that changes in weather conditions play a major role in determining when rapid growth of Pss occurs under field conditions. Nearly all increases in pathogen population sizes greater than about 10-fold which occurred within a span of 1–3 days were associated with rainstorms during which peak rainfall rates met or exceeded 1 mm/min (Hirano *et al.*, 1994, 1996). In experiments in which pathogen population sizes were determined every two hours within 24-hour periods randomly selected regardless of weather conditions, we measured a large net increase (28-fold) within one of the 24-hour periods but not in the other two (Hirano & Upper, 1989). Coincidentally, 22 mm of rain was recorded immediately before the start of the experiment in which the large increase was measured. No rain fell on the day before or during the other two 24-hour sampling periods in which net changes in pathogen population sizes increased ~5-fold in one case and none in the other.

Bursts in net growth of Pss were not observed in the absence of intense rains even when leaves were wet with dew. We examined other weather variables such as temperature, humidity, solar radiation, duration of leaf wetness, and so forth. None were strongly correlated with bursts in rapid growth of Pss. At this point, our working hypothesis was that intense rains trigger the onset of rapid growth of Pss in association with bean plants in the field. The hypothesis, however, was based merely on correlative analysis of changes in weather conditions and changes in pathogen population sizes.

Direct evidence that blooms in pathogen population sizes are effected by intense rains was obtained by manipulating the environment in the field (Hirano *et al.*, 1996). Bursts in pathogen population sizes were prevented when bean plants were shielded from natural rain in the field. Water temperature, pH, and many other variables were ruled out as explanations for how rain triggers growth of Pss. Simply having leaves wet for a duration equivalent to that encountered following intense rains did not lead to substantial growth. Some property of intense rain other than mere wetness was required. Eventually, we found that the momentum of intense rain is required for the growth-triggering process (Hirano *et al.*, 1996). Absorbing the momentum of intense rains with inert screens placed over bean canopies, which allowed the same amount of water of comparable quality to fall gently onto bean leaves, prevented the growth-triggering process.

At this point, we know what aspect of the environment triggers the onset of rapid growth of Pss in the field, but we do not know how the momentum of raindrops triggers growth of the bacterium. It is clear that understanding the phenomenon is central to understanding the interactions of Pss with its host and environment. In order to unravel the regulatory network that controls growth of Pss on leaves in the field, we must first determine how growth of the bacteria is triggered by rain. How does one undertake such a task?

The literature describes several conditions under which Pss will grow on bean leaves without intense rain. For example, under laboratory conditions it is sufficient to spray the bacteria on bean leaves and maintain them under moderately high humidity and the bacteria will grow (e.g. Willis *et al.*, 1990). Pss infiltrated into bean leaves in either the laboratory or the field will grow (e.g. Halberg & Hirano, unpublished; Willis *et al.*, 1990; Wilson *et al.*, 1999). Several lines of evidence suggest that either the apoplast or the substomatal cavities, or both are the preferred sites for growth of some strains of *P. syringae* under laboratory conditions (e.g. *P. syringae* pv. *tomato* DC3000 (Boureau *et al.*, 2002; Romantschuk *et al.*, 2002); see also Beattie & Lindow, 1999).

Thus, we know how to get Pss to grow in association with bean leaves, and we have some indication where they grow. Does this not mean that we know what intense rain must be doing? If they drive the bacteria into the apoplast of leaves, is it not obvious that this is how growth is triggered? We do not think so. Suppose we want to understand what it is that makes an automobile accelerate. If we head an automobile down a steep hill, take it out of gear and release the brake, it will accelerate. Such an experiment would tell us one way to make an automobile accelerate. Does it, however, tell us what makes it accelerate on a level road – a different environment?

The long periods without substantial growth that we normally see in the field have not been successfully reproduced under laboratory conditions. Lindow (1993) has described conditions of low humidity, high temperature, and intense light under which Pss will not grow. However, as soon as the plants are returned to cooler, more normal conditions, the bacteria resume growth. In the field, the bacteria do not resume growth after a hot sunny day, after the sun goes down, the air cools, and the leaves become wet with dew. Is it reasonable to study how a switch gets turned *on* under conditions where turning it on appears to be unnecessary?

At this stage, many possible alternative hypotheses cannot be ruled out to explain how rain drop momentum triggers growth of Pss in the field. Some possible mechanisms act only on the plant. For example, intense rain may remove inhibitory compounds from leaves or stimulate nutrient release by the plant, or otherwise modify leaves to make them more favorable habitats. Others may require the bacteria to be present. Such possibilities include transporting the bacteria either into or out of the leaves, facilitating bacterial movement either within or outside the leaves or disrupting bacterial aggregates. Recently, Sabaratnam and Beattie (2003) found that both leaf surface and internal population sizes of Pss strain B728a increased following an intense rainfall. These results suggest that mechanisms that require growth to occur in a single class of sites, such as the apoplast or the leaf surface, are unlikely.

A systematic series of experiments to attempt to rule out each of these possibilities are needed to get us where we want to be – to understand exactly what it is about intense rain that switches growth of Pss from *off* to *on* in association with bean leaves in the field. When we have that knowledge, we will be able to proceed to look for the regulatory system and what it does.

We have dealt with rain and its role in pathogen population development at some length to emphasize the importance of the environment in pathogen–host interactions. In the absence of any one of the three corners of the disease triangle, susceptible host, virulent pathogen, and appropriate environment (in this case, intense rain), the interactions that eventually lead to disease simply do not occur. Despite this, nearly all molecular studies of interactions of bacterial pathogens with their hosts have been carried out under conditions where the need for intense rain is avoided. Should we be surprised if, when all the answers are in, it turns out that such studies missed key aspects of the plant–bacterial interaction in the field? Only when the question is asked under conditions where the system responds to the trigger will we know the answer to the intriguing question, "What is the nature of the genetic system regulating bacterial–plant response to intense rains?"

5.4.5   Enter the host

We know very little about host factors that affect rates of the bacterial population processes. At the scale of whole plants, we know that host genotype has a significant effect on relative abundance of Pss (Hirano *et al.*, 1996). Some cultivars and breeding lines support larger numbers of Pss than others. When two cultivars that differed in susceptibility to bacterial brown spot were planted at the same time, changes in pathogen population sizes occurred in parallel. The magnitude of the changes, particularly increases, however, differed. Relative pathogen population sizes appear to be controlled by quantitative traits at multiple loci in the host (Jung *et al.*, 2003).

At a scale smaller than individual leaves, we know that leaves are structurally heterogeneous. The surface is dotted with stomata, trichomes, and other appendages. The cuticle layer separates the leaf proper from the surrounding air. Below the epidermis is the realm of the apoplast and plant cells that form the leaf. The fate of immigrants arriving on leaves, the likelihood that bacterial cells will multiply or die or emigrate, is surely affected by the combined effects of factors related to the leaf, the environment, and the bacterium at the micro-scale. Studies using bioreporter constructs of *P. syringae* and other phyllosphere bacterial species that sense, for example, nutrient and water availability are beginning to provide some information on interactions that occur at the scale of individual bacterial cells (Axtell & Beattie, 2002; Leveau & Lindow, 2001, 2002; Lindow & Brandl, 2003). To date, the vast majority of studies on populations at a scale smaller than an intact leaf have been done under laboratory conditions that avoid many of the plant–bacterial–environmental interactions that regulate population sizes in the field. At some future time, we expect that linking findings from micro-scale or within-leaf studies to those dealing with populations of leaves will provide a better understanding of plant–microbe interactions in the natural environment.

5.5   Pss population sizes and the likelihood of disease development

Pss is often found in association with asymptomatic bean plants (Ercolani *et al.*, 1974; Lindemann *et al.*, 1984a). This finding has been well documented not only for Pss but for other pathovars of *P. syringae* and other foliar bacterial pathogens (Hirano & Upper, 1983). We now ask: How do temporal and leaf-to-leaf variabilities in pathogen population sizes relate to the development of brown spot disease in the field? To address this question, doses (pathogen population sizes) were varied and responses (amounts of disease) were measured in two different sets of field experiments (Lindemann *et al.*, 1984b; Rouse *et al.*, 1985). In both, pathogen population sizes on individual bean leaflets and amounts of disease were measured at various times during the growing season and the experiments were repeated in different growing seasons.

In the experiments of Lindemann *et al.* (1984b), mean pathogen population sizes were not quantitatively related to subsequent amounts of brown spot disease. However, the frequencies with which epiphytic population sizes of Pss were equal to or greater than about 10^4 CFU on asymptomatic individual bean leaflets were, indeed, predictive of new brown spot disease incidence a week later. This quantitative relationship was described in terms of a threshold model; i.e. the amount of new disease that develops between some time, t_0 and time t_0+t, is proportional to the frequency of pathogen population sizes greater than or equal to a threshold of about 10^4 CFU per asymptomatic bean leaflet times the proportion of leaves that are asymptomatic at t_0. The frequencies were estimated from the lognormal distribution of pathogen population sizes.

Rouse *et al.* (1985) modified the threshold model into a stochastic model by including the notion that there is some probability of disease occurring at any given pathogen population size, and that the probability increases with population size. Lesions may develop on leaflets with population sizes smaller than 10^4 CFU. However, the probability of such an occurrence is smaller than that for leaves with larger pathogen population sizes. The probit-lognormal model of Rouse *et al.* (1985) can be stated as: the probability of disease in a canopy is equal to the probability that a given leaflet will have a particular number of bacteria (described by the lognormal model for bacterial numbers) multiplied by the probability of disease occurring on a leaflet, given that particular number of bacteria (described by the probit function) summed over all leaves in a canopy. In this way, the model uses information available in the variability in population sizes among individual leaves to estimate the likelihood of future disease. The model was used to estimate the dose of inoculum that corresponded to disease incidence of 50% (i.e. ED_{50}). These values were approximately $1–5 \times 10^5$ CFU per bean leaflet. Epidemics of brown spot disease were always preceded by pathogen population build-up. When population sizes of Pss were compared to amounts of disease, disease correlated well with population sizes measured 4–7 days before disease was measured. This correlation was much stronger than correlation of disease and population size measured on the same day. The amount of disease did not correlate with numbers of Pss measured at any time after disease was estimated.

The model, however, does not tell the whole story. Not all increases in pathogen population sizes lead to increased amounts of disease. This was frequently the case when pathogen population increases were relatively transient as exemplified by the ~270-fold increase that occurred between 29 and 30 days after planting (Fig. 5.2B). The model predicted that disease should develop but it did not. We do not know why. What we do know is that Pss can multiply, in some cases to rather large numbers, without causing disease. Somehow, the regulatory machinery in plant and bacteria that controls growth of Pss in association with leaves that determines when lesions will be formed, also allows pathogen growth that does not result in lesions. How (or why) is it that each Pss cell has the genetic machinery sufficient to cause a brown spot lesion, although the likelihood of lesion development per cell is so very small and stochastically linked to pathogen population sizes? A thorough understanding of the way this bacterial–plant system is regulated will not be complete until these phenomena can be explained.

The likelihood of disease is related to the numbers of bacteria associated with leaves. Thus, the processes that give rise to large population sizes of the bacteria in association with leaves are the ones important for determining the amount of subsequent disease. For the Pss–bean system, intense rain is a major force driving rapid and sustained pathogen multiplication. We have come full circle to return to the basic, simple life cycle of bacteria – they divide. However, the process involves a complex regulatory network that controls timing and extent of replication of the bacteria in their natural habitat.

5.6 How does Pss cause brown spot lesions?

Causation of brown spot lesions on bean is unique to some strains within the pathovar *syringae*. Should we, therefore, expect the genes required for brown spot lesion formation to be unique to brown spot strains of Pss? Which genes are the ones really important for disease causation? – The ones that actually code for the functional molecules that cause the lesions? The regulatory systems that allow the bacteria to flourish in association with their hosts? Housekeeping genes such as DNA polymerase? In a broad context, all are. In a more specific context, the only ones really important for lesion causation are those involved directly in causing lesions. Although these are interesting questions from an academic perspective, they are all more or less irrelevant in the practical world of empirical science. In that world, the genes that are recognized as being required for disease causation are the ones for which we can detect some sort of phenotype associated with loss of pathogenicity or virulence. The genes we discuss here have been identified on the basis of phenotype exhibited in some sort of assay. They affect disease causation in that assay, and are therefore virulence or pathogenicity genes under the conditions of the assay from which they came, in the minds of their discoverers and in the literature. What remains is to sort out the way they affect virulence or pathogenicity, be it through a direct effect on lesion causation, on growth and survival in association with the host, or through housekeeping functions within the total package within which all of these functions are integrated to form the whole – the bacterium.

Most (perhaps all) of the genes known to be associated with pathogenicity in the Pss–bean system fall into one of the two major regulons: *hrp* (hypersensitivity reaction and pathogenicity) and *gac* (global activator) (reviewed in Hirano & Upper, 2002). The *hrp* regulon encompasses all regulatory and structural genes that code for components of a Type III secretion system (TTSS) and a suite of effector molecules that traverse the system. Current models describe the system as one in which bacterial effector molecules are delivered directly into plant cells. Interactions of bacterial and plant molecules are said to lead to resistance in the case of incompatible interactions and disease in the case of compatible interactions. *hrp* genes are present in clusters spanning roughly 25 kb in the genome of various pathovars of Ps (Lindgren, 1997). The region has been described as a pathogenicity island with a tripartite mosaic structure (Alfano *et al.*, 2000). The regulatory and structural genes that code for the secretory apparatus itself are centrally located within the region. The organization of the genes in this core region is generally similar in all strains of Ps for which there are data. The regions bordering the core *hrp/hrc* genes have been referred to as the conserved effector locus (CEL), to indicate the similarity in genes found in the region to the right of the core genes, and exchangeable effector locus (EEL) due to the variable nature of genes on the left (Alfano *et al.*, 2000; Charity *et al.*, 2003; Deng *et al.*, 2003). Many reviews are available for greater detail regarding the Hrp TTSS and the ongoing search for effector molecules that traverse the system (e.g. He, 1998; Hueck, 1998; Collmer *et al.*, 2002; Greenberg & Vinatzer, 2003).

The *gac* regulon is controlled by the master two-component regulators, GacS (sensor kinase) and GacA (response regulator) (Hrabak & Willis, 1992; Rich *et al.*, 1994). Typically, two-component regulatory systems consist of a kinase which senses environmental signals and relays the signals via phosphorylation of the cognate response regulator which, in turn, mediates changes in transcription either directly or indirectly (Hoch & Silhavy, 1995). The *gacS* and *gacA* genes have been found in a number of Gram-negative bacterial species including non-pathogens and pathogens of plants and animals. They regulate genetic systems responsible for a variety of phenotypes that differ among bacterial species (reviewed in Heeb & Haas, 2001). The phenotypes affected in *gacS* (and *gacA*) mutants of Pss strain B728a are listed in Table 5.1. Hence, some of the genes necessary for each of these phenotypes are within the *gac* regulon. A gene of particular interest is the regulatory gene *salA* (syringomycin and lesion formation) (Kitten *et al.*, 1998). Pss B728a *salA* mutants are impaired in lesion formation and toxin production but wild type for other phenotypes affected in a *gacS* or *gacA* mutant. Expression of *salA* is regulated by GacS/GacA. SalA, in turn, regulates expression of only a subset of genes in the regulon – those required for lesion formation and toxin production. Syringomycin has long been recognized as a virulence factor in Pss strains (Bender *et al.*, 1999). In strains such as Pss B301D (isolated from pear), syringomycin and a second toxin, syringopeptin, together contribute significantly to disease severity in immature cherry fruit assays (Scholz-Schroeder *et al.*, 2001a). The toxins are cyclic lipo-peptide molecules that are biosynthesized by non-ribosomal peptide synthetases (Bender *et al.*, 1999). The genes coding for both toxins span a contiguous ~140 kb in the Pss genome (Scholz-Schroeder *et al.*, 2001b, 2003; Lu *et al.*, 2002). Interestingly, *salA* is located at the right end of this large gene cluster. The genes for toxin production including *salA* are not widespread among bacterial species as are *gacS/gacA* and the core *hrp/hrc* genes in the *hrp* regulon.

The *hrp* and *gac* regulons have been studied as two separate regulons that affect interactions of Pss with plants. Within the intricate and complex regulatory networks of each of these regulons lies part of what makes Pss do what it does in association with plants. Let us compare behaviors of a few selected mutants (Table 5.1) under different environmental conditions to try to sort out if these genes are involved in lesion formation *per se*, in enabling growth in association with the plant, and/or in some other function(s) in the plant environment. The mutants we discuss were derived from wild type B728a, the model strain for the Pss–brown spot system. B728a has proven to be a good model strain (amenable to genetic manipulations, highly virulent and field competent) whose genome is currently available in draft form (http://www.jgi.doe.gov/JGI_microbial/html/pseudomonas_syr/pseudo_syr_homepage .html). For ease of discussion we refer to the mutants of interest by the names provided in Table 5.1.

5.6.1 Growth chamber assays

When infiltrated into bean leaves, the *hrp* secretion mutants BHrcC and BHrpJ and the *gac* regulon mutants BGacS and BSalA all fail to cause a pathogenic reaction. The mutants, however, differ with respect to *in planta* growth. BGacS and BSalA grow well and are indistinguishable from the wild type strain. BHrcC and BHrpJ do not grow well or achieve wild type population sizes. They (as well as many other *hrp* mutants of phytopathogenic bacteria) are impaired in their ability to grow when infiltrated into leaves. The findings with the BGacS and BSalA mutants demonstrate that *in planta* growth and lesion formation are genetically separate phenomena. At this point, the *in planta*

Table 5.1 Laboratory phenotypes of *hrp* and *gac* regulon mutants derived from *P. syringae* pv. *syringae* wild type strain B728a[1]

Mutant[2]	BHrcC	BHrpJ	BHrpZ	BGacS	BSalA	BSyrB
gene/function	*hrcC* TTSS[3]	*hrpJ* operon TTSS	*hrpZ* harpin$_{Pss}$ TTSS secreted protein	*gacS* sensor kinase, two-component regulatory system	*salA* regulator	*syrB* subunit of syringomycin peptide synthetase
Brown spot lesion formation	–	–	+	–	–	+
In planta growth	reduced	reduced	+	+	+	+
Hypersensitive reaction	–	–	+	+	+	+
Syringomycin				–	–	–
Syringopeptin				–	unknown	+
Extracellular protease				–	+	+
Homoserine lactone				–	+	+
Swarming				–	+	+

[1] For more information, see Charkowski *et al.* (1997) and Hirano *et al.* (1999) for *hrp* genes and mutants; Willis *et al.* (1990) and Hrabak & Willis (1992) for *gacS*, plant assays; Hrabak & Willis (1993) for syringomycin and protease assays; Kitten *et al.* (1998) for *salA*, homoserine lactone assay; Kinscherf & Willis (1999) for swarming phenotype; Zhang *et al.* (1995) for *syrB* gene; Grgurina *et al.* (2002) for toxins produced by B728a.

[2] Original names in the literature are NPS3136 for BGacS (Willis *et al.* 1990), BSal1 for BSalA (Kitten *et al.* 1998), and KW329 for BSyrB (Hrabak & Willis, 1993).

[3] Type III secretion system.

growth defect of the *hrp* secretion mutants is a plausible explanation for their lesion-minus phenotype.

BHrpZ is deficient in production of harpin$_{Pss}$, one of the first proteins demonstrated to be secreted by the Hrp TTSS (He *et al.*, 1993). BHrpZ and the syringomycin-deficient mutant BSyrB were indistinguishable from wild type with respect to *in planta* growth and lesion formation. Thus, at least under conditions of growth chamber assays, lack of syringomycin and harpin$_{Pss}$ production appear not to affect bacteria–leaf interactions in the leaf apoplast in the Pss–bean system.

5.6.2 Field experiments

In the field, we have the opportunity to examine the behavior of mutants relative to wild type within the context of the overall dynamics of populations of bacteria on populations of leaves. We introduce the bacteria onto bean seeds at the time of planting. In moist soil, Pss B728a multiplies rapidly on the germinating seeds and is well adapted to life under field conditions by the time the seedlings emerge. Bacterial population sizes are measured over a 5–9-week period; first on germinating seeds, then leaves as they are produced. There is ample time for many iterations of bacterial immigration, multiplication, emigration, and death. Except for the initial inoculation of seeds with laboratory-grown bacterial cells, the system develops naturally under variable weather conditions. Because we homogenize all samples prior to dilution plating, what we measure are numbers of bacteria associated with entire individual leaflets (or germinating seeds). We do not know where they are in or on leaves. Many individual samples are collected at frequent intervals to capture the inherent leaf-to-leaf and temporal variability in population sizes. The following summarizes what we have learned from experiments conducted over the past decade with a number of *hrp* and *gac* regulon mutants (Table 5.1) (Hirano *et al.*, 1997, 1999; Hirano & Upper, 2002, unpublished).

5.6.2.1 Germinating bean seeds
Within 3–5 days after planting (depending on experiment), all bacterial strains increased in numbers from an average initial dose of ~10^2–10^3 CFU per seed to ~10^6–10^8 CFU per pre-emergent seedling. Population sizes of the mutants and wild type were indistinguishable on germinating seeds. The unimpaired growth of the *hrp* secretion mutants was unexpected, given their inability to grow well in leaves in growth chamber assays. Genes in the *hrp* and *gac* regulons appear not to play major roles in that part of the life style of *P. syringae* which requires multiplication of bacteria and immigration to get from seed to pre-emergent seedlings.

5.6.2.2 Leaves
Population sizes differed among the mutants and with one exception between mutant and wild type as summarized below. The strains are ordered based on population sizes pooled across all samplings and experiments. The order can be viewed as depicting the relative fitness of the various bacterial strains and with some caveats

the relative effects of the genes on fitness. The order from the growth chamber assays is presented for comparative purposes.

Relative overall population sizes
Field:
    BHrcC = BHrpJ < BGacS < BSalA < BSyrB < BHrpZ = B728a wild type

Growth chamber:
    BHrcC = BHrpJ < BGacS = BSalA = BSyrB = BHrpZ = B728a wild type

Population sizes of the *hrp* secretion mutants BHrcC and BHrpJ decreased with time in association with leaves. Following intense rain events, B728a multiplied as expected. BHrcC and BHrpJ did not. Differences in population sizes between the mutants and wild type were as large as 1000–10 000-fold. Similar findings from two very different environments (growth chamber and field) tell us rather clearly that the TTSS is an essential component of the machinery that regulates growth of Pss. *hrpZ*, on the other hand, has no effect on bacterial population sizes in either environment – laboratory or field.

In contrast to the *hrp* secretion mutants, the *gac* regulon mutants behaved differently in the field compared to growth chamber. In the field, relative population sizes of BGacS, BSalA, and BSyrB were not only different from the wild type but they differed among themselves. The order of relative field fitness of the mutants paralleled the tier at which the genes (or proteins) are thought to participate in the regulon. Estimates of population differences between the mutants and wild type are roughly in the order of 100–1000-fold for BGacS, 10–100-fold for BSalA, and 3–10-fold for BSyrB. The findings tell us that in addition to the *hrp* regulon, genes in the *gac* regulon are also essential components of the machinery that regulates multiplication of Pss in the phyllosphere in the field but not under normal laboratory conditions. The relatively small but reproducible reduction in population sizes exhibited by BSyrB relative to wild type suggests that the *syrB* gene affects fitness of B728a and that at least a part of the reduced fitness of BSalA may be attributed to syringomycin production.

5.6.2.3   Brown spot disease
There also were some interesting differences in lesion formation under field versus growth chamber conditions. BHrcC and BHrpJ caused brown spot lesions in the field. This may seem paradoxical since we previously noted that population sizes of the mutants decreased in association with leaves. However, in one experiment, population sizes of the mutants on emerging primary leaves were unusually large, presumably because of unusually cool, wet conditions between planting and emergence. A few young primary leaves harbored population sizes greater than 10^6 CFU per leaf. Fourteen days after planting, disease incidence on primary leaves was about 4% in plots inoculated with BHrcC and BHrpJ (~63% for B728a). With time, population sizes of the mutants decreased on leaves and no additional disease

was found. Consistent with findings from growth chamber assays, BHrpZ and BSyrB caused disease in the field. Disease incidence was similar for BHrpZ relative to wild type and slightly less, though not statistically different, for BSyrB relative to wild type.

The amounts of disease caused by the *hrp* secretion mutants, BHrpZ and BSyrB, in the field were as expected based on the established quantitative relationship between pathogen population sizes and likelihood of disease as discussed in a previous section. We isolated BSalA from occasional lesions. However, the incidence of disease was very much less than predicted from the population sizes established by this mutant. This might be expected for a gene that is directly involved in brown spot lesion formation, albeit in a regulatory role, and provided that we accept the premise that bacterial multiplication and lesion formation are under separate genetic control, at least to some extent.

5.6.3   Growth chamber assays revisited

With the finding that the *hrp* secretion mutants are able to cause lesions under certain field conditions, we returned to the growth chamber to determine whether we could reproduce the field phenomenon. We inoculated the mutants onto seeds as was done in the field experiments. We manipulated conditions in the growth chamber to better simulate the weather conditions that led to substantial growth of the mutants on emerging seedlings. Under these conditions, BHrcC and BHrpJ caused disease in the growth chamber. The amounts were similar to those measured in the field.

We reasoned that if *hrp* secretion mutants are unable to cause disease largely because of their growth defect, we may be able to overcome the defect by merely increasing the dose of inoculum in leaf infiltration assays. Indeed, when this was done, BHrcC and BHrpJ caused pathogenic reactions, although the reactions were not as strong as those of the wild type. In a different set of experiments, we found that the growth defect of BHrcC and BHrpJ could be rescued by co-inoculating the secretion mutants with BGacS. In the presence of BGacS, the *hrp* secretion mutants grew to wild type levels. Moreover, pathogenic reactions developed. Neither BGacS nor the secretion mutants alone can cause the reaction when inoculated at low doses. When co-inoculated, both types of mutants may have directly contributed to the pathogenic reaction. Alternatively, it may be that BGacS with its TTSS intact provided the effector molecules necessary to enable growth. Rescued growth of the *hrp* secretion mutants, in turn, resulted in population sizes that were sufficient for the mutants to cause disease. Neither hypothesis can be ruled out at this time. However, the latter is consistent with findings from the other experiments described here.

5.6.4   Reconciliation

Elevated population sizes of the *hrp* secretion mutants developed or were established under three different conditions: (1) seed inoculation with conditions favorable for growth of the mutants on emerging seedlings, (2) large doses infiltrated into leaves, and (3) co-inoculation of *hrp* mutant with BGacS. In all cases, disease developed. Is the Hrp TTSS directly involved in lesion formation? A subset of molecules that traverse the pathway have been described as pathogenicity or virulence factors. However, none have been identified for which effects on lesion formation *per se* have been demonstrated to be distinct from effects on growth in or on leaves. At the time that *hrp* genes were first discovered, the suggestion was made that the loss of pathogenicity in *hrp* mutants may be due to their growth defect (Lindgren *et al.*, 1986). Since then, no evidence has been found to refute this hypothesis. Indeed, findings from laboratory and field experiments demonstrate clearly that the requirement of *hrp* genes for pathogenicity is largely mediated through the important role that the TTSS plays in enabling pathogen growth in association with leaves.

Interestingly, deletion of the EEL (~8 kb coding for three putative effector proteins) in Pss B728a had an assay-dependent effect on growth of the mutant in laboratory experiments (Deng *et al.*, 2003). Growth of the EEL deletion mutant and wild type was similar when the strains were infiltrated into bean leaves. When leaves were inoculated by dipping them into bacterial suspensions, the EEL mutant was significantly reduced in population sizes and caused smaller lesions than the wild type. The findings suggest that some molecules coded by the EEL region may participate in early interactions between pathogen and plant (on the leaf surface, perhaps) which are circumvented when bacteria are injected into the apoplast. If this is the case for B728a, it apparently is not for other *P. syringae* strains (e.g. *P. syringae* pv. *tomato* DC3000 (Boureau *et al.*, 2002)).

If *hrp* genes are not directly responsible for lesion formation, what genes are? We think the likely candidates may lie in the *salA*-regulated branch of the *gac* regulon. The genes appear not to be those required for syringomycin production as syringomycin mutants of strain B728a cause lesions in the laboratory and field. What about syringopeptin? We do not know. What about other genes that are not involved in toxin production but that code for proteins that either directly or indirectly act in concert with the toxins? All are possible until proven otherwise.

We do not know the full story of how Pss causes brown spot disease. We have learned that genes in the *hrp* and *gac* regulons are part of the complex genetic machinery that regulates bacterial growth and survival in the phyllosphere. Even when equipped with all of the genes necessary for growth on susceptible plants, however, the bacterium does not flourish in the field in the absence of frequent intense rains. For a bacterium such as Pss, for which growth is so closely coupled to weather, are not the systems regulating the bacterial response to weather at least as important for the likelihood of disease as those plant genes that control the suitability of the plant as habitat for the bacterium? This seems to us an excellent prospect for future research.

5.7 Summary

Immigration delivers founding populations of Pss to bean leaves. Growth of the pathogen, the frequency with which cells divide, is the most important determinant of pathogen population sizes on individual bean leaves. Intense rain triggers growth of Pss in association with bean leaves. The likelihood that disease will develop increases with increasing population sizes. Thus, bacterial growth triggered by intense rain is the key to development of population sizes large enough to cause epidemics. This picture is summarized in Fig. 5.4.

With the advent of each new growing season, new crops of *Phaseolus* beans are planted. The crops may become diseased. They may not. Amounts of disease may be insignificantly small or of epidemic proportions. It all depends on the components at the vertices of the Disease Triangle: pathogen, host, and environment. The amount of disease depends on the virulence of the pathogen, the susceptibility of

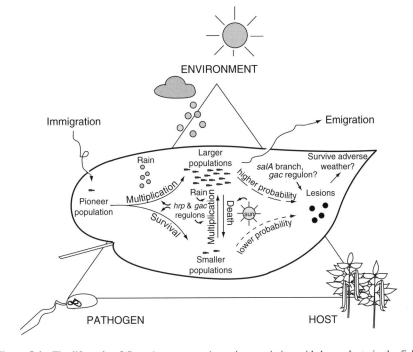

Figure 5.4 The life style of *P. syringae* pv. *syringae* in association with bean plants in the field. Immigrants arrive on leaves, either by growth on germinating seeds or through the air. The fate of the immigrants depends on interactions among pathogen, host, and the environment. On susceptible bean cultivars, frequent intense rains are required for rapid multiplication of the pathogen that leads to the establishment of large bacterial population sizes, which, in turn, increases the probability of brown spot disease development. During dry, sunny weather, some portion of the population dies, others may survive (e.g. in lesions if present). Bacteria may emigrate during dry (e.g. aerosol dispersal during dry, sunny weather) and wet (e.g. wash-off by rain) conditions. Genes in the *hrp* and *gac* regulons impact on pathogen multiplication and survival. Genes in the *salA* branch of the *gac* regulon are likely candidates directly responsible for brown spot lesion formation. The diagram illustrates processes that occur on a single bean leaflet. These processes are iterated on each of the millions of leaflets within a hectare of the bean crop. Pathogen and host genomes and the environment interact to regulate the rates with which these processes occur. Modified from Hirano and Upper (2000); reprinted with permission.

the host, and the suitability of the environment for disease development. The more susceptible the host, the more virulent the pathogen, and the more conducive the environment, the greater will be the amount of disease. Host cultivars vary quantitatively with regard to the sizes of pathogen populations they carry. Fitness of the pathogen determines the sizes of the population that will arise, and thus the amount of disease. And finally, weather conditions, principally the amount of intense rain for the *P. syringae* – bean system, drive dynamics of Pss population sizes. All three, host, pathogen, and environment, are key players.

By selecting conditions that minimize the effect of the environment, we essentially rotate the disease triangle out of the plane of the paper until the corner representing "environment" is completely obscured by the line connecting host and pathogen. When this happens, the important effects of environmental factors on plant–pathogen systems are removed from consideration. Most of our understanding of bacterial–plant interactions has been gathered in a growth chamber environment, where bacterial cells are often injected directly into the apoplast and there are no environmental constraints on pathogen growth or lesion formation. The entire process is compressed in time, and separating effects on growth from those on lesion formation becomes difficult if not impossible. Thus, the genetic systems we know about that affect lesion development in one way or another may lack any or all environmental responsive elements, and may also be poorly resolved with regard to where in the system they function, be it in lesion formation *per se*, in regulation of plant-associated growth, or in some housekeeping function critical to growth or lesion formation. Such limitations need to be corrected for complete illumination of the interior of the black box of plant–bacterial interactions.

Where do we stand along the path to understanding how Pss causes disease? Is the proverbial glass half full or half empty? In this case, the answer depends on the size of the glass. If our objective is to identify and understand the genetic systems related to disease causation under laboratory conditions, we have excellent cause to be optimistic and declare the glass half full, or maybe more! On the other hand, if we are trying to understand the nature and function of the complex matrix of structural and regulatory genes necessary for epidemic causation under field conditions, it is probably appropriate to declare the glass half empty – or less. The difference between these two views can be found at the neglected corner of the disease triangle – the environment. The genetic systems that regulate growth of Pss in response to intense rain are completely essential to development of epidemics of bacterial brown spot disease, and beg to be elucidated.

References

Alfano, J.R., Charkowski, A.O., Deng, W.L., Badel, J.L., Petnicki-Ocwieja, T., van Dijk, K. & Collmer, A. (2000) The *Pseudomonas syringae* Hrp pathogenicity island has a tripartite mosaic structure composed of a cluster of Type III secretion genes bounded by exchangeable effector and conserved effector loci that contribute to parasitic fitness and pathogenicity in plants. *Proc. Natl. Acad. Sci. USA*, **97**, 4856–4861.

Axtell, C.A. & Beattie, G.A. (2002) Construction and characterization of a *proU-gfp* transcriptional fusion that measures water availability in a microbial habitat. *Appl. Environ. Microbiol.*, **68**, 4604–4612.

Beattie, G.A. & Lindow, S.E. (1999) Bacterial colonization of leaves: a spectrum of strategies. *Phytopathology*, **89**, 353–359.

Bender, C.L., Alarcón-Chaidez, F. & Gross, D.C. (1999) *Pseudomonas syringae* phytotoxins: mode of action, regulation, and biosynthesis by peptide and polyketide synthetases. *Microbiol. Mol. Biol. Rev.*, **63**, 266–292.

Boureau, T., Routtu, J., Roine, E., Taira, S. & Romantschuk, M. (2002) Localization of *hrpA*-induced *Pseudomonas syringae* pv. *tomato* DC3000 in infected tomato leaves. *Mol. Plant Pathol.*, **3**, 451–460.

Charity, J.C., Pak, K., Delwiche, C.F. & Hutcheson, S.W. (2003) Novel exchangeable effector loci associated with the *Pseudomonas syringae hrp* pathogenicity island: evidence for integron-like assembly from transposed gene cassettes. *Mol. Plant Microbe Interact.*, **16**, 495–507.

Charkowski, A.O., Huang, H.-C. & Collmer, A. (1997) Altered localization of HrpZ in *Pseudomonas syringae* pv. syringae *hrp* mutants suggests that different components of the Type III secretion pathway control protein translocation across the inner and outer membranes of Gram-negative bacteria. *J. Bacteriol.*, **179**, 3866–3874.

Collmer, A., Lindeberg, M., Petnicki-Ocwieja, T., Schneider, D.J. & Alfano, J.R. (2002) Genomic mining Type III secretion system effectors in *Pseudomonas syringae* yields new picks for all TTSS prospectors. *Trends Microbiol.*, **10**, 462–469.

Deng, W.L., Rehm, A.H., Charkowski, A.O., Rojas, C.M. & Collmer, A. (2003) *Pseudomonas syringae* exchangeable effector loci: sequence diversity in representative pathovars and virulence function in *P. syringae* pv. syringae B728a. *J. Bacteriol.*, **185**, 2592–2602.

Ercolani, G.L., Hagedorn, D.J., Kelman, A. & Rand, R.E. (1974) Epiphytic survival of *Pseudomonas syringae* on hairy vetch in relation to epidemiology of bacterial brown spot of bean in Wisconsin. *Phytopathology*, **64**, 1330–1339.

Greenberg, J.T. & Vinatzer, B.A. (2003) Identifying Type III effectors of plant pathogens and analyzing their interaction with plant cells. *Curr. Opin. Microbiol.*, **6**, 20–28.

Grgurina, I., Mariotti, F., Fogliano, V., Gallo, M., Scaloni, A., Iacobellis, N.S., Cantore, P.L., Mannina, L., van Axel Castelli, V., Greco, M.L. & Graniti, A. (2002) A new syringopeptin produced by bean strains of *Pseudomonas syringae* pv. *syringae*. *Biochim. Biophys. Acta*, **1597**, 81–89.

He, S.Y. (1998) Type III protein secretion systems in plant and animal pathogenic bacteria. *Annu. Rev. Phytopathol.*, **36**, 363–392.

He, S.Y., Huang, H.-C. & Collmer, A. (1993) *Pseudomonas syringae* pv. *syringae* harpin$_{Pss}$: a protein that is secreted via the hrp pathway and elicits the hypersensitive response in plants. *Cell*, **73**, 1255–1266.

Heeb, S. & Haas, D. (2001) Regulatory roles of the GacS/GacA two-component system in plant-associated and other Gram negative bacteria. *Mol. Plant Microbe Interact.*, **14**, 1351–1363.

Hirano, S.S. & Upper, C.D. (1983) Ecology and epidemiology of foliar bacterial plant pathogens. *Annu. Rev. Phytopathol.*, **21**, 243–269.

Hirano, S.S. & Upper, C.D. (1989) Diel variation in population size and ice nucleation activity of *Pseudomonas syringae* on snap bean leaflets. *Appl. Environ. Microbiol.*, **55**, 623–630.

Hirano, S.S. & Upper, C.D. (2000) Bacteria in the leaf ecosystem with emphasis on *Pseudomonas syringae* – a pathogen, ice nucleus, and epiphyte. *Microbiol. Mol. Biol. Rev.*, **64**, 624–653.

Hirano, S.S. & Upper, C.D. (2002) Effects of pathogenicity-associated genes on field fitness of *Pseudomonas syringae* pv. *syringae*. In: *Phyllosphere Microbiology* (eds S.E. Lindow E.I. Hecht-Poinar & V.J. Elliott), pp. 81–99. American Phytopathological Society, St Paul, MN.

Hirano, S.S., Baker, L.S. & Upper, C.D. (1996) Raindrop momentum triggers growth of leaf-associated populations of *Pseudomonas syringae* on field-grown snap bean plants. *Appl. Environ. Microbiol.*, **62**, 2560–2566.

Hirano, S.S., Nordheim, E.V., Arny, D.C. & Upper, C.D. (1982) Lognormal distribution of epiphytic bacterial populations on leaf surfaces. *Appl. Environ. Microbiol.*, **44**, 695–700.

Hirano, S.S., Clayton, M.K. & Upper, C.D. (1994) Estimation of and temporal changes in means and variances of populations of *Pseudomonas syringae* on snap bean leaflets. *Phytopathology*, **84**, 934–940.

Hirano, S.S., Rouse, D.I., Clayton, M.K. & Upper, C.D. (1995) *Pseudomonas syringae* pv. *syringae* and bacterial brown spot of snap bean: a study of epiphytic phytopathogenic bacteria and associated disease. *Plant Dis.*, **79**, 1085–1093.

Hirano, S.S., Charkowski, A.O., Collmer, A., Willis, D.K. & Upper, C.D. (1999) Role of the Hrp Type III protein secretion system in growth of *Pseudomonas syringae* pv. *syringae* B728a on host plants in the field. *Proc. Natl. Acad. Sci. USA*, **96**, 9851–9856.

Hirano, S.S., Ostertag, E.M., Savage, S.A., Baker, L.S., Willis, D.K. & Upper, C.D. (1997) Contribution of the regulatory gene *lemA* to field fitness of *Pseudomonas syringae* pv. *syringae*. *Appl. Environ. Microbiol.*, **63**, 4304–4312.

Hoch, J.A. & Silhavy, T.J. (eds) (1995) *Two-component Signal Transduction*, American Society for Microbiology, Washington, DC.

Hrabak, E.M. & Willis, D.K. (1992) The *lemA* gene required for pathogenicity of *Pseudomonas syringae* pv. *syringae* on bean is a member of a family of two-component regulators. *J. Bacteriol.*, **174**, 3011–3020.

Hrabak, E.M. & Willis, D.K. (1993) Involvement of the *lemA* gene in production of syringomycin and protease by *Pseudomonas syringae* pv. *syringae*. *Mol. Plant Microbe Interact.*, **6**, 368–375.

Hueck, C.J. (1998) Type III protein secretion systems in bacterial pathogens of animals and plants. *Microbiol. Mol. Biol. Rev.*, **62**, 379–433.

Jung, G., Ariyarathne, H.M., Coyne, D.P. & Nienhuis, J. (2003) Mapping QTL for bacterial brown spot resistance under natural infection in field and seedling stem inoculation in growth chamber in common bean. *Crop Sci.*, **43**, 350–357.

Kinscherf, T.G. & Willis, D.K. (1999) Swarming by *Pseudomonas syringae* B728a requires *gacS* (*lemA*) and *gacA* but not the acyl-homoserine lactone biosynthetic gene *ahlI*. *J. Bacteriol.*, **181**, 4133–4136.

Kitten, T., Kinscherf, T.G., McEvoy, J.L. & Willis, D.K. (1998) A newly-identified regulator is required for virulence and toxin production in *Pseudomonas syringae*. *Mol. Microbiol.*, **28**, 917–929.

Leveau, J.H. & Lindow, S.E. (2001) Appetite of an epiphyte: quantitative monitoring of bacterial sugar consumption in the phyllosphere. *Proc. Natl. Acad. Sci. USA*, **98**, 3446–3453.

Leveau, J.H. & Lindow, S.E. (2002) Bioreporters in microbial ecology. *Curr. Opin. Microbiol.*, **5**, 259–265.

Lindemann, J. & Upper, C.D. (1985) Aerial dispersal of epiphytic bacteria over bean plants. *Appl. Environ. Microbiol.*, **50**, 1229–1232.

Lindemann, J., Arny, D.C. & Upper, C.D. (1984a) Epiphytic populations of *Pseudomonas syringae* pv. *syringae* on snap bean and nonhost plants and the incidence of bacterial brown spot disease in relation to cropping patterns. *Phytopathology*, **74**, 1329–1333.

Lindemann, J., Arny, D.C. & Upper, C.D. (1984b) Use of an apparent infection threshold population of *Pseudomonas syringae* to predict incidence and severity of brown spot of bean. *Phytopathology*, **74**, 1334–1339.

Lindgren, P.B. (1997) The role of *hrp* genes during plant–bacterial interactions. *Annu. Rev. Phytopathol.*, **35**, 129–152.

Lindgren, P.B., Peet, R.C. & Panopoulos, N.J. (1986) Gene cluster of *Pseudomonas syringae* pv. "*phaseolicola*" controls pathogenicity on bean plants and hypersensitivity on nonhost plants. *J. Bacteriol.*, **168**, 512–522.

Lindow, S.E. (1993) Novel method for identifying bacterial mutants with reduced epiphytic fitness. *Appl. Environ. Microbiol.*, **59**, 1586–1592.

Lindow, S.E. & Brandl, M.T. (2003) Microbiology of the phyllosphere. *Appl. Environ. Microbiol.*, **69**, 1875–1883.

Lu, S.E., Scholz-Schroeder, B.K. & Gross, D.C. (2002) Characterization of the *salA, syrF*, and *syrG* regulatory genes located at the right border of the syringomycin gene cluster of *Pseudomonas syringae* pv. *syringae. Mol. Plant Microbe Interact.*, **15**, 43–53.

Rich, J.J., Kinscherf, T.G., Kitten, T. & Willis, D.K. (1994) Genetic evidence that the *gacA* gene encodes the cognate response regulator for the *lemA* sensor in *Pseudomonas syringae. J. Bacteriol.*, **176**, 7468–7475.

Romantschuk, M., Boureau, T., Roine, E., Haapalainen, M. & Taira, S. (2002) The role of pili and flagella in leaf colonization by *Pseudomonas syringae*. In: *Phyllosphere Microbiology* (eds S.E. Lindow, E.I. Hecht-Poinar & V.J. Elliott), pp. 101–113. American Phytopathological Society, St Paul, MN.

Rouse, D.I., Nordheim, E.V., Hirano, S.S. & Upper, C.D. (1985) A model relating the probability of foliar disease incidence to the population frequencies of bacterial plant pathogens. *Phytopathology*, **75**, 505–509.

Sabaratnam, S. & Beattie, G.A. (2003) Differences between *Pseudomonas syringae* pv. *syringae* B728a and *Pantoea agglomerans* BRT98 in epiphytic and endophytic colonization of leaves. *Appl. Environ. Microbiol.*, **69**, 1220–1228.

Scholz-Schroeder, B.K., Hutchison, M.L., Grgurina, I. & Gross, D.C. (2001a) The contribution of syringopeptin and syringomycin to virulence of *Pseudomonas syringae* pv. *syringae* strain B301D on the basis of *sypA* and *syrB1* biosynthesis mutant analysis. *Mol. Plant Microbe Interact.*, **14**, 336–348.

Scholz-Schroeder, B.K., Soule, J.D., Lu, S.E., Grgurina, I. & Gross, D.C. (2001b) A physical map of the syringomycin and syringopeptin gene clusters located to an approximately 145-kb DNA region of *Pseudomonas syringae* pv. *syringae* strain B301D. *Mol. Plant Microbe Interact.*, **14**, 1426–1435.

Scholz-Schroeder, B.K., Soule, J.D. & Gross, D.C. (2003) The *sypA, sypB*, and *sypC* synthetase genes encode twenty-two modules involved in the nonribosomal peptide synthesis of syringopeptin by *Pseudomonas syringae* pv. *syringae* B301D. *Mol. Plant Microbe Interact.*, **16**, 271–280.

Upper, C.D. & Hirano, S.S. (2002) Revisiting the roles of immigration and growth in the development of *Pseudomonas syringae* in the phyllosphere. In: *Phyllosphere Microbiology* (eds S.E. Lindow, E.I. Hecht-Poinar & V.J. Elliott), pp. 69–79. American Phytopathological Society, St Paul, MN.

Upper, C.D., Hirano, S.S., Dodd, K.K. & Clayton, M.K. (2003) Factors that affect spread of *Pseudomonas syringae* in the phyllosphere. *Phytopathology*, **93**, 1082–1092.

Willis, D.K., Hrabak, E.M., Rich, J.J., Barta, T.M., Lindow, S.E. & Panopoulos, N.J. (1990) Isolation and characterization of a *Pseudomonas syringae* pv. *syringae* mutant deficient in lesion formation on bean. *Mol. Plant Microbe Interact.*, **3**, 149–156.

Wilson, M., Hirano, S.S. & Lindow, S.E. (1999) Location and survival of leaf-associated bacteria in relation to pathogenicity and potential for growth within the leaf. *Appl. Environ. Microbiol.*, **65**, 1435–1443.

Zhang, J.-H., Quigley, N.B. & Gross, D.C. (1995) Analysis of the *syrB* and *syrC* genes of *Pseudomonas syringae* pv. *syringae* indicates that syringomycin is synthesized by a thiotemplate mechanism. *J. Bacteriol.*, **177**, 4009–4020.

6 Fungal pathogenesis in the rice blast fungus *Magnaporthe grisea*

Chaoyang Xue, Lei Li, Kyeyong Seong and Jin-Rong Xu

6.1 Introduction

Rice blast, caused by *Magnaporthe grisea* (Hebert) Barr (anamorph *Pyricularia grisea* Sacc) (Rossman *et al.*, 1990), is one of the most severe diseases of rice throughout the world (Ou, 1985). *M. grisea* also causes disease on many other grass species, including economically important crops such as barley, wheat, and millet (Valent & Chumley, 1991). It was proposed that the rice blast fungus be changed to *Magnaporthe oryzae* (Couch & Kohn, 2002), but we will use *M. grisea* in this chapter because the proposed name change has not yet been adopted by the rice blast community. The rice blast fungus attacks all above-ground parts of the rice plant, and seedlings can be killed during epidemics. In mature plants, the fungus attacks the emerging seed panicle, resulting in the complete loss of rice grains (Ou, 1985). As rice remains the major food crop for more than a third of the world's population (Ford *et al.*, 1994), rice blast is recognized as one of the main pathological threats to world food supplies.

M. grisea is amenable to classical and molecular genetic manipulations and has been proposed as a model system to study fungal–plant interactions (Valent, 1990). Several efficient protocols and selectable markers are available for fungal transformation. Various genetic resources, including genetic maps, and numerous genomic and cDNA libraries have been developed in the past decade. Significant progress also has been made in genetic and cell biological studies of the infection-related morphogenesis (Talbot & Foster, 2001). Recently, the rice and *M. grisea* genomes have been sequenced, which enables genomic studies of rice–rice blast interactions and pathogenesis.

6.2 Life cycle of *Magnaporthe grisea*

M. grisea can be cultured on a number of artificial media, and produces three-celled conidia by holoblastic conidiogenesis. Conidiation is induced by light on oatmeal plates but inhibited in submerged cultures. On nutritionally rich media such as complete medium, conidiation is significantly reduced. It is a heterothallic haploid ascomycete with two mating types, *MAT1-1* and *MAT1-2* (Kang *et al.*, 1994). Mating between compatible strains produces fertile perithecia that are usually buried in substrates. Eight ascospores arranged unordered inside each ascus are suitable for tetrad analysis. Although perithecia have not been observed in nature and most field isolates are female sterile, genetic recombination likely occurs in some areas, and self-fertile strains have been isolated in mountainous regions of South and East Asia (Zeigler, 1998; Kumar *et al.*, 1999). Under laboratory conditions, female fertility rapidly degenerates after a few generations of subculture (Valent & Chumley, 1991).

The infection cycle is initiated by the attachment of the three-celled conidia to the rice leaf surface. Germ tubes produced from conidia can recognize hydrophobic surfaces and differentiate into specialized infection structures called appressoria (Fig. 6.1). *M. grisea* uses the enormous turgor pressure developed in appressoria to physically penetrate the underlying plant surface with a penetration peg. Mutants blocked during appressorium formation or appressorial turgor generation fail to infect healthy rice plants (Tucker & Talbot, 2001). Once inside plants, infectious hyphae grow in and between plant cells. Eventually, lesions develop and the fungus release conidia to reinitiate the infection cycle (Fig. 6.1). *M. grisea* infects rice plants throughout the season and causes severe yield losses throughout the world. In the past decade, many aspects of infection-related morphogenesis and fungal–plant interaction have been extensively studied with *M. grisea* (Howard & Valent, 1996; Hamer & Talbot, 1998; Talbot & Foster, 2001; Tucker & Talbot, 2001). A list of virulence-associated genes and their potential biological functions is given in Tables 6.1 and 6.2.

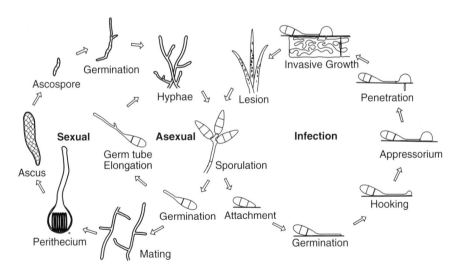

Figure 6.1 Life cycle of *Magnaporthe grisea*. For asexual reproduction, pyriform conidia are produced sympodially on conidiophores. Sexual reproduction is initiated after mating between *MAT1-1* and *MAT1-2* strains. Ascospores are hyaline and four-celled. The infection cycle is initiated by attaching conidia to rice leaf surface. Germ tubes produced from conidia differentiate into melanized appressoria for penetration. Once inside plants, infectious hyphae grow in and between plant cells. Eventually, lesions develop on rice leaves, and the fungus produces conidia on infected area to reinitiate infection.

Table 6.1 Pathogenicity genes characterized in *Magnaporthe grisea*

Gene	Protein	Major mutant phenotype	Reference
ABC1	ABC-transporter	Reduced in infectious growth	Urban et al. (1999)
ACE1	PKS/NRPS	Defective in interacting with rice R gene *Pi33*	Böhnert *et al.* (2003)
ACR1	Transcription factor	Acropetal conidia, reduced in virulence	Lau & Hamer (1998)
ALB1	Polyketide synthase	Melanin-deficient	Romao & Hamer (1992)
AVR-CO39	No known homolog	Virulent on rice cultivar CO39	Farman & Leong (1998)
AVR-PITA	Neutral zinc metalloprotease	Defective in interacting with rice R gene *Pi-ta*	Orbach *et al.* (2000)
BUF1	Polyhydroxynaphthalene reductase	Melanin-deficient	Chumley & Valent (1990)
CBP1	Chitin-binding protein	Altered appressorium morphology	Kamakura *et al.* (2002)
CON7	Transcription factor	Altered appressorium morphology	Shi & Leung (1995)
CPKA	Catalytic subunit of PKA	Delayed appressorium formation, defective in appressorial penetration	Mitchell & Dean (1995)
CYP1	Cyclophilin	Reduced in penetration and virulence	Viaud *et al.* (2002)
GAS1	Virulence factor	Reduced in penetration and virulence	Xue *et al.* (2002)
GAS2	Virulence factor	Reduced in penetration and virulence	Xue *et al.* (2002)
ICL1	Isocitrate lyase	Reduced in penetration and virulence	Wang *et al.* (2003)
MAC1	Adenylate cyclase	No appressorium formation, non-pathogenic, reduced in growth and conidiation	Choi & Dean (1997)
MAGB	Gα subunit	Reduced in conidiation, appressorium formation, and virulence	Liu & Dean (1997)
MPG1	Class I hydrophobin	Reduced in appressorium formation and virulence	Talbot *et al.* (1993)
MPS1	Homolog of yeast Slt2 MAP kinase	No appressorial penetration, reduced in conidiation	Talbot *et al.* (1993)
MST12	Transcription factor	No appressorial penetration and infectious hyphal growth	Park *et al.* (2002)
NTH1	Neutral trehalase	Reduced virulence	Foster *et al.* (2003)
ORP1	No known homolog	Reduced virulence	Villalba *et al.* (2001)
PAT531	Transmembrane protein	Penetration and colonization deficiency	Fujimoto *et al.* (2002)
PDE1	P-type ATPase	Reduced penetration and virulence on rice	Balhadere & Talbot (2001)
PLS1	Tetraspanin	No appressorial penetration	Clergeot *et al.* (2001)
PMK1	Homolog of yeast Fus3/Kss1 MAP kinase	No appressorium formation and infectious hyphal growth	Xu & Hamer (1996)
PTH1	Yeast Grr1 homolog	Reduced virulence	Sweigard *et al.* (1998)
PTH2	Carnitine acetyltransferase	Reduced virulence	Sweigard *et al.* (1998)
PTH3	Imidazole glycerol phosphate dehydratase	Histidine auxotrophic, reduced in virulence	Sweigard *et al.* (1998)
PTH8	UDP-glucose:sterol glucosyltransferase	Normal appressoria, reduced in virulence	Sweigard *et al.* (1998)

PTH11	Transmembrane protein	Reduced in appressorium formation and plant infection	DeZwaan *et al.* (1999)
PWL2	No known homolog	Avirulence gene	Sweigard *et al.* (1995)
RSY1	Scytalone dehydratase	Melanin-deficient	Motoyama *et al.* (1998)
TPS1	Trehalose-6-phosphate synthase	No trehalose synthesis, reduced in virulence	Foster *et al.* (2003)

Table 6.2 Defects of *Magnaporthe grisea* mutants in different infection stages

Infection stage	Mutants
Attachment and germination	*acr1, gde1, icl1, smo*
Appressorium morphogenesis	*apf1, app1, app2, app3, app5, alb1, buf1, cbp1, con1, con2, con4, con7, cpkA, gde1, igd1, mac1, magB, met1, mpg1, pig1, pmk1, pth4, pth11, rsy1, smo, sum1*
Penetration	*alb1, buf1, cpkA, cyp1, gas1, gas2, mps1, mst12, pat531, pde1, pde2, pls1, pmk1, orp1, rsy1, tps1*
Infectious growth	*abc1, ace1, avr-Pita, avr-CO39, con1, cpkA, icl1, mac1, magB, mst12, npr1, npr2, nth1, orp1, pal144, pat531, pmk1, pth1, pth2, pth3 pth8, pwl1, pwl2, tps1*
Conidium production	*acr1, apf1, ccn1, con1, con2, con3, con4, con5, con6, con7, gde1, hsp1, igd1, mac1, magB, magC, met1, pth10, smo*

6.3 Conidium attachment and germination

6.3.1 Attachment

Adhesion of dispersing propagules to plant surfaces is commonly the first step of infection by fungal pathogens. In *M. grisea*, the adhesive material known as the spore tip mucilage (STM) is pre-formed and stored in the periplasmic regions of the apical conidial cell (Hamer *et al.*, 1988). STM is released upon contact with plant or artificial surfaces, and forms a viscous pad for attaching the conidium to the substrate (Hamer *et al.*, 1988). The adhesive strength of *M. grisea* conidia has been determined by measuring the capillary force needed to detach conidia from the surface (Gerbeaud *et al.*, 2001).

The plant lectin concanavalin A binds to STM and inhibits conidium adhesion and appressorium formation, indicating that STM contains α-linked mannosyl and/ or glucosyl residues (Bourett *et al.*, 1993). However, the exact chemical composition of STM is not clear. It has been reported that STM contains protein, carbohydrate, and lipid components (Howard & Valent, 1996). To date, no genes involved in biosynthesis of STM components have been identified, and no adhesive compound has been experimentally proved by directed mutagenesis to mediate attachment in *M. grisea*. Since the production of STM seems to be affected by culture age and incubation conditions and may vary among different isolates (Howard & Valent, 1996), it will be interesting to determine the importance of STM in disease development under field conditions.

6.3.2   Germination

After attachment, conidia germinate and produce germination tubes or germ tubes. In *M. grisea*, germ tube emergence is not proceeded by obvious conidial swelling, but endocytosis can be detected within 2–3 min of conidial hydration (Atkinson *et al.*, 2002). For conidia attached to solid surfaces, nutrients are not required for germination, and a single germ tube usually emerges from the apical and/or basal cell of the conidia. The middle cell rarely germinates and may function as an energy reservoir for appressorial turgor generation. These germ tubes are usually unbranched and surrounded by an extracellular matrix material (Jelitto *et al.*, 1994; Xiao *et al.*, 1994a). The germ tube mucilage mediates germ tube adhesion and possibly provides protection against harmful environmental factors. Nutrients can stimulate germination in conidia attached to solid surfaces or suspended in liquid solutions. However, there are contradicting reports about the essentiality of nutrients for conidium germination in liquid suspensions (Lee & Dean, 1993; Xiao *et al.*, 1994b).

Conidium germination requires active metabolic activities. In several filamentous fungi such as *Colletotrichum* species and *Aspergillus nidulans*, cAMP signaling and a mitogen-activated protein (MAP) kinase pathway have been implicated in regulating spore germination (Osherov & May, 2001). In *M. grisea*, however, conidium germination is not affected in mutants deleted of the catalytic subunit of protein kinase A (PKA) (*CPKA*), or MAP kinase genes *PMK1, MPS1*, or *OSM1* (Xu, 2000), indicating that the cAMP signaling and these MAP kinase signaling pathways are dispensable for conidial germination. To our knowledge, no mutants blocked in conidium germination have been identified in *M. grisea*, but a few mutants are known to be delayed in conidium germination, such as the *icl1* and *gde1* mutants (Balhadere *et al.*, 1999; Wang *et al.*, 2003). It has been reported that *M. grisea* conidia possess lipophilic self-inhibitors that prevent conidia from germination and appressorium differentiation on clean glass surfaces when the spore density is above 10^5–10^6 conidia/ml (Hegde & Kolattukudy, 1997). Although the nature of these conidial surface lipids are not clear, their inhibitory effect on germination and appressorium formation can be relieved by nutrients or rice leaf wax components (Hegde & Kolattukudy, 1997).

6.4   Appressorium morphogenesis

6.4.1   Surface recognition and appressorium initiation

Many foliar fungal pathogens rely on thigmotropic and/or chemotropic sensing mechanisms to decipher the topographic or chemical features of plant surfaces and regulate the infection process. For example, the germ tubes of urediniospores of *Uromyces appendiculatus* recognize the guard cell lips to form appressoria directly over stomata. *M. grisea* normally penetrates rice cuticles directly instead of entering through natural openings. Germ tubes of *M. grisea* can recognize the hydrophobicity of the substrate and form appressoria on hydrophobic surfaces that mimic the rice leaf surface (Lee & Dean, 1993; Jelitto *et al.*, 1994; Xiao *et al.*, 1994b). Appressorium formation is initiated by cessation of the tip growth of germ tubes attached to hydrophobic surfaces. The germ tube tips then begin to bend and swell, a process known as germ tube tip deformation or hooking (Bourett & Howard, 1990). In normal situations, one appressorium is formed at the tip of each germ tube. It has been reported that surface hardness also affects appressorium formation, and well-differentiated appressoria are formed only on hard surfaces (Xiao *et al.*, 1994b).

In addition to physical attachment, germ tubes of *M. grisea* are capable of recognizing certain chemical cues on the host surface (Uchiyama & Okuyama, 1990). Germ tubes differentiate into appressoria at higher percentages on rice leaves than on artificial substrata (Jelitto *et al.*, 1994). More direct evidence comes from studies showing that some plant cutin monomers and wax compounds, such as 1,16-hexadecanedial and 1,16-hexadecanediol, induce appressorium formation effectively in *M. grisea* (Gilbert *et al.*, 1996). However, these chemical factors on the host surface are not essential for stimulating appressorium formation, and likely play no role in host-specific recognition because *M. grisea* forms appressoria efficiently even on non-host plants and inert artificial surfaces (Gilbert *et al.*, 1996).

6.4.2   Appressorium maturation

After the initial stage of appressorium formation, appressoria progress through the maturation process that involves many structural modifications and physiological activities, such as the deposition of additional cell wall layers, melanization, and turgor generation. Mature appressoria contain one nucleus that is assumed to be arrested in the G1 stage. One septum separates the appressorium from the rest of the germ tube. Conidial and germ tube cells usually collapse when appressoria mature.

The melanin layer located between the appressorium cell wall and cytoplasm membrane is essential for lowering the porosity of the appressorial wall and turgor generation (Bourett & Howard, 1990). Blocking melanin synthesis by mutations or inhibitors prevents appressorial penetration of the plant cuticle. Melanin in *M. grisea* is polymerized from the polyketide precursor 1,8-dihydroxynaphthalene (DHN). DHN is synthesized via the pentaketide pathway (Valent & Chumley, 1991). Three major genes involved in melanin synthesis, *ALB1*, *RSY1*, and *BUF1*, are named after the albino, rosy, and buff colony phenotypes of the corresponding melanin-deficient mutants. The *ALB1* and *RSY1* genes encode a putative polyketide synthase and a scytalone dehydratase respectively (Howard & Valent, 1996). *BUF1* encodes a polyhydroxynaphthalene reductase that is the molecular target for several blast fungicides such as tricyclazole. The *alb1*, *buf1*, and *rsy1* mutants fail to infect intact host plants but are able to colonize wounded or surface-abraded rice leaves (Chumley & Valent, 1990). Interestingly, melanin synthesis appears to be regulated by different mechanisms in mycelia and appressoria, because deletion of the homolog of the *PIG1* transcription factor in *Colletotrichum lagenarium* affects melanin synthesis only in mycelia but not in appressoria (Tsuji *et al.*, 2000).

As in other fungi, the appressorial pore located at the contact area between the appressorium and the substratum is where the penetration peg emerges (Deising *et al.*, 2000). In *M. grisea*, the appressorium pore has a specialized single-layered cell wall that apparently lacks chitin and is much thinner than other areas of appressorium (Bourett & Howard, 1990). The melanin layer is absent from the appressorial pore. The tight adhesion of appressoria to the plant surface is mediated by a ring of appressorium mucilage (Howard *et al.*, 1991). Preliminary analysis indicated that the appressorium mucilage contains lipids, proteins, and sugars (Ebata *et al.*, 1998). The bonding strength of appressoria on hydrophobic surfaces is weakened by treating with protease and inhibitors of lipid or glycoprotein synthesis, suggesting that lipid and glycoprotein components are closely associated with appressorium adhesion (Ohtake *et al.*, 1999).

6.4.3   The role of cAMP signaling in surface recognition and appressorium initiation

On hydrophilic surfaces, conidia produce long germ tubes that eventually branch and grow as vegetative hyphae. Appressorium formation can be induced in germ tubes on hydrophilic surfaces with exogenous cyclic AMP (cAMP) or 3-iso-butyl-1-methylxanthine (IBMX), suggesting that cAMP signaling is involved in surface recognition (Lee & Dean, 1993). In fungal cells, cAMP is synthesized by adenylate cyclases. The *M. grisea mac1* (adenylate cyclase gene) disruption mutants fail to form appressoria, and are reduced in growth and conidiation (Choi & Dean, 1997; Adachi & Hamer, 1998). Defects in *mac1* mutants can be complemented by exogenous cAMP or suppressed by a mutation in *SUM1*, the regulatory subunit of PKA (Adachi & Hamer, 1998). Under normal growth conditions, *mac1* mutants are unstable and produce many sectors carrying various spontaneous suppressor mutations. Several suppressor mutants form appressoria on either hydrophobic or hydrophilic surfaces (Adachi & Hamer, 1998).

Activation of adenylate cyclase is mediated likely by heterotrimeric GTP-binding proteins. Three Gα subunits, *MAGA*, *MAGB*, and *MAGC*, have been identified in *M. grisea* (Liu & Dean, 1997). While deletion of *MAGA* has no effect on vegetative growth, conidiation, or appressorium formation, *magC* mutants are reduced in conidiation, but normal in vegetative growth or appressorium formation. Different from *magA* and *magC* mutants, *magB* deletion mutants are significantly reduced in vegetative growth, conidiation, appressorium formation, and plant infection (Choi & Dean, 1997; Liu & Dean, 1997). *MAGB* is highly homologous to *CPG-1*, which plays an important role in growth and pathogenesis in *Cryphonectria parasitica*

(Chen *et al.*, 1996). Transformants expressing the dominant active $MAGB^{G42R}$ allele form appressoria on both hydrophobic and hydrophilic surfaces (Fang & Dean, 2000). These $MAGB^{G42R}$ transformants are reduced in plant infection as well as sexual or asexual reproduction, and exhibit a colony autolysis phenotype in aged cultures and mis-scheduled melanization of hyphal tips (Fang & Dean, 2000). Transformants expressing a putative, dominant, negative $MAGB^{G203R}$ allele have no obvious phenotypes in appressorium formation or plant infection (Fang & Dean, 2000). However, whether this G203R mutation has any effect on disassociation of MagB from Gβ is not clear. Mutants deleted of the *GBM1* Gβ subunit gene are non-pathogenic and fail to form appressoria on hydrophobic surfaces. Exogenous cAMP stimulates appressorium formation in *gbm1* mutants but these appressoria are morphologically abnormal and non-functional for penetration (Nishimura & Xu, unpublished).

Interestingly, gene replacement mutants of *CPKA*, a gene encoding the catalytic subunit of PKA, are delayed only in appressorium formation and have normal growth and conidiation (Mitchell & Dean, 1995; Xu *et al.*, 1997). Appressoria formed by *cpkA* mutants are defective in penetration and colonization of rice leaf tissues. While non-pathogenic on healthy rice plants, *cpkA* mutants can still infect through wounds and cause rare lesions (Xu *et al.*, 1997). On hydrophilic surfaces, *cpkA* mutants still respond to exogenous cAMP, indicating that there is an additional catalytic subunit of PKA in *M. grisea*. We have identified a second catalytic subunit of PKA (*CPK2*) in the recently available *M. grisea* 70–15 genome sequence. Mutants deleted of *CPK2* have no obvious phenotype, we have no success in generating *cpkA cpk2* double mutants (Kim & Xu, unpublished).

6.4.4 PMK1 *regulates appressorium formation and maturation*

The *PMK1* MAP kinase gene is the only homolog of yeast Fus3 and Kss1 in *M. grisea*, and it can functionally complement the mating defect of a yeast *fus3 kss1* double mutant (Xu & Hamer, 1996). The *pmk1* gene replacement mutants produce normal mycelia and conidia, but are non-pathogenic. Conidia from *pmk1* mutants can attach to and germinate on artificial surfaces as efficiently as wild-type conidia. Germ tubes of *pmk1* mutants still recognize hydrophobic surfaces and form sub-apical swollen bodies, but fail to arrest the germ tube tip growth and form appressoria on Teflon membranes or on onion epidermal cells. In addition, *pmk1* mutants fail to grow invasively in rice plants when inoculated through wound sites (Xu & Hamer, 1996). However, *pmk1* mutants remain responsive to cAMP for germ tube tip deformation, although appressoria are not formed (Xu & Hamer, 1996), indicating that *PMK1* may act downstream from the cAMP signaling for regulating appressorium formation and infectious growth (Fig. 6.2).

Recent studies on several phytopathogenic fungi, including *C. lagenarium*, *Pyrenophora teres*, *Fusarium oxysporum*, *Fusarium graminearum*, and *Claviceps purpurea*, have indicated that the *PMK1* pathway may be well conserved for regulating

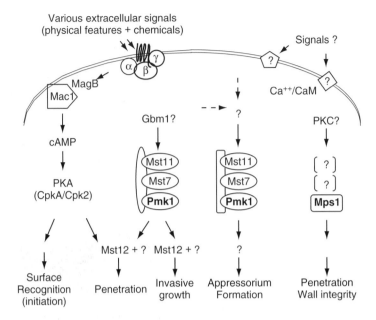

Figure 6.2 A model of signaling pathways involved in appressorium morphogenesis. Various surface attachment and chemical signals can stimulate appressorium formation in *M. grisea*. Some of these signals may be recognized by G-protein-coupled receptors and activate trimeric G-proteins. The activated Gα subunit (presumably MagB) will initiate appressorium formation by activating the Mac1-cAMP-PKA pathway. Some external stimuli may be recognized by other receptors, and function through signaling systems such as the Ca⁺⁺/CaM signaling. Late stages of appressorium formation and maturation are regulated by *PMK1*, presumably through a MAP kinase cascade *MST11-MST7-PMK1*. *PMK1* is also necessary for penetration and infectious hyphal growth, possibly by regulating the *MST12* transcription factor. The *PMK1* pathway may directly respond to external stimuli via *GBM1* (Gβ) or some internal secondary signals. Different scaffold proteins and transcription factors may be involved in regulating specific functions of the *PMK1* pathway during appressorium maturation, penetration, and invasive growth. *MPS1* is essential for appressorial penetration and it also controls cell wall integrity and conidiation. In yeast, the Slt2 (a homolog of *MPS1*) MAP kinase cascade is regulated by protein kinase C (PKC). The *OSM1* MAP kinase is dispensable for appressorium formation and penetration.

appressorium formation and other plant infection processes (Xu, 2000). The *M. grisea* genome contains distinct homologs of components of the yeast pheromone and filamentation pathways, including Ste2, Ste3, Ste5, Ste20, Ste11, Ste7, Ste12, and Far1 (Gustin *et al.*, 1998). *MST12* encodes a transcription factor homologous to yeast Ste12 in *M. grisea*. Mutants deleted of *MST12* are normal in vegetative growth and conidiation but are non-pathogenic on rice and barley leaves. In contrast to *pmk1* mutants, *mst12* mutants produce typical dome-shaped and melanized appressoria. However, the appressoria formed by *mst12* mutants fail to penetrate onion epidermal cells. Similar to *pmk1* mutants, *mst12* mutants are defective in infecting wounded or abraded leaves, indicating that the *MST12* may be defective in infectious growth. *MST12* interacts weakly with *PMK1* in yeast two-hybrid assays. It is likely that *MST12* functions downstream of *PMK1* to regulate genes involved in appressorial penetration and infectious hyphal growth. A subtraction library enriched with genes regulated by *PMK1* has also been constructed and sequenced (Xue *et al.*, 2002). Two genes identified in this library, *GAS1* and *GAS2*, encode small proteins that are specific to filamentous fungi. Both are expressed specifically during appressorium formation and are localized in the cytoplasm. Mutants deleted of *GAS1*, *GAS2*, or both *GAS1* and *GAS2* have no defect in growth, conidiation, or appressoria formation, but are reduced in appressorial penetration and lesion development (Xue *et al.*, 2002).

6.4.5   Hydrophobin MPG1 *and surface recognition*

MPG1 is the first well-characterized *M. grisea* gene to play an important role during appressorium formation (Talbot *et al.*, 1993). It encodes a small, cysteine-rich, secreted protein with characteristics of a class I fungal hydrophobin. Targeted deletion of *MPG1* reduces the efficiency of appressorium formation and virulence. Germ tubes of *mpg1* mutants are normal in tip deformation on rice leaves or highly hydrophobic artificial substrates but are defective in appressorium differentiation. Interestingly, several artificial substrates were found to support efficient appressorium formation of *mpg1* mutants (Beckerman & Ebbole, 1996), suggesting that *MPG1* is involved in the interaction with and recognition of the host surface. Mpg1 proteins can self-assemble *in vitro* and on the conidial surface, to form the hydrophobic rodlet layer (Kershaw *et al.*, 1998). During appressorium formation, Mpg1 hydrophobin proteins may assemble into an amphipathic layer over the rice leaf surface to assist germ tube adhesion and acts as a signal for appressorium formation.

Although *MPG1* was originally isolated as a gene differentially expressed during plant infection, it is also highly expressed during appressorium formation and conidiation and in mycelia that starved for nitrogen (Talbot *et al.*, 1993). The expression of *MPG1* is affected by nutritional conditions and mutations in genes involved in different signal transduction pathways. Yeast extract represses *MPG1* expression *in vitro*. *PMK1*, *NPR1*, and *NUT1* are required for full expression of *MPG1* in response to starvation stress, while *CPKA* is required for repression of *MPG1* during growth under nutrient-rich conditions. During appressorium formation, induced expression of *MPG1* requires the *CPKA* and *NPR1* genes (Soanes *et al.*, 2002).

6.4.6   Other factors affecting appressorium formation

In addition to the physical and chemical features of the attached surface, pharmacological studies have identified many compounds that stimulate or inhibit appressorium formation in *M. grisea*, such as polyamines, ceramide, neobulgarones, glisoprenins, sphingomyelin, zosteric acid, oleic acid, and various respiratory inhibitors (Lee & Dean, 1993; Gilbert *et al.*, 1996; Thines *et al.*, 1997, 1998; Eilbert *et al.*, 1999a; Stanley *et al.*, 2002). Certain nutrients such as 2% yeast extract also inhibit appressorium formation. Interestingly, yeast pheromones inhibit appressorium formation in a mating type-dependent manner (Beckerman *et al.*, 1997), but this observation has not been confirmed or further investigated.

Besides cAMP signaling, Ca^{++}/calmodulin (CaM) signaling also has been implicated by pharmacological and molecular studies to be involved in appressorium formation in several fungi (Deising *et al.*, 2000). In *M. grisea*, several Ca^{++} modulators and CaM antagonists inhibit appressorium formation at µM concentrations (Lee & Lee, 1998). During appressorium formation, the expression of the CaM gene is induced (Liu & Kolattukudy, 1999). Interestingly, diacylglycerol (DAG), a known activator of protein kinase C (PKC), also can induce appressorium formation in *M. grisea* (Thines *et al.*, 1997) and reverse the inhibition by glisoprenins (Thines *et al.*, 1998). Oleic acid inhibits appressorium formation induced by 1,16-hexadecanediol, but it has no effect on appressorium formation induced by cAMP (Thines *et al.*, 1998; Eilbert *et al.*, 1999b). Cyclosporin A, a calcineurin inhibitor, also inhibits appressorium development in *M. grisea* (Viaud *et al.*, 2002). These observations suggest that appressorium formation is regulated by multiple signals and signaling pathways in *M. grisea*.

6.4.7   Other mutants defective in appressorium morphogenesis

In the past decade, several mutants defective in appressorium formation have been isolated by different mutagenesis approaches. The first genetically analyzed appressorium mutant is the *smo* mutant, which forms morphologically abnormal conidia and appressoria (Hamer *et al.*, 1989). The *smo* mutant was isolated in an UV mutagenesis designed to screen for mutants reduced in adhesion. Appressorium formation is completely absent in the spontaneous mutants *apf1* and *app5* (Silue *et al.*, 1998; Chun & Lee, 1999). The *app1*, *app2*, and *app3* mutants also were isolated by UV mutagenesis (Zhu *et al.*, 1996). Appressorium formation on hydrophobic surfaces is greatly reduced in *app1* and *app2* mutants. The *app3* mutant forms appressoria on hydrophilic surfaces. Unfortunately, none of the mutations responsible for phenotypes observed in these mutants identified by UV mutagenesis or spontaneous mutations have been identified.

In addition to targeted deletion of genes involved in the cAMP, PMK1, and G-protein signaling pathways, insertional mutagenesis also has been applied to identify genes involved in appressorium formation. The *con1* and *con7* mutants generated by restriction enzyme mediated integration (REMI) mutagenesis are blocked in appressorium formation (Shi & Leung, 1995). Appressorium formation is reduced by 70 and 22% in *con2* and *con4* mutants respectively (Shi & Leung, 1995). The *acr1* mutant disrupted in a gene homologous to *medA* of *A. nidulans* is reduced in appressorium formation and plant infection (Lau & Hamer, 1998; Nishimura *et al.*, 2000). The *con1*, *con2*, *con4*, *con7*, and *acr1* mutants all produce conidia with altered morphology. REMI mutants *igd1* and *met1* also are defective in both conidiation and appressorium function (Balhadere *et al.*, 1999).

A few mutants, such as the *gde1* and *pth11* mutants, produce normal conidia but are defective in appressorium formation. The *gde1* mutant is delayed in conidial germination and appressorium formation and reduced in virulence on rice but not on barley (Balhadere *et al.*, 1999). *PTH11* encodes a transmembrane protein localized to the cell membrane and vacuoles. Germ tubes of *pth11* mutants undergo hooking and apical swelling, but only 10–15% of them form morphologically normal appressoria (DeZwaan *et al.*, 1999). *pth11* mutants are non-pathogenic on healthy leaves but can colonize wounded plant tissues. Exogenous cAMP and DAG both restore appressorium formation, but only cAMP restores pathogenicity in *pth11* mutants (DeZwaan *et al.*, 1999). The germ tube-specific *CBP1* gene encoding a putative extracellular protein with two chitin-binding domains also is likely involved in the recognition of physical factors acting on the host surface. Mutants deleted of *CBP1* have abnormal appressorium differentiation on artificial surfaces but produce appressoria normally on the leaf surface (Kamakura *et al.*, 2002).

6.5   Penetration

6.5.1   Penetration peg

Once formed, mature appressoria develop penetration pegs to pierce the host surface and enter cells of the leaf epidermis. The penetration peg is usually much narrower than somatic hyphae and germ tubes (Bourett & Howard, 1990). At early stages of penetration, the peg cytoplasm contains a few apical vesicles and consists primarily of a zone-of-exclusion that is continuous with a similar region in the appressorium (Bourett & Howard, 1992). Apical vesicles are, however, observed in elongating penetration pegs at later penetration stages. The penetration peg contains high concentrations of actin filaments (Bourett & Howard, 1992) that may be necessary to stabilize the tip of the penetration peg and compensate for the difference in osmotic pressure between appressorial and host cell protoplasts (Howard & Valent, 1996). Actin and cytoskeletal elements also may be involved in the selection of peg emergence site and re-establishment of polarized growth.

6.5.2   Forces of penetration

Fungal pathogens have evolved distinct strategies for penetrating various natural barriers on plant surfaces. In *M. grisea*, elevated osmotic pressures within melanized appressoria are used to directly puncture plant cuticles and synthetic membranes. The appressorial turgor is estimated to be as high as 8 MPa or 80 bars by cytorrhysis assays (Howard *et al.*, 1991). In other fungi forming melanized appressoria, it is likely that appressorial turgor pressure also plays a major role in plant penetration (Bastmeyer *et al.*, 2002). In *C. graminicola*, the force exerted by appressoria was estimated to be about 17 μN by a microscopic method using elastic optical waveguides (Bechinger et al., 1999).

Phytopathogenic fungi are known to produce various hydrolytic enzymes that are able to erode the plant cuticle and degrade plant cell walls. The microfibrils of host cell walls around the penetration pegs are disorganized, indicating that localized plant cell wall maceration may facilitate penetration in M. grisea (Koga, 1995). However, deletion of a cutinase gene CUT1 has no detectable effect on plant infection and sporulation in M. grisea (Sweigard et al., 1992). Two distinct xylanases encoded by XYL1 and XYL2 are also dispensable for fungal pathogenicity (Wu et al., 1997). It is well known that determining the importance of individual fungal cutin- or cell wall-degrading enzymes in plant pathogenic fungi is complicated by the genetic redundancy and variable regulation of these enzymes (Tonukari et al., 2000). M. grisea is probably similar because cutinase activity is still detectable in cut1 mutants, and four additional xylanases are expressed in xyl1 xyl2 double mutants. Further studies are necessary to clarify the role of these hydrolytic enzymes during penetration by the blast fungus.

6.5.3   Turgor generation

In M. grisea, appressorial turgor is generated by the accumulation of an enormous amount of glycerol, which draws water into appressoria by osmosis (deJong et al., 1997). Since appressoria of M. grisea can develop turgor pressure in water droplets without any exogenous energy sources, the fungus apparently uses the conidial storage products, mainly glycogen and lipid, to synthesize glycerol. During appressorium formation, both glycogen and lipid droplets are mobilized from conidial compartments to the young appressorium and degraded before turgor generation (Thines et al., 2000). Glycogen degradation occurs during conidium germination and appressorium maturation. Glycerol-3-phosphate dehydrogenase and glycerol dehydrogenase are present in appressoria (Thines et al., 2000), but their enzymatic activities are not induced during appressorium maturation. In contrast, triacylglycerol lipase activity is strongly induced during appressorium maturation. Lipid droplets accumulated in the central vacuoles in developing appressoria are totally degraded in appressoria formed on inert artificial surfaces. However, appressorial penetration of softer surfaces (e.g. onion epidermis) can occur before the complete degradation of lipid droplets (Weber et al., 2001).

The M. grisea OSM1 gene encodes a MAP kinase homologous to yeast HOG1. The osm1 deletion mutant has no defect in turgor generation and plant infection, indicating that glycerol accumulation in appressoria is not regulated by OSM1 (Dixon et al., 1999). PMK1 regulates the transfer of lipid bodies and glycogen deposits into the developing appressorium. The mobilization of glycogen and lipid was not observed in pmk1 mutants and significantly retarded in cpkA mutants (Thines et al., 2000). Lipid and glycogen degradation are severely delayed in cpkA mutants but occur very rapidly in a mac1 sum1-99 mutant (Thines et al., 2000). Therefore, cAMP signaling is likely involved in regulating the mobilization and degradation of conidial storage products. These data indicate that the cAMP signaling and *PMK1* pathways coordinate in the generation of appressorial turgor.

Trehalose is a non-reducing disaccharide often functioning as a storage carbohydrate or a stress-tolerance factor in eukaryotic cells. In *M. grisea*, mutants deleted of the trehalose-6-phosphate synthase gene *TPS1* fail to synthesize trehalose and conidiate poorly. Appressoria produced by *tps1* mutants do not develop full turgor or elaborate penetration hyphae efficiently. Interestingly, neither of the two trehalase genes that have been characterized in *M. grisea* is essential for appressorial penetration (Foster *et al.*, 2003). *NTH1* (=*PTH9*) encodes a neutral trehalase that is expressed during conidiogenesis, plant infection, and in response to hyperosmotic stress. *nth1* disruption mutants have no defect in appressorial penetration, but are reduced in lesion formation (Sweigard *et al.*, 1998; Foster *et al.*, 2003). The other trehalase gene *TRE1* has characteristics of both neutral and acidic trehalases. *TRE1* is required for growth on trehalose and mobilization of intracellular trehalose during appressorium formation, but it is dispensable for pathogenesis. It is likely that trehalose degradation is important only for infectious growth but not for appressorial penetration.

6.5.4 Other genes involved in appressorial penetration

In addition to genes involved in turgor generation, several *M. grisea* genes have been identified by insertional or targeted deletion mutagenesis to play important roles in appressorial penetration. *MPS1* is highly homologous to the *Saccharomyces cerevisiae* MAP kinase *SLT2* (Xu *et al.*, 1998). The *mps1* deletion mutant is non-pathogenic on healthy rice leaves but capable of infecting rice plants through wound sites. Appressoria formed by *mps1* mutants are melanized but fail to penetrate underlying plant cells and fail to form infectious hyphae. Interestingly, *mps1* appressoria still elicit autofluorescence and papilla formation in the underlying onion epidermal cells (Xu *et al.*, 1998). Similar to *mps1* mutants, *pls1* deletion mutants form appressoria but fail to penetrate and infect rice plants. *PLS1* encodes a putative membrane protein structurally related to the tetraspanins. It is specifically expressed in appressoria and localized in plasma membranes and vacuoles (Clergeot *et al.*, 2001). Mutants deleted of the *MST12* transcription factor also form non-functional, melanized appressoria. Different from *mps1* mutants, *mst12* mutants are capable of colonizing plant tissues through wounds, and appressoria formed by *mst12* mutants do not elicit plant defense responses in the underlying plant cells (Park *et al.*, 2002). In addition to these mutants completely blocked in penetration, a few are known to be reduced in appressorial penetration, such as the *pde1* and *pde2* mutants (Balhadere *et al.*, 1999). *PDE1* encodes a P-type ATPase that is expressed in germinating conidia and developing appressoria. Mutants deleted of *PDE1* are reduced in virulence on rice but normal on barley leaves (Balhadere & Talbot, 2001). Deletion of the *CYP1* cyclophilin gene also reduces appressorial penetration and plant infection (Viaud *et al.*, 2002). Reduced penetration in the *cyp1* mutant may be related to its defect in turgor pressure generation.

6.6   Infectious growth and lesion formation

After penetrating the host cell, the penetration peg develops into an unbranched, thickened primary infection hypha or vesicle, which then further differentiates into branched, bulbous, lobed secondary infectious hyphae. The secondary infectious hyphae, often simply known as infectious hyphae, are morphologically distinct from germ tubes and somatic hyphae. Lesions will develop on plant leaves due to invasive fungal growth. Typical rice blast lesions are diamond-shaped with a gray or white central area with brown margins. Long periods of high humidity and moderate temperatures (63–73°F) are favorable for blast disease development. The blast disease is also favored by excessive nitrogen fertilization, aerobic soils, and drought stress (Bonman, 1992).

6.6.1   Infectious hyphae

M. grisea has been described as a hemibiotrophic pathogen (Morosov, 1992). During the initial biotrophic stage, the infectious hyphae can grow inter- and intracellularly without damaging plant cells (Heath *et al.*, 1992). However, it is not known whether an intact plant cytoplasm membrane surrounds the primary and secondary infectious hyphae at the early infection stages. Switch of infectious hyphae from the biotrophic phase to the necrotrophic phase in later infection stages results in plant cell death. It is likely that cell wall-degrading enzymes, proteinases, and other hydrolytic enzymes play important roles during infectious hyphal growth, to break down plant cell wall and other defense-related structures. Although *CUT1*, *XYL1*, and *XYL2* are dispensable for plant infection, additional cutinases, xylanases, and other cell-wall-degrading enzymes exist in *M. grisea*. It may be necessary to characterize the global regulatory mechanisms, such as the *SNF1* homolog of *C. carbonum* (Tonukari *et al.*, 2000), of these degradative enzymes, to clarify their role in infectious growth in *M. grisea*.

For infectious hyphae to grow successfully in plant tissue, the fungus must use different strategies to overcome constitutive and induced plant defenses, including degradation of preformed or induced antimicrobial compounds. A variety of compounds toxic or inhibitory to *M. grisea* have been isolated from rice (Kumar & Sridhar, 1993; Koga *et al.*, 1995). However, no specific enzymes degrading these compounds have been identified in *M. grisea*. Interestingly, the *abc1* deletion mutant is dramatically reduced in lesion formation. *ABC1* encodes a protein belonging to the ATP-binding cassette (ABC) superfamily of membrane transporters. After penetrating rice or barley epidermal cells, the *abc1* mutant arrests growth and dies shortly thereafter. Transcription of *ABC1* is induced by several toxic drugs and a rice phytoalexin (Urban *et al.*, 1999). The up-regulation of *ABC1* is critical for pathogenesis because an insertion in its promoter has the same effect as deletion of *ABC1* on pathogenicity. It is likely that *ABC1* is involved in the tolerance of *M. grisea* to certain toxic plant defense compounds or exporting fungal metabolites suppressive to host defense responses.

6.6.2 Phytotoxins produced by M. grisea

Numerous phytotoxic metabolites, including dihydropyriculol, pyrichalasin H, terrestric acid, pyriculol, pyriculariol, pyriculone, pyricuol, and other pyriculol-related metabolites, have been isolated from culture filtrates of *M. grisea* (Nukina, 1999). However, none of them are host-specific toxins, and their role in blast disease development is not clear. Pyriculol-related toxins, such as pyriculol and pyriculone, induce blast-like necrotic lesions on rice leaves but their toxic effect is not race specific (Kim *et al.*, 1998). Partially purified toxins inhibit root growth of rice seedlings and induce necrosis on the leaves of cultivar Sekiguchi-asashi in a light-dependent manner (Iedome *et al.*, 1995). Some phytotoxic factors of *M. grisea* are proteinous, such as the 30 kDa rice protoplast-disrupting protein (Nomura & Kiyosawa, 1992), leaf senescence-promoting factors in filtrates of cultures that starved for nitrogen (Talbot *et al.*, 1997), and heat-labile molecules capable of releasing cell-wall fragments toxic to maize cells (Bucheli *et al.*, 1990).

Interestingly, toxins isolated from germinating conidia have been shown to induce susceptibility of rice leaves to infection by *Alternaria alternata*, a pathogen normally non-pathogenic on rice (Fujita *et al.*, 1994). The induced susceptibility is independent of the rice cultivars and the pathogen races. Toxins from rice blast isolates also induce susceptibility in other plant species that are susceptible to rice isolates (e.g. barley), but not in non-host plant species (Fujita *et al.*, 1994). In an independent study, the concentrated conidium germination fluid of rice blast isolates were shown to disrupt protoplasts of different rice cultivars but had little effect on protoplasts of two non-host plant species (Rathour *et al.*, 2002). These results indicate that toxins in germination fluids of *M. grisea* are host-selective at the plant species level and play an important role in determining basic compatibility.

6.6.3 Avirulence genes

As in other pathosystems, the rice–*M. grisea* interaction is governed by specific interactions between fungal avirulence (*AVR*) genes and their corresponding plant resistance *R* genes. There are over 40 known major rice blast resistance genes (Sallaud *et al.*, 2003). Many corresponding avirulence genes have been mapped in *M. grisea* (Valent and Chumley, 1991; Dioh *et al.*, 2000). A few of them that have been cloned, including *PWL2*, *AVR-Pita*, and *AVR*-CO39, do not share common structural features or conserved domains. *AVR-Pita* (=*AVR2-YAMO*) encodes a neutral zinc metalloprotease with an N-terminal secretory signal and pro-protein sequences (Orbach *et al.*, 2000). Diverse mutations in *AVR-Pita*, including point mutation, insertion, and deletion, permit the fungus to avoid triggering resistance responses mediated by *Pi-ta*. The *PWL2* gene encodes a glycine-rich, hydrophilic protein with a putative secretion signal sequence (Sweigard *et al.*, 1995). It is an avirulence gene isolated from a rice pathogen that prevents this fungus from infecting weeping lovegrass. Multiple *PWL2* homologs with varying degrees of sequence homology and different chromosome locations, including *PWL1*, *PWL3*, and *PWL4*, have been isolated (Kang *et al.*, 1995). The *AVR1*-CO39 gene was cloned from a weeping lovegrass pathogen (Farman & Leong, 1998). *AVR1-CO39* confers avirulence on the rice cultivar CO39, but its biological function during plant infection is not clear (Farman *et al.*, 2002). Recently, another *AVR* gene *ACE1* has been cloned (Böhnert *et al.*, 2003). *ACE1* encodes a protein with combined polyketide synthase and non-ribosomal peptide synthetase domains. It is specifically expressed in penetrating appressoria and may be responsible for the synthesis of an unidentified secondary metabolite recognized by resistance gene *Pi33*. Interestingly, sequence analysis of the genomic region adjacent to *ACE1* has revealed a cluster of genes potentially involved in secondary metabolism (Böhnert *et al.*, 2003).

To date, *AVR-Pita* is the only *M. grisea AVR* gene whose corresponding rice resistance gene *Pita* has been cloned (Orbach *et al.*, 2000), and it is the first fungal avirulence gene shown to directly interact with its corresponding resistance gene (Jia *et al.*, 2000). *Pi-ta* encodes a putative cytoplasmic receptor with a centrally localized nucleotide-binding site (NBS) and leucine-rich domain (LRD) at the C-terminus (Bryan *et al.*, 2000). Transient expression of *AVR-Pita* inside plant cells results in a *Pi-ta*-dependent resistance response. Single amino acid mutations in the *Pi-ta* LRD or in the *AVR-Pita* protease motif disrupt the physical interaction between Pi-ta and Avr-Pita proteins and result in loss of resistance responses in rice plants. It is likely that the direct interaction between *AVR-Pita* and *Pi-ta* inside the plant cell initiates a *Pi-ta*-mediated defense response. The only other blast resistance gene that has been cloned in rice is *Pib*, which is also a member of the NBS-LRD class of plant disease resistance genes. Interestingly, the expression of *Pib* is up-regulated by environmental conditions favorable for *M. grisea* infection, suggesting the evolution of anticipatory control of *R* gene expression (Wang *et al.*, 2001).

6.6.4   *Nutritional requirements and metabolic activities during infectious growth*

The infectious hyphae must physiologically adapt to the plant environment and produce enzymes necessary for nutrient absorption. Unfortunately, only a few mutants defective in specific metabolic activities are known in *M. grisea*. While the sulfate non-utilizing (*sub*) mutant has no defects in plant infection (Harp & Correll, 1998), different amino acid auxotrophic mutations exhibit various effects on pathogenicity. The arginine auxotrophic mutant *argB* is fully pathogenic (Sweigard *et al.*, 1992), but the histidine and methionine auxotrophic REMI mutants, *pth3* and *met1*, are significantly reduced in plant colonization and lesion development (Sweigard *et al.*, 1998; Balhadere *et al.*, 1999). The gene disrupted in *pth3* encodes the imidazole glycerol phosphate dehydratase (Sweigard *et al.*, 1998). For the *met1* mutant, a contradictory report suggests that the same mutant is normal in infecting leaf and root tissues (Dufresne & Osbourn, 2001).

In *M. grisea*, nitrogen limitation induces the expression of many genes expressed during infectious growth (Talbot *et al.*, 1993), and filtrates of cultures under nitrogen starvation are able to cause senescence of rice leaves (Talbot *et al.*, 1997), indicating that nitrogen starvation may, to a certain degree, mimic the *in planta* growth condition of infectious hyphae. Unfortunately, nitrogen metabolism is not well studied in *M. grisea*. Deletion of *NUT1* results in the failure to grow on a variety of nitrogen sources, but it has no effect on pathogenicity (Froeliger & Carpenter, 1996). *NUT1* encodes a gene homologous to the major nitrogen-regulatory genes *areA* of *A. nidulans*. However, different from the *areA* mutant that fails to utilize nitrogen sources other than ammonium and glutamine, *nut1* mutants are able to use proline and alanine as the nitrogen source (Froeliger & Carpenter, 1996). Two non-allelic mutants, *npr1* and *npr2*, defective in utilization of a wide range of nitrogen sources (e.g. nitrate or amino acids) are also defective in plant infection. It is likely that *NPR1* and *NPR2* encode novel regulators of nitrogen metabolism and play a critical role in pathogenesis (Lau & Hamer, 1996).

Carbon starvation also is known to induce the expression of many genes expressed during plant infection. Interestingly, the most abundant fungal clone identified in Expressed sequence tags (ESTs) from a cDNA library of infected rice leaves is a homolog of the *Neurospora crassa* glucose-repressible gene *GRG1* (Kim *et al.*, 2001). Mutants *pth1* and *pth2* show reduced pathogenicity (Sweigard *et al.*, 1998). *PTH1* encodes a protein with homology to yeast *GRR1*. Yeast *grr1* mutants are defective in high-affinity glucose transport and glucose repression. *PTH2* encodes a protein homologous to yeast Cat2, the carnitine acetyltransferase associated with peroxisomes and mitochondria. *PTH2* is likely involved in carnitine metabolism in *M. grisea*. Recently, the isocitrate lyase gene *ICL1*, a key component of the glyoxylate cycle, was characterized in *M. grisea* (Wang *et al.*, 2003). The expression of *ICL1* is up-regulated during plant penetration. The *icl1* deletion mutant is delayed in conidium germination and appressorium formation and reduced in lesion development (Wang *et al.*, 2003). In *Leptosphaeria maculans*, the isocitrate lyase gene is also essential for the successful colonization of canola and growth on fatty acids such as monolaurate (Idnurm & Howlett, 2002). Interestingly, 2.5% glucose can restore the lesion size defect of the *L. maculans icl1* mutant (Idnurm & Howlett, 2002). These findings indicate that, similar to its role in human pathogen *Candida albicans*, the glyoxylate cycle also is important for disease development in plant pathogenic fungi.

6.6.5 *Other mutants defective in infectious growth and lesion formation*

The *pmk1*, *mst12*, and *mac1* mutants are non-pathogenic on healthy or abraded leaves (Xu & Hamer, 1996; Choi & Dean, 1997; Adachi & Hamer, 1998; Park *et al.*, 2002), indicating that both the *PMK1* and the cAMP signaling pathways are essential for regulating infectious hyphal growth and colonization of plant tissues. Although the relationship between these two pathways is not clear in *M. grisea*, it is likely that they interact with each other and are coordinated in regulating different infection processes. Studies in other fungi also have shown that both the cAMP signaling and the *PMK1*-like MAP kinase pathways are involved in fungal differentiation and pathogenesis (Kronstad *et al.*, 1998). In *U. maydis*, both the cAMP signaling and the *ubc3* (*kpp2*) MAP kinase pathways are involved in mating, dimorphic switching, and virulence (Kronstad *et al.*, 1998).

Many mutants defective in appressorium formation or penetration (see Table 6.2) are also known to be reduced in infectious hyphal growth and lesion development. This may reflect that appressorium morphogenesis is a complicated process, and that many genes are involved in both appressorium morphogenesis and plant colonization. For some mutants, such as the *mpg1*, *pls1*, *gas1*, and *gas2*, reduction in lesion formation results primarily from reduction in appressorial penetration. For some mutants, such as *nth1* and *con1*, defects in colonizing plant tissue are more than the reduction in appressorial penetration. The *con1* mutant fails to cause lesions on wounded leaves (Shi & Leung, 1995). The *nth1* mutant is reduced in appressorial turgor and colonization of plant tissue (Foster *et al.*, 2003). These genes may be important for infectious hyphal development or growth, but their exact function during infectious growth is not clear.

In addition, several mutants are defective in fungal pathogenicity but not in appressorium morphogenesis, such as *pth8*, *orp1*, *pal144*, and *pat531* (Sweigard *et al.*, 1998; Balhadere *et al.*, 1999; Fujimoto *et al.*, 2002). According to the sequence deposited in GenBank, *PTH8* encodes a putative UDP-glucose:sterol glucosyltransferase homologous to *UGT51*. In *S. cerevisiae*, *ugt51* deletion mutants fail to synthesize sterol glucoside but grow normally under various culture conditions. Two mutants, *pal144* and *pat531*, are reduced in the efficiency of penetration and colonization of host tissue, and require 10-fold higher inocula to achieve wild-type levels of infection (Fujimoto *et al.*, 2002). In mutant *pat531*, a putative transmembrane protein gene with unknown function is disrupted (Fujimoto *et al.*, 2002). A mutant generated by transposon, *impala*-mediated insertional mutagenesis is also known to be defective in plant infection. The corresponding gene disrupted in this mutant, *ORP1*, has no sequence homology to known genes (Villalba *et al.*, 2001).

6.6.6   Genes specifically or highly expressed during infectious growth

Several approaches, including differential screening and ESTs profiling, have been used to identify genes highly or specifically expressed during the colonization of rice leaves. While the expression of *MPG1* is induced at the onset of lesion formation (Talbot *et al.*, 1993), the ubiquitin extension protein gene *UEP1* is highly expressed 48 hours after inoculation (McCafferty & Talbot, 1998). *UEP1* is down-regulated after this time, despite further extensive growth of infectious hyphae. High-level expression during plant infection is not observed for another ubiquitin extension protein gene *UEP3* and a polyubiquitin gene *PUB4* (McCafferty & Talbot, 1998).

The EST sequencing approach has been used to identify fungal genes expressed in infected rice leaves (Shi & Leung, 1995; Kim *et al.*, 2001). Among

511 clones sequenced from a cDNA library of infected rice leaves collected at 84, 96, and 120 hours after inoculation, 72 clones were fungal genes (Kim *et al.*, 2001). In an independent study, only 20 of 619 random clones sequenced from a 48-hour post-inoculation infected rice leaf cDNA library have significant fungal homologs in GenBank (Rauyaree *et al.*, 2001). Since both libraries were constructed with rice cultivar Nipponbare infected with 70–15 genome sequence, the difference in the abundance of fungal genes between these two libraries likely resulted from different amounts of fungal biomass accumulated in leaves collected at different stages of disease development. After sequencing an additional 460 clones that were identified as putative fungal genes by differential hybridization, 40 known *M. grisea* genes, including *MPG1* and *PUB4*, were found (Rauyaree *et al.*, 2001). A proteomics approach has also been used to study proteins expressed in rice leaves infected by *M. grisea*, but none of the proteins sequenced were fungal genes (Konishi *et al.*, 2001).

6.7 Conidiation

Conidiation occurs during the later stage of the rice blast disease cycle. Many environmental cues, including moisture, temperature, and light, are known to affect conidiation in *M. grisea* (Pinnschmidt *et al.*, 1995). To date, several mutants with altered conidia morphology, including *smo*, *con1*, *con2*, *con4*, *con7*, *acr1*, *apf1*, *gde1*, and *met1*, have been identified by chemical or insertional mutagenesis. The *smo* mutant produces normal amounts of conidia that are aberrantly shaped (Hamer *et al.*, 1989). Conidiation is significantly reduced (>90% reduction) in both *con1* and *con2* mutants. The *con1* mutant produces a terminal elongated conidium, but the *con2* mutant produces mostly non-septate or two-celled conidia (Shi & Leung, 1995). The *con4* and *con7* mutants produce conidia of abnormal cell shape, and are reduced in sporulation by approximately 35% (Shi & Leung, 1995). According to the sequence deposited in GenBank, *CON7* encodes a transcription factor with a zinc finger motif and nuclear localization signal. The *acr1* mutant was first isolated as a REMI mutant reduced in pathogenicity (Lau & Hamer, 1998; Nishimura *et al.*, 2000). Deletion or disruption of *ACR1* results in the production of head-to-tail (acropetal) arrays of elongated conidia. The *ACR1* gene seems to be a stage-specific negative regulator of conidiation, required to establish a sympodial pattern of conidiation (Lau & Hamer, 1998). One of the two spontaneous mutants producing abnormal cylindrical conidia with more than two septa (Arase *et al.*, 1994) is phenotypically similar to the *acr1* mutant.

A few *M. grisea* mutants produce altered levels of morphologically normal conidia. In the *mac1*, *magB*, and *magC* deletion mutants, reduction in conidiation can be suppressed by exogenous cAMP (Choi & Dean, 1997; Liu & Dean, 1997), suggesting that cAMP signaling is involved in conidiation. The *mps1* mutant is also significantly reduced in conidiation, presumably due to its defect in aerial hyphal development (Xu *et al.*, 1998). The *pth10* mutant produces approximately

50-fold fewer conidia than the wild-type strain on oatmeal plates, but it is fully pathogenic and conidiates at a wild-type level on infected leaves (Sweigard *et al.*, 1998). In contrast, conidiation in the *hsp1* mutant is 10-fold higher on artificial medium but 3-fold lower on infected leaves than the wild-type strain (Tharreau *et al.*, 1997). The *ccn1* mutant produces abundant conidia on complete medium (Zhu *et al.*, 1996). Conidial production is completely absent in the *con5* and *con6* mutants on oatmeal agar (Shi & Leung, 1995). Although pairwise crosses have been used to establish epistatic relationships between *con1* and *con7* mutations (Shi & Leung, 1995), only *CON7* has been further characterized. In summary, many mutants with either altered conidium morphology or conidiation levels are known, but the regulatory mechanism of conidiation and genes involved in conidiogenesis are not well characterized in *M. grisea*.

6.8 Genomics studies

M. grisea has seven chromosomes and an estimated genome size of approximately ~40 Mb. A high-density genetic map consisting of 203 markers spanning approximately 900 cM was constructed by integrating markers from several labs (Nitta *et al.*, 1997). A physical map consisting of bacterial artificial chromosome (BAC) contigs assembled by fingerprinting analysis has been constructed for all seven chromosomes and integrated with the genetic map of *M. grisea*. The BAC end sequences and physical and genetic map data are incorporated in the MagnaportheDB database (Martin *et al.*, 2002).

Currently, over 28 000 *M. grisea* ESTs are deposited in GenBank. The largest *M. grisea* EST dataset was generated in Dr Ebbole's lab at Texas A & M University by sequencing cDNA libraries constructed with RNAs isolated from appressoria, conidia, mating cultures, *pmk1* mutant germlings, and mycelia grown on minimal, complete, nitrogen-deficient, and rice cell wall media. In 2003, the Whitehead Institute/ Massachusetts Institute of Technology Center for Genome Research generated over 6X coverage of the *M. grisea* 70–15 genome sequence by the shotgun approach. Information about sequencing, assembly, and annotation is available to the public at www-genome.wi.mit.edu/annotation/fungi/magnaporthe. It is worth noting here that the 70–15 genome sequence (Chao & Ellingboe, 1991) is a strain generated by backcrossing three times to Guy11, which is the first hermaphroditic *M. grisea* isolate pathogenic to rice and which was widely used in earlier genetic studies. Microarrays of genes predicted from the rice blast genome sequence are under development and will be very useful for generating expression profiles of *M. grisea* genes in different mutants and plant infection stages.

Several labs have generated or are generating a large number of random insertion mutants by using REMI or *Agrobacterium tumefaciens*-mediated transformation (ATMT) or transposon hopping approaches (Sweigard *et al.*, 1998; Balhadere *et al.*, 1999). These collections of transformants are very useful to identify mutants defective in different stages of infection. However, the conventional gene replacement or disruption approaches are time-consuming. The transposon-arrayed gene knockout (TAGKO) approach is efficient in generating gene disruption constructs and sequencing templates at the same time (Hamer *et al.*, 2001), but it is not suitable for large-scale disruption of genes identified in the *M. grisea* genome sequence. More efficient functional analysis approaches need to be developed for *M. grisea* and other filamentous fungi.

6.9   Future perspectives

In the past few years, both forward and reverse genetics approaches have been used to characterize genes important for infection-related morphogenesis. However, our knowledge about molecular mechanisms involved in the intimate fungal–plant interactions is still very limited, particularly the events after penetration (Kahmann & Basse, 2001). Functional and comparative genomics studies are likely to be the focus for the next few years and will lead to identification of genes important for different developmental and plant infection stages as well as genes specific for fungal pathogens or those specific for different infection mechanisms. Recent advances in imaging techniques will also be very helpful to further characterize fungal infection structures and cell-related changes during fungal pathogenesis.

References

Adachi, K. & Hamer, J.E. (1998) Divergent cAMP signaling pathways regulate growth and pathogenesis in the rice blast fungus *Magnaporthe grisea*. *Plant Cell*, **10**, 1361–1373.

Arase, S., Katano, Y., Li, X., Honda, Y. & Nozu, M. (1994) Morphological variation in spores of *Pyricularia oryzae* Cavara. *J. Phytopathol.*, **142**, 253–257.

Atkinson, H.A., Daniels, A. & Read, N.D. (2002) Live-cell imaging of endocytosis during conidial germination in the rice blast fungus *Magnaporthe grisea*. *Fungal Genet. Biol.*, **37**, 233–244.

Balhadere, P.V. & Talbot, N.J. (2001) *PDE1* encodes a P-type ATPase involved in appressorium-mediated plant infection by the rice blast fungus *Magnaporthe grisea*. *Plant Cell*, **13**, 1987–2004.

Balhadere, P.V., Foster, A.J. & Talbot, N.J. (1999) Identification of pathogenicity mutants of the rice blast fungus *Magnaporthe grisea* by insertional mutagenesis. *Mol. Plant Microbe Interact.*, **12**, 129–142.

Bastmeyer, M., Deising, H.B. & Bechinger, C. (2002) Force exertion in fungal infection. *Annu. Rev. Biophys. Biomol. Struct.*, **31**, 321–341.

Bechinger, C., Giebel, K.F., Schnell, M., Leiderer, P., Deising, H.B. & Bastmeyer, M. (1999) Optical measurements of invasive forces exerted by appressoria of a plant pathogenic fungus. *Science*, **285**, 1896–1899.

Beckerman, J.L. & Ebbole, D.J. (1996) *MPG1*, a gene encoding a fungal hydrophobin of *Magnaporthe grisea*, is involved in surface recognition. *Mol. Plant Microbe Interact.*, **9**, 450–456.

Beckerman, J.L., Naider, F. & Ebbole, D.J. (1997) Inhibition of pathogenicity of the rice blast fungus by *Saccharomyces cerevisiae* alpha-factor. *Science*, **276**, 1116–1119.

Böhnert, H.U., Fudal, I., Houlle, A. & Lebrun, M. (2003) Secondary metabolism and avirulence in *Magnaporthe grisea*. *Fungal Genetics Newsletter* 50S: abstract 346.

Bonman, J.M. (1992) Durable resistance to rice blast disease – environmental influences. *Euphytica*, **63**, 115–123.

Bourett, T.M. & Howard, R.J. (1990) *In vitro* development of penetration structures in the rice blast fungus *Magnaporthe grisea. Can. J. Bot.*, **68**, 329–342.

Bourett, T.M. & Howard, R.J. (1992) Actin in penetration pegs of the fungal rice blast pathogen *Magnaporthe grisea. Protoplasma*, **168**, 20–26.

Bourett, T.M., Picollelli, M.A. & Howard, R.J. (1993) Postembedment labeling of intracellular concanavalin A-binding sites in freeze-substituted fungal cells. *Exp. Mycol.*, **17**, 223–235.

Bryan, G.T., Wu, K.S., Farrall, L., Jia, Y.L., Hershey, H.P., McAdams, S.A., Faulk, K.N., Donaldson, G.K., Tarchini, R. & Valent, B. (2000) A single amino acid difference distinguishes resistant and susceptible alleles of the rice blast resistance gene *Pi-ta. Plant Cell*, **12**, 2033–2045.

Bucheli, P., Doares, S.H., Albersheim, P. & Darvill, A. (1990) Host pathogen interactions 36. Partial purification and characterization of heat-labile molecules secreted by the rice blast pathogen that solubilize plant cell wall fragments that kill plant cells. *Physiol. Mol. Plant Pathol.*, **36**, 159–173.

Chao, C.C.T. & Ellingboe, A.H. (1991) Selection for mating competence in *Magnaporthe grisea* pathogenic to rice. *Can. J. Bot.*, **69**, 2130–2134.

Chen, B.S., Gao, S.J., Choi, G.H. & Nuss, D.L. (1996) Extensive alteration of fungal gene transcript accumulation and elevation of G-protein-regulated cAMP levels by a virulence-attenuating hypovirus. *Proc. Natl. Acad. Sci. USA*, **93**, 7996–8000.

Choi, W.B. & Dean, R.A. (1997) The adenylate cyclase gene *MAC1* of *Magnaporthe grisea* controls appressorium formation and other aspects of growth and development. *Plant Cell*, **9**, 1973–1983.

Chumley, F.G. & Valent, B. (1990) Genetic analysis of melanin-deficient, nonpathogenic mutants of *Magnaporthe grisea. Mol. Plant Microbe Interact.*, **3**, 135–143.

Chun, S.J. & Lee, Y.H. (1999) Genetic analysis of a mutation on appressorium formation in *Magnaporthe grisea. FEMS Microbiol. Lett.*, **173**, 133–137.

Clergeot, P.H., Gourgues, M., Cots, J., Laurans, F., Latorse, M.P., Pepin, R., Tharreau, D., Notteghem, J.L. & Lebrun, M.H. (2001) PLS1, a gene encoding a tetraspanin-like protein, is required for penetration of rice leaf by the fungal pathogen *Magnaporthe grisea. Proc. Natl. Acad. Sci. USA*, **98**, 6963–6968.

Couch, B.C. & Kohn, L.M. (2002) A multilocus gene genealogy concordant with host preference indicates segregation of a new species, *Magnaporthe oryzae*, from *M. grisea. Mycologia*, **94**, 683–693.

Deising, H.B., Werner, S. & Wernitz, M. (2000) The role of fungal appressoria in plant infection. *Microbes Infect.*, **2**, 1631–1641.

deJong, J.C., McCormack, B.J., Smirnoff, N. & Talbot, N.J. (1997) Glycerol generates turgor in rice blast. *Nature*, **389**, 244–245.

DeZwaan, T.M., Carroll, A.M., Valent, B. & Sweigard, J.A. (1999) *Magnaporthe grisea* Pth11p is a novel plasma membrane protein that mediates appressorium differentiation in response to inductive substrate cues. *Plant Cell*, **11**, 2013–2030.

Dioh, W., Tharreau, D., Notteghem, J.L., Orbach, M. & Lebrun, M.H. (2000) Mapping of avirulence genes in the rice blast fungus, *Magnaporthe grisea*, with RFLP and RAPD markers. *Mol. Plant Microbe Interact.*, **13**, 217–227.

Dixon, K.P., Xu, J.R., Smirnoff, N. & Talbot, N.J. (1999) Independent signaling pathways regulate cellular turgor during hyperosmotic stress and appressorium-mediated plant infection by *Magnaporthe grisea. Plant Cell*, **11**, 2045–2058.

Dufresne, M. & Osbourn, A.E. (2001) Definition of tissue-specific and general requirements for plant infection in a phytopathogenic fungus. *Mol. Plant Microbe Interact.*, **14**, 300–307.

Ebata, Y., Yamamoto, H. & Uchiyama, T. (1998) Chemical composition of the glue from appressoria of *Magnaporthe grisea. Biosci. Biotechnol. Biochem.*, **62**, 672–674.

Eilbert, F., Thines, E. & Anke, H. (1999a) Effects of antifungal compounds on conidial germination and on the induction of appressorium formation of *Magnaporthe grisea. Z.Naturforsch. (C)*, **54**, 903–908.

Eilbert, F., Thines, E., Sterner, O. & Anke, H. (1999b) Fatty acids and their derivatives as modulators of appressorium formation in *Magnaporthe grisea. Biosci. Biotechnol. Biochem.*, **63**, 879–883.

Fang, E.G.C. & Dean, R.A. (2000) Site-directed mutagenesis of the *MAGB* gene affects growth and development in *Magnaporthe grisea*. *Mol. Plant Microbe Interact.*, **13**, 1214–1227.

Farman, M.L. & Leong, S.A. (1998) Chromosome walking to the *AVR1-CO39* avirulence gene of *Magnaporthe grisea*: discrepancy between the physical and genetic maps. *Genetics*, **150**, 1049–1058.

Farman, M.L., Eto, Y., Nakao, T., Tosa, Y., Nakayashiki, H., Mayama, S. & Leong, S.A. (2002) Analysis of the structure of the *AVR1-CO39* avirulence locus in virulent rice-infecting isolates of *Magnaporthe grisea*. *Mol. Plant Microbe Interact.*, **15**, 6–16.

Ford, T.L., Cooley, J.T. & Christou, P. (1994) Current status for gene transfer into rice utilizing variety-independent delivery systems. In: *Rice Blast Disease* (eds R.S. Zeigler, S.A. Leong & P.S. Teng), CAB International, Wallingford, UK.

Foster, A.J., Jenkinson, J.M. & Talbot, N.J. (2003) Trehalose synthesis and metabolism are required at different stages of plant infection by *Magnaporthe grisea*. *Embo J.*, **22**, 225–235.

Froeliger, E.H. & Carpenter, B.E. (1996) *NUT1*, a major nitrogen regulatory gene in *Magnaporthe grisea*, is dispensable for pathogenicity. *Mol. Gen. Genet.*, **251**, 647–656.

Fujimoto, D., Shi, Y., Christian, D., Mantanguihan, J.B. & Leung, H. (2002) Tagging quantitative loci controlling pathogenicity in *Magnaporthe grisea* by insertional mutagenesis. *Physiol. Mol. Plant Pathol.*, **61**, 77–88.

Fujita, K., Arase, S., Hiratsuka, H., Honda, Y. & Nozu, M. (1994) The role of toxins produced by germinating spores of *Pyricularia oryzae* in pathogenesis. *J. Phytopathol.*, **142**, 245–252.

Gerbeaud, C., Giermanska-Kahn, J., Meleard, P., Pouligny, B. & Latorse, M.P. (2001) Using capillary forces, to estimate the adhesion, strength of *Magnaporthe grisea*, spores on glass. *C.R. Acad. Sci. Ser. IV-Phys. Astrophys.*, **2**, 1235–1240.

Gilbert, R.D., Johnson, A.M. & Dean, R.A. (1996) Chemical signals responsible for appressorium formation in the rice blast fungus *Magnaporthe grisea*. *Physiol. Mol. Plant Pathol.*, **48**, 335–346.

Gustin, M.C., Albertyn, J., Alexander, M. & Davenport, K. (1998) MAP kinase pathways in the yeast *Saccharomyces cerevisiae*. *Microbiol. Mol. Biol. Rev.*, **62**, 1264–1300.

Hamer, J.E. & Talbot, N.J. (1998) Infection-related development in the rice blast fungus *Magnaporthe grisea*. *Curr. Opin. Microbiol.*, **1**, 693–697.

Hamer, J.E., Chumley, F.G., Howard, R.J. & Valent, B. (1988) A mechanism for surface attachment in spores of a plant pathogenic fungus. *Science*, **239**, 288–290.

Hamer, J.E., Valent, B. & Chumley, F.G. (1989) Mutations at the *smo* genetic-locus affect the shape of diverse cell types in the rice blast fungus. *Genetics*, **122**, 351–361.

Hamer, L., Adachi, K., Montenegro-Chamorro, M.V., Tanzer, M.M., Mahanty, S.K., Lo, C., Tarpey, R.W., Skalchunes, A.R., Heiniger, R.W., Frank, S.A., Darveaux, B.A., Lampe, D.J., Slater, T.M., Ramamurthy, L., DeZwaan, T.M., Nelson, G.H., Shuster, J.R., Woessner, J. & Hamer, J.E. (2001) Gene discovery and gene function assignment in filamentous fungi. *Proc. Natl. Acad. Sci. USA*, **98**, 5110–5115.

Harp, T.L. & Correll, J.C. (1998) Recovery and characterization of spontaneous, selenate-resistant mutants of *Magnaporthe grisea*, the rice blast pathogen. *Mycologia*, **90**, 954–963.

Heath, M.C., Howard, R.J., Valent, B. & Chumley, F.G. (1992) Ultrastructural interactions of one strain of *Magnaporthe grisea* with goosegrass and weeping lovegrass. *Can. J. Bot.*, **70**, 779–787.

Hegde, Y. & Kolattukudy, P.E. (1997) Cuticular waxes relieve self-inhibition of germination and appressorium formation by the conidia of *Magnaporthe grisea*. *Physiol. Mol. Plant Pathol.*, **51**, 75–84.

Howard, R.J. & Valent, B. (1996) Breaking and entering: host penetration by the fungal rice blast pathogen *Magnaporthe grisea*. *Annu. Rev. Microbiol.*, **50**, 491–512.

Howard, R.J., Ferrari, M.A., Roach, D.H. & Money, N.P. (1991) Penetration of hard substrates by a fungus employing enormous turgor pressures. *Proc. Natl. Acad. Sci. USA*, **88**, 11281–11284.

Idnurm, A. & Howlett, B.J. (2002) Isocitrate lyase is essential for pathogenicity of the fungus *Leptosphaeria maculans* to canola *Brassica napus*. *Eukaryot. Cell*, **1**, 719–724.

Iedome, M., Arase, S., Honda, Y. & Nozu, M. (1995) Studies on host-selective infection mechanism of *Magnaporthe grisea* 10. Light-dependent necrosis formation by *Magnaporthe grisea* toxins in rice cv. Sekiguchi-Asahi. *J. Phytopathol.*, **143**, 325–328.

Jelitto, T.C., Page, H.A. & Read, N.D. (1994) Role of external signals in regulating the pre-penetration phase of infection by the rice blast fungus *Magnaporthe grisea. Planta*, **194**, 471–477.

Jia, Y., McAdams, S.A., Bryan, G.T., Hershey, H.P. & Valent, B. (2000) Direct interaction of resistance gene and avirulence gene products confers rice blast resistance. *Embo J.*, **19**, 4004–4014.

Kahmann, R. & Basse, C. (2001) Fungal gene expression during pathogenesis-related development and host plant colonization. *Curr. Opin. Microbiol.*, **4**, 374–380.

Kamakura, T., Yamaguchi, S., Saitoh, K., Teraoka, T. & Yamaguchi, I. (2002) A novel gene, *CBP1*, encoding a putative extracellular chitin-binding protein, may play an important role in the hydrophobic surface sensing of *Magnaporthe grisea* during appressorium differentiation. *Mol. Plant Microbe Interact.*, **15**, 437–444.

Kang, S.C., Chumley, F.G. & Valent, B. (1994) Isolation of the mating-type genes of the phytopathogenic fungus *Magnaporthe grisea* using genomic subtraction. *Genetics*, **138**, 289–296.

Kang, S.C., Sweigard, J.A. & Valent, B. (1995) The *PWL* host specificity gene family in the blast fungus *Magnaporthe grisea. Mol. Plant Microbe Interact.*, **8**, 939–948.

Kershaw, M.J., Wakley, G. & Talbot, N.J. (1998) Complementation of the Mpg1 mutant phenotype in *Magnaporthe grisea* reveals functional relationships between fungal hydrophobins. *Embo J.*, **17**, 3838–3849.

Kim, J.C., Min, J.Y., Kim, H.T., Cho, K.Y. & Yu, S.H. (1998) Pyricuol, a new phytotoxin from *Magnaporthe grisea. Biosci. Biotechnol. Biochem.*, **62**, 173–174.

Kim, S., Ahn, I.P. & Lee, Y.H. (2001) Analysis of genes expressed during rice–*Magnaporthe grisea* interactions. *Mol. Plant Microbe Interact.*, **14**, 1340–1346.

Koga, H. (1995) An electron-microscopic study of the infection of spikelets of rice by *Pyricularia oryzae. J. Phytopathol.*, **143**, 439–445.

Koga, J., Shimura, M., Oshima, K., Ogawa, N., Yamauchi, T. & Ogasawara, N. (1995) Phytocassane A, phytocassane B, phytocassane C and phytocassane D, novel diterpene phytoalexins from rice *Oryza sativa. Tetrahedron*, **51**, 7907–7918.

Konishi, H., Ishiguro, K. & Komatsu, S. (2001) A proteomics approach towards understanding blast fungus infection of rice grown under different levels of nitrogen fertilization. *Proteomics*, **1**, 1162–1171.

Kronstad, J., De Maria, A., Funnell, D., Laidlaw, R.D., Lee, N. & Ramesh, M. (1998) Signaling via cAMP in fungi: interconnections with mitogen-activated protein kinase pathways. *Arch. Microbiol.*, **170**, 395–404.

Kumar, S. & Sridhar, R. (1993) Phytoalexins in rice–*Pyricularia oryzae* interaction – factors affecting phytoalexin production. *Acta Phytopathol. Entomol. Hung.*, **28**, 59–69.

Kumar, J., Nelson, R.J. & Zeigler, R.S. (1999) Population structure and dynamics of *Magnaporthe grisea* in the Indian Himalayas. *Genetics*, **152**, 971–984.

Lau, G. & Hamer, J.E. (1996) Regulatory genes controlling *MPG1* expression and pathogenicity in the rice blast fungus *Magnaporthe grisea. Plant Cell*, **8**, 771–781.

Lau, G.W. & Hamer, J.E. (1998) Acropetal: a genetic locus required for conidiophore architecture and pathogenicity in the rice blast fungus. *Fungal Genet. Biol.*, **24**, 228–239.

Lee, S.C. & Lee, Y.H. (1998) Calcium/calmodulin-dependent signaling for appressorium formation in the plant pathogenic fungus *Magnaporthe grisea. Mol. Cells*, **8**, 698–704.

Lee, Y.H. & Dean, R.A. (1993) cAMP regulates infection structure formation in the plant pathogenic fungus *Magnaporthe grisea. Plant Cell*, **5**, 693–700.

Liu, S.H. & Dean, R.A. (1997) G protein alpha subunit genes control growth, development, and pathogenicity of *Magnaporthe grisea. Mol. Plant Microbe Interact.*, **10**, 1075–1086.

Liu, Z.M. & Kolattukudy, P.E. (1999) Early expression of the calmodulin gene, which precedes appressorium formation in *Magnaporthe grisea*, is inhibited by self-inhibitors and requires surface attachment. *J. Bacteriol.*, **181**, 3571–3577.

Martin, S.L., Blackmon, B.P., Rajagopalan, R., Houfek, T.D., Sceeles, R.G., Denn, S.O., Mitchell, T.K., Brown, D.E., Wing, R.A. & Dean, R.A. (2002) MagnaportheDB: a federated solution for integrating physical and genetic map data with BAC end derived sequences for the rice blast fungus *Magnaporthe grisea. Nucleic Acids Res.*, **30**, 121–124.

McCafferty, H.R.K. & Talbot, N.J. (1998) Identification of three ubiquitin genes of the rice blast fungus *Magnaporthe grisea*, one of which is highly expressed during initial stages of plant colonisation. *Curr. Genet.*, **33**, 352–361.

Mitchell, T.K. & Dean, R.A. (1995) The cAMP-dependent protein kinase catalytic subunit is required for appressorium formation and pathogenesis by the rice blast pathogen *Magnaporthe grisea*. *Plant Cell*, **7**, 1869–1878.

Morosov, Y.M. (1992) The ultrastructure of parasite and host cells during rice blast disease *Pyricularia oryzae* Cav. *Izvestiya Akademii Nauk Sssr Seriya Biologicheskaya*, 519–530.

Motoyama, T., Imanishi, K. & Yamaguchi, I. (1998) cDNA cloning, expression, and mutagenesis of scytalone dehydratase needed for pathogenicity of the rice blast fungus *Pyricularia oryzae*. *Biosci. Biotechnol. Biochem.*, **62**, 564–566.

Nishimura, M., Hayashi, N., Jwa, N.S., Lau, G.W., Hamer, J.E. & Hasebe, A. (2000) Insertion of the LINE retrotransposon MGL causes a conidiophore pattern mutation in *Magnaporthe grisea*. *Mol. Plant Microbe Interact.*, **13**, 892–894.

Nitta, N., Farman, M.L. & Leong, S.A. (1997) Genome organization of *Magnaporthe grisea*: integration of genetic maps, clustering of transposable elements and identification of genome duplications and rearrangements. *Theor. Appl. Genet.*, **95**, 20–32.

Nomura, K. & Kiyosawa, S. (1992) Differences in hypersensitive reaction among rice cultivars carrying various resistance genes to the blast fungus *Pyricularia oryzae*. *Jpn. J. Breeding*, **42**, 213–225.

Nukina, M. (1999) The blast disease fungi and their metabolic products. *J. Pestic. Sci.*, **24**, 293–298.

Ohtake, M., Yamamoto, H. & Uchiyama, T. (1999) Influences of metabolic inhibitors and hydrolytic enzymes on the adhesion of appressoria of *Pyricularia oryzae* to wax-coated cover-glasses. *Biosci. Biotechnol. Biochem.*, **63**, 978–982.

Orbach, M.J., Farrall, L., Sweigard, J.A., Chumley, F.G. & Valent, B. (2000) A telomeric avirulence gene determines efficacy for the rice blast resistance gene *Pi-ta*. *Plant Cell*, **12**, 2019–2032.

Osherov, N. & May, G.S. (2001) The molecular mechanisms of conidial germination. *FEMS Microbiol. Lett.*, **199**, 153–160.

Ou, S.H. (1985) *Rice Diseases*. Commonwealth Mycological Institute, Surrey, UK.

Park, G., Xue, C.Y., Zheng, L., Lam, S. & Xu, J.R. (2002) *MST12* regulates infectious growth but not appressorium formation in the rice blast fungus *Magnaporthe grisea*. *Mol. Plant Microbe Interact.*, **15**, 183–192.

Pinnschmidt, H.O., Bonman, J.M. & Kranz, J. (1995) Lesion development and sporulation of rice blast. *J. Plant Dis. Prot.*, **102**, 299–306.

Rathour, R., Singh, B.M. & Plaha, P. (2002) Host species-specific protoplast damaging activity of spore germination fluids of blast pathogen *Magnaporthe grisea*. *J. Phytopathol.*, **150**, 576–578.

Rauyaree, R., Choi, W., Fang, E., Blackmon, B. & Dean, R.A. (2001) Genes expressed during early stages of rice infection with the rice blast fungus *Magnaporthe grisea*. *Mol. Plant Pathol.*, **2**, 347–354.

Romao, J. & Hamer, J.E. (1992) Genetic organization of a repeated DNA sequence family in the rice blast fungus. *Proc. Natl. Acad. Sci. USA*, **89**, 5316–5320.

Rossman, A.Y., Howard, R.J. & Valent, B. (1990) *Pyricularia grisea*, the correct name for the rice blast disease fungus. *Mycologia*, **82**, 509–512.

Sallaud, C., Lorieux, M., Roumen, E., Tharreau, D., Berruyer, R., Svestasrani, P., Garsmeur, O., Ghesquiere, A. & Notteghem, J.L. (2003) Identification of five new blast resistance genes in the highly blast-resistant rice variety IR64 using a QTL mapping strategy. *Theor. Appl. Genet.*, **106**, 794–803.

Shi, Z.X. & Leung, H. (1995) Genetic analysis of sporulation in *Magnaporthe grisea* by chemical and insertional mutagenesis. *Mol. Plant Microbe Interact.*, **8**, 949–959.

Silue, D., Tharreau, D., Talbot, N.J., Clergeot, P.H., Notteghem, J.L. & Lebrun, M.H. (1998) Identification and characterization of *apf1* in a non-pathogenic mutant of the rice blast fungus *Magnaporthe grisea* which is unable to differentiate appressoria. *Physiol. Mol. Plant Pathol.*, **53**, 239–251.

Soanes, D.M., Kershaw, M.J., Cooley, R.N. & Talbot, N.J. (2002) Regulation of the MPG1 hydrophobin gene in the rice blast fungus *Magnaporthe grisea*. *Mol. Plant Microbe Interact.*, **15**, 1253–1267.

Stanley, M.S., Callow, M.E., Perry, R., Alberte, R.S., Smith, R. & Callow, J.A. (2002) Inhibition of fungal spore adhesion by zosteric acid as the basis for a novel, nontoxic crop protection technology. *Phytopathology*, **92**, 378–383.

Sweigard, J.A., Chumley, F.G. & Valent, B. (1992) Disruption of a *Magnaporthe grisea* cutinase gene. *Mol. Gen. Genet.*, **232**, 183–190.

Sweigard, J.A., Carroll, A.M., Farrall, L., Chumley, F.G. & Valent, B. (1998) *Magnaporthe grisea* pathogenicity genes obtained through insertional mutagenesis. *Mol. Plant Microbe Interact.*, **11**, 404–412.

Sweigard, J.A., Carroll, A.M., Kang, S., Farrall, L., Chumley, F.G. & Valent, B. (1995) Identification, cloning, and characterization of *PWL2*, a gene for host species specificity in the rice blast fungus. *Plant Cell*, **7**, 1221–1233.

Talbot, N.J. & Foster, A.J. (2001) Genetics and genomics of the rice blast fungus *Magnaporthe grisea*: developing an experimental model for understanding fungal diseases of cereals. In: *Advances in Botanical Research Incorporating Advances in Plant Pathology*, **34**, 263–287.

Talbot, N.J., Ebbole, D.J. & Hamer, J.E. (1993) Identification and characterization of *MPG1*, a gene involved in pathogenicity from the rice blast fungus *Magnaporthe grisea*. *Plant Cell*, **5**, 1575–1590.

Talbot, N.J., McCafferty, H.R.K., Ma, M., Moore, K. & Hamer, J. E. (1997) Nitrogen starvation of the rice blast fungus *Magnaporthe grisea* may act as an environmental cue for disease symptom expression. *Physiol. Mol. Plant Pathol.*, **50**, 179–195.

Tharreau, D., Notteghem, J.L. & Lebrun, M.H. (1997) Mutations affecting perithecium development and sporulation in *Magnaporthe grisea*. *Fungal Genet. Biol.*, **21**, 206–213.

Thines, E., Eilbert, F., Sterner, O. & Anke, H. (1997) Signal transduction leading to appressorium formation in germinating conidia of *Magnaporthe grisea*: effects of second messengers diacylglycerols, ceramides and sphingomyelin. *FEMS Microbiol. Lett.*, **156**, 91–94.

Thines, E., Eilbert, F., Sterner, O. & Anke, H. (1998) Inhibitors of appressorium formation in *Magnaporthe grisea*: a new approach to control rice blast disease. *Pestic. Sci.*, **54**, 314–316.

Thines, E., Weber, R.W.S. & Talbot, N.J. (2000) MAP kinase and protein kinase A-dependent mobilization of triacylglycerol and glycogen during appressorium turgor generation by *Magnaporthe grisea*. *Plant Cell*, **12**, 1703–1718.

Tonukari, N.J., Scott-Craig, J.S. & Walton, J.D. (2000) The *Cochliobolus carbonum SNF1* gene is required for cell wall-degrading enzyme expression and virulence on maize. *Plant Cell*, **12**, 237–247.

Tsuji, G., Kenmochi, Y., Takano, Y., Sweigard, J., Farrall, L., Furusawa, I., Horino, O. & Kubo, Y. (2000) Novel fungal transcriptional activators, Cmr1p of *Colletotrichum lagenarium* and Pig1p of *Magnaporthe grisea*, contain Cys2His2 zinc finger and Zn(II)2Cys6 binuclear cluster DNA-binding motifs and regulate transcription of melanin biosynthesis genes in a developmentally specific manner. *Mol. Microbiol.*, **38**, 940–954.

Tucker, S.L. & Talbot, N.J. (2001) Surface attachment and pre-penetration stage development by plant pathogenic fungi. *Annu. Rev. Phytopathol.*, **39**, 385–417.

Uchiyama, T. & Okuyama, K. (1990) Participation of *Oryza sativa* leaf wax in appressorium formation by *Pyricularia oryzae*. *Phytochemistry*, **29**, 91–92.

Urban, M., Bhargava, T. & Hamer, J.E. (1999) An ATP-driven efflux pump is a novel pathogenicity factor in rice blast disease. *Embo J.*, **18**, 512–521.

Valent, B. (1990) Rice blast as a model system for plant pathology. *Phytopathology*, **80**, 33–36.

Valent, B. & Chumley, F.G. (1991) Molecular genetic analysis of the rice rlast fungus *Magnaporthe grisea*. *Annu. Rev. Phytopathol.*, **29**, 443–467.

Viaud, M.C., Balhadere, P.V. & Talbot, N.J. (2002) A *Magnaporthe grisea* cyclophilin acts as a virulence determinant during plant infection. *Plant Cell*, **14**, 917–930.

Villalba, F., Lebrun, M.H., Hua-Van, A., Daboussi, M.J. & Grosjean-Cournoyer, M.C. (2001) Transposon impala, a novel tool for gene tagging in the rice blast fungus *Magnaporthe grisea*. *Mol. Plant Microbe Interact.*, **14**, 308–315.

Wang, Z.X., Yamanouchi, U., Katayose, Y., Sasaki, T. & Yano, M. (2001) Expression of the *Pib* rice-blast-resistance gene family is up-regulated by environmental conditions favouring infection and by chemical signals that trigger secondary plant defences. *Plant Mol. Biol.*, **47**, 653–661.

Wang, Z.Y., Thornton, C.R., Kershaw, M.J., Li, D.B. & Talbot, N.J. (2003) The glyoxylate cycle is required for temporal regulation of virulence by the plant pathogenic fungus *Magnaporthe grisea*. *Mol. Microbiol.*, **47**, 1601–1612.

Weber, R.W.S., Wakley, G.E., Thines, E. & Talbot, N. J. (2001) The vacuole as central element of the lytic system and sink for lipid droplets in maturing appressoria of *Magnaporthe grisea*. *Protoplasma*, **216**, 101–112.

Wu, S.C., Ham, K.S., Darvill, A.G. & Albersheim, P. (1997) Deletion of two endo-beta-1,4-xylanase genes reveals additional isozymes secreted by the rice blast fungus. *Mol. Plant Microbe Interact.*, **10**, 700–708.

Xiao, J.Z., Ohshima, A., Kamakura, T., Ishiyama, T. & Yamaguchi, I. (1994a) Extracellular glycoprotein(s) associated with cellular differentiation in *Magnaporthe grisea*. *Mol. Plant Microbe Interact.*, **7**, 639–644.

Xiao, J.Z., Watanabe, T., Kamakura, T., Ohshima, A. & Yamaguchi, I. (1994b) Studies on cellular-differentiation of *Magnaporthe grisea* – physicochemical aspects of substratum surfaces in relation to appressorium formation. *Physiol. Mol. Plant Pathol.*, **44**, 227–236.

Xu, J.R. (2000) MAP kinases in fungal pathogens. *Fungal Genet. Biol.*, **31**, 137–152.

Xu, J.R. & Hamer, J.E. (1996) MAP kinase and cAMP signaling regulate infection structure formation and pathogenic growth in the rice blast fungus *Magnaporthe grisea*. *Genes Dev.*, **10**, 2696–2706.

Xu, J.R., Urban, M., Sweigard, J.A. & Hamer, J.E. (1997) The *CPKA* gene of *Magnaporthe grisea* is essential for appressorial penetration. *Mol. Plant Microbe Interact.*, **10**, 187–194.

Xu, J.R., Staiger, C.J. & Hamer, J.E. (1998) Inactivation of the mitogen-activated protein kinase *MPS1* from the rice blast fungus prevents penetration of host cells but allows activation of plant defense responses. *Proc. Natl. Acad. Sci. USA*, **95**, 12713–12718.

Xue, C.Y., Park, G., Choi, W.B., Zheng, L., Dean, R.A. & Xu, J.R. (2002) Two novel fungal virulence genes specifically expressed in appressoria of the rice blast fungus. *Plant Cell*, **14**, 2107–2119.

Zeigler, R.S. (1998) Recombination in *Magnaporthe grisea*. *Annu. Rev. Phytopathol.*, **36**, 249–275.

Zhu, H., Whitehead, D.S., Lee, Y.H. & Dean, R.A. (1996) Genetic analysis of developmental mutants and rapid chromosome mapping of *APP1*, a gene required for appressorium formation in *Magnaporthe grisea*. *Mol. Plant Microbe Interact.*, **9**, 767–774.

7   The *Ustilago maydis*–maize interaction

Maria D. Garcia-Pedrajas, Steven J. Klosterman,
David L. Andrews and Scott E. Gold

7.1   Introduction

In the introduction of a now half-century-old landmark paper (Rowell & Devay, 1954), several fascinating unanswered questions about the biology of *Ustilago maydis* (then called *U. zeae*) were raised: "There are, therefore, a number of unanswered questions on the pathogenicity of *U. zeae*. Why does fusion of the haploid sporidia cause a change from a saprophytic to a parasitic habit? Are the processes of sex and pathogenicity governed by the same or separate genic factors? What are the pathogenic mechanisms causing gall formation?" Today, these questions are partially answered and studies are in progress that should help to explain these processes more fully, but nonetheless these same questions still remain at the heart of much of the current research with this organism.

The earliest recordings of *U. maydis*, the causal agent of corn smut, date back to the 1750s and 1760s (Christensen, 1963). One of the earliest great experimental scientists, M. Tillet of Bordeux, France, proved by using dusting experiments that wheat smut was a transmissible disease. His similar experiments with teliospores of *U. maydis* proved in vain, leading him to conclude erroneously that corn smut was not a contagious disease (Tillet, 1766). It has since been established that *U. maydis* is a heterothallic fungus with a tetrapolar mating system and a dimorphic life cycle consisting of a saprophytic asexual phase and a parasitic sexual stage (Fig. 7.1). Unlike other smut fungi, the life cycle of *U. maydis* is shorter, because it can produce galls in sexually immature plants.

Resting sexual spores (teliospores) germinate and undergo meiosis, producing haploid sporidia. These are saprophytic and grow by budding. Fusion of compatible haploid sporidia produces a filamentous dikaryon that is an obligate parasite. The dikaryon colonizes the plant inducing tumors (galls) within which nuclear fusion takes place and diploid teliospores are produced. Upon dispersal from the host plant, the teliospores germinate in this way completing the life cycle. In this review, we will focus on the biology of *U. maydis*, employing the life and disease cycle as a guide to describe the various topics of study. We will start with the germination and consequent meiosis in the resting teliospores and end full circle back at their production and survival. We will also discuss a few aspects of the host's biology in response to the pathogen and how this relates to disease control. Finally, a brief note on future directions of research on this model basidiomycete plant pathogen is included.

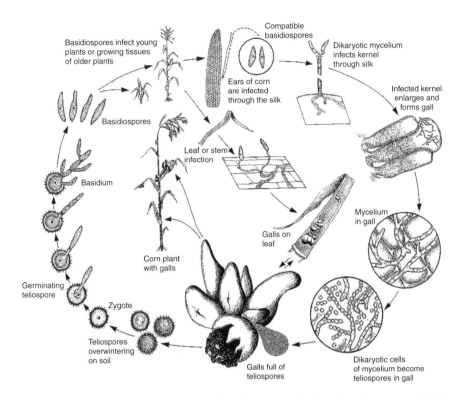

Figure 7.1 Disease cycle of corn smut, caused by *Ustilago maydis*. Reproduced from Agrios (1997) with the permission of Academic Press.

A note on nomenclature: The common corn smut fungus, *Ustilago maydis* (DC.) Corda, has been studied for two and a half centuries and has consequently had several synonyms recorded in the literature (reviewed in Christensen, 1963). The most common synonym causing confusion is *Ustilago zeae* which overtook *maydis* in use but became less favored due to an argument based on International Rules of Botanical Nomenclature (Stevenson & Johnson, 1944). Since about 1955, *U. zeae* has become rare in mainstream literature and nearly all more recent papers refer to *U. maydis* instead. Additionally, the darkly pigmented resting spores found in galls have variously been referred to as, brand-spores, chlamydospores, and teliospores. Currently, the term teliospore is seen in the vast majority of the literature. A final note regarding symbolism: it is convention in the *U. maydis* literature to employ italicized lowercase designations for genes (e.g. *ubc1*) and non-italicized designations with the first letter capitalized in reference to the protein encoded by that gene (e.g. Ubc1).

7.2   Teliospore germination and meiosis

In *U. maydis*, meiosis occurs in germinating teliospores (Christensen, 1963; O'Donnell & McLaughlin, 1984). As the teliospore germinates, the nucleus migrates to the center of the developing promycelium and the first meiotic division occurs in the midregion of the promycelium (Christensen, 1963; O'Donnell & McLaughlin, 1984). The resultant daughter nuclei divide again to produce four meiotic products. One of these nuclei generally remains in the teliospore and the other three are typically observed in the promycelium (Christensen, 1963; O'Donnell & McLaughlin, 1984). Septa are formed between these nuclei. In general, the behavior of the nucleus and the spindle pole body during meiosis is comparable to that of other basidiomycetes (O'Donnell & McLaughlin, 1984).

Typically, although not always, in the four-celled promycelium, each cell gives rise to a single primary sporidium. While the above description of teliospore germination and sporidial formation is the *norm*, Christensen (1963) stated that it is not the most frequent. A single promycelium may give rise to five to seven primary sporidia (see Fig. 7.2) and some of these may be diploid (Christensen, 1963). Fibrillar

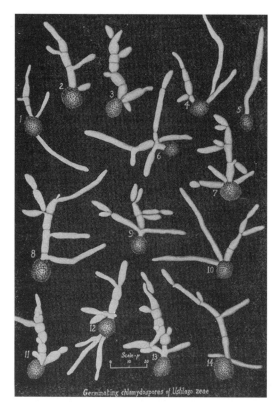

Figure 7.2 Germinating teliospores of *Ustilago maydis*, showing diverse types of germination. Reproduced from Christensen (1963) with the permission of APS press.

outgrowths on sporidia have been termed fimbriae and may aid in the attachment to host tissues (Snetselaar & Mims, 1993). It should be noted that, unlike many other basidiomycetes, the sporidia (basidiospores) of *U. maydis* do not develop on sterigmata and are not forcibly discharged. Rather, the primary sporidia develop on the surface of promycelia (Alexopoulus *et al.*, 1996).

Meiosis in the promycelium leads to the production of sporidia, which are normally not capable of causing infection singly. Occasionally, meiosis may not occur, giving rise to diploid sporidia (Christensen, 1932). Likewise, it is conceivable that two haploid nuclei from a promycelium may fuse to form the diploid (Banuett & Herskowitz, 1989). In either case, this may result in the production in nature of solopathogenic strains (Christensen, 1932). Christensen (1932) conducted extensive analyses of these types of solopathogens. Of 275 teliospores from 25 different galls, seven gave rise to diploid solopathogenic strains.

DNA synthesis during spore formation in *U. maydis* has also been investigated. Using microdensitometry, a technique to measure DNA content of nuclei, Snetselaar and McCann (1997) correlated nuclear density with sporidial cell morphology. DNA synthesis occurs in sporidia before bud formation (Snetselaar & McCann, 1997) and production of secondary sporidia. This study also revealed that older cultures of uniformly budding sporidia sometimes produce large numbers of binucleate cells when transferred to fresh culture medium. Septa were formed between these nuclei and the cells bud at both ends. Thus, it is important to observe and maintain *U. maydis* cultures appropriately for morphological observations.

Little is known about the genetic regulation of teliospore germination. Nevertheless, several mutants have been identified that affect spore germination. Banuett (1991) identified a mutant strain, *fuz2⁻*, in which teliospore germination is abolished, and suggested that *fuz2* may encode an enzyme necessary for the breakdown of the thick teliospore cell walls. Mayorga and Gold (1999) also reported a 100-fold reduction in the germination of teliospores formed by strains carrying a deletion mutation of the MAP kinase *ubc3* (*kpp2*), suggesting a role for the MAP kinase cascade that includes *ubc3* in regulating spore germination. Table 7.1 summarizes the roles of these and many other genes in morphogenesis and pathogenicity.

7.3 Mating and dikaryon formation

Genetic analysis of the sexual system has been a major focus of work on *U. maydis*. Most species of *Ustilago* are heterothallic and possess a bipolar mating pattern in which only two sexes are encountered. In this regard, *U. maydis* is unusual and more complex, possessing a tetrapolar mating system. Its sexual system is a paradigm for the higher basidiomycetes (Casselton & Olesnicky, 1998; Casselton, 2002). *U. maydis* was recognized as early as 1927 to be heterothallic (Stakman & Christensen, 1927). Shortly later, multiple sex types due to the activity of (at least) two separate mating factors were suggested (Hana, 1929). As discussed by Hana (1929), in 1912 Rawitscher had established that *U. maydis* had an alternation of generations between haploid and diploid. Employing tetrad analysis Hana showed conclusively

Table 7.1 Roles of *U. maydis* genes in morphogenesis and pathogenicity. This table focuses on genes with an effect on morphology and/or pathogenicity. It is not meant to be an exhaustive list of the genes that have been analyzed in *U. maydis* to date. A few genes with no effect on these processes have been included because their differential expression in budding versus filamentous cells or in a particular step of disease development, made them good candidates for analysis of their contribution to morphogenesis and pathogenesis. Genes with similarities to other genes shown to have an effect on these processes, i.e. kinases, have also been included

Gene	Process/function	ID Method[1]	Effect on morphology[2]	Effect on pathogenicity[2]	Reference
adr1	cAMP signalling major PKA catalytic subunit	Resistance to vinclozolin/ candidate gene approach	Filamentous	Non-pathogenic	Orth *et al.* (1995), Durrenberger *et al.* (1998)
bE (*bW*)	Mating/ homeodomain transcription factor	Functional assay	Required for postfusion filamentation	Non-pathogenic	Kronstad & Leong (1990), Schulz *et al.* (1990)
crk1	Putative protein kinase	Candidate gene approach	Defect in morphological response to environmental factors *Overexpressing strain (by fusion to inducible promoter): filamentous	Not tested	Garrido & Perez-Martin (2003)
don1	Cell separation/ guanine nucleotide exchange factor	Complementation of mutants defective in cell separation	Defect in cell separation, tree-like structures	No effect	Weinzierl *et al.* (2002)
don3	Cell separation/ Ste-20 like protein kinase	Complementation of mutants defective in cell separation	Defect in cell separation, tree-like structures	No effect	Weinzierl *et al.* (2002)
egl1	(β-1,4) endoglucanase	Subtractive cDNA hybridization	No effect	No effect	Schauwecker *et al.* (1995)
fuz1	Unknown (gene has not been cloned)	UV-mutagenesis	*Uncharacterized mutation: long thin cells, required for filamentation	*Uncharacterized mutation: tumors, no teliospores	Banuett (1991)
fuz2	Unknown (gene has not been cloned)	UV-mutagenesis	*Uncharacterized mutation: long thin cells, required for filamentous growth	*Uncharacterized mutation: abolished teliospore germination	Banuett (1991)
fuz7/ ubc5	MAPK signaling/ MAPK kinase	Candidate gene approach/ suppressor of *uac1⁻* filamentous phenotype	Required for filamentation	Non-pathogenic	Banuett & Herskowitz (1994), Andrews *et al.* (2000), Brachmann *et al.* (2003)

γ-ada	Protein transport/ γ-adaptin	Complementation of a osmotic-remedial thermosensitive mutant	Uncharacterized mutation: required for apical extension	Not tested	Keon *et al.* (1995)
gpa3	cAMP signaling/ trimeric G-protein α-subunit	Candidate gene approach	Filamentous	Non-pathogenic	Regenfelder *et al.* (1997), Kruger *et al.* (2000)
			*Constitutively active form ($gpa3_{Q206L}$ allele): glossy colonies	*Constitutively active form ($gpa3_{Q206L}$ allele): tumors, no teliospores	
hda1	*b*-regulated gene expression/ histone deacetylase	UV mutagenesis screening for mutants leading to *b*-independent expression of *egl1*	Dark-pigmented colonies	Tumors, no teliospores	Reichmann *et al.* (2002)
hgl1	Putative target of PKA regulatory subunit	Suppressor of *adr1⁻* filamentous phenotype	Reduced filamentation, darkly pigmented cultures	Tumors, no teliospores	Durrenberger (2001)
kin1	Cytoskeleton/ motor protein	Candidate gene approach	No effect	No effect	Lehmler *et al.* (1997)
kin2	Cytoskeleton/ heavy chain of kinesin	Candidate gene approach	Required for vacuole formation and hyphal extension	Severely reduced virulence	Lehmler *et al.* (1997), Steinberg *et al.* (1998)
kpp6	MAPK signaling/ *b*-regulated MAPK kinase	RNA fingerprinting	No effect	Reduced virulence; lack of anthocyanin	Brachmann *et al.* (2003)
			*Mutation in the putative phosphorylation motif ($kpp6^{T355AY357F}$ allele): no effect	*Mutation in the putative phosphorylation motif ($kpp6^{T355AY357F}$ allele): defect in plant penetration	
mfa1 (2)	Mating/ pheromone	Complementation of the closely linked gene *pan1*/ functional assay	Mating with either mating types abolished but able to mate with diploid *a1a2/b1b1*	No effect on solopathogenic strains	Froeliger & Leong (1991), Bolker *et al.* (1992)
mig1	Maize induced gene/ unknown	Differential display	No effect	No effect	Basse *et al.* (2000)
mig2-1 through mig2-5 gene cluster	Maize induced genes/ unknown	Differential display	No effect	Deletion of the entire cluster has no effect on pathogenicity	Basse *et al.* (2002)

Table 7.1 (Continued)

Gene	Process/function	ID Method[1]	Effect on morphology[2]	Effect on pathogenicity[2]	Reference
myp1	Possible role in hyphal elongation/ unknown	Insertional mutagenesis of mycelial haploid	Reduced filamentation	Reduced virulence	Giasson & Kronstad (1995)
pra1(2)	Mating/ pheromone receptor	Complementation of the closely linked gene *pan1*/ functional assay	Mating completely abolished with both haploid and diploid strains	No effect on solopathogenic strains	Froeliger & Leong (1991), Bolker *et al.* (1992)
prf1	Mating/ transcription	Amplification of HMG-conserved domain with degenerate primers	Required for filamentation	Non-pathogenic	Hartmann *et al.* (1996)
ras1	Presumably cAMP signaling/ small GTP-binding protein	Candidate gene approach	*Dominant, active allele (ras1$_{Q67L}$): no effect	*Dominant, active allele (ras1$_{Q67L}$): not tested	Muller *et al.* (2003)
ras2	Crosstalk between cAMP and MAP kinase pathways/small GTP-binding protein	Suppressor of *adr1⁻* filamentous phenotype/ candidate gene approach	Cells shorter and rounder than wild-type budding cells	Non-pathogenic	Lee & Kronstad (2002), Muller *et al.* (2003)
			*Point mutation (activated alleles *ras2*^{Val19} and *ras2*$_{Q65L}$): filamentous	*Point mutation (activated alleles *ras2*^{Val19} and *ras2*$_{Q65L}$): increased pathogenicity in solopathogenic haploid	
rep1	Cell wall protein filament-specific	Purification of an abundant cell wall protein specific to filaments	Required for aerial filament formation	No effect	Wosten *et al.* (1996)
rtf1	Unknown (gene has not been cloned)	UV-mutagenesis	*Uncharacterized mutation: required for filamentation.	*Uncharacterized mutation: induction of tumors in inoculations with strains with the same b	Banuett (1991)
rum1	b-regulated gene expression/ transcriptional repressor	UV mutagenesis screening for mutants leading to b-independent expression of *egl1*	No effect	Tumors, no teliospores	Quadbeck-Seeger *et al.* (2000)

sql1	cAMP signaling/ trancriptional repressor	Suppressor of glossy colony phenotype of strain $gpa3_{Q206L}$	Null mutants grow poorly and are unstable, truncated versions of Sql1 induce filamentation	Not tested	Loubradou *et al.* (2001)
sql2	cAMP signaling/ guanyl nucleotide exchange factor	Suppressor of glossy colony phenotype of strain $gpa3_{Q206L}$	Glossy colonies	Reduced virulence	Muller *et al.* (2003)
			*Overexpressing strain (by fusion to constitutive promoter): filamentous	*Overexpressing strain (by fusion to constitutive promoter): no effect	
ssp1	Putative dioxygenase	Purification of Ssp1 (abundant in teliospores)	No effect	No effect	Huber *et al.* (2002)
uac1	cAMP signaling/ adenylate cyclase	Complementation of constitutive filamentous mutant	Filamentous	Non-pathogenic	Gold *et al.* (1994)
ubc1	cAMP signaling/ PKA regulatory subunit	Suppressor of *uac1⁻* filamentous phenotype	Multiple budding, glossy colonies	No tumors	Gold *et al.* (1994, 1997), Kruger *et al.* (2000)
			*Single point mutation ($ubc1_{R321Q}$ allele): multiple budding, glossy colonies	*Single point mutation ($ubc1_{R321Q}$ allele): tumors, no teliospores	
ubc2	MAPK signaling/ adaptor	Suppressor of *uac1⁻* filamentous phenotype	Required for filamentation	Non-pathogenic	Mayorga & Gold (2001)
ubc3/ kpp2	MAPK signaling/MAP kinase	Suppressor of *uac1⁻* filamentous phenotype/ candidate gene approach	Required for filamentation	Reduced virulence, reduced teliospore germination	Mayorga & Gold (1999), Muller *et al.* (1999)
ubc4	MAPK signaling/MAPKK kinase	Suppressor of *uac1⁻* filamentous phenotype	Required for filamentation	Non-pathogenic	Andrews *et al.* (2000), Mueller *et al.* (2001)
uka1	cAMP signaling/PKA catalytic subunit	Candidate gene approach	No effect	No effect	Durrenberger *et al.* (1998)
ukb1	Serine/ threonine protein kinase	Candidate gene approach	Lateral buds, aerial mycelium in the absence of mating partner	No tumors	Abramovitch *et al.* (2002)

Table 7.1 (Continued)

Gene	Process/function	ID Method[1]	Effect on morphology[2]	Effect on pathogenicity[2]	Reference
ukc1	Protein kinase involves in morphogenesis	Candidate gene approach	Highly distorted pigmented cells resembling chlamydospores, required for filamentation	No tumors	Durrenberger & Kronstad (1999)
umodc	Polyamine biosynthesis/ ornithine decarboxylase	Candidate gene approach	Defect in dimorphic transition at low polyamine concentration	Not tested	Guevara-Olvera *et al.* (1997)

[1]Method used for gene identification.

[2]The effect on morphology and pathogenicity is referred to the null mutant unless noted otherwise (*).

that there were at least two pairs of segregating sex factors because in some cases four distinct mating types were derived from a single tetrad. He designated the diploid nucleus as possessing an AaBb genotype and the progeny as AB, Ab, aB, and ab. This symbolism was not far from the currently used designations. No distinction was made for the function of A or B in Hana's work. Rowell and DeVay (1954) designated these two mating factors as *a* and *b* and determined that two specificities of *a* and multiple specificities of *b* existed. In this work, they clearly demonstrated that different specificities were necessary at both the *a* and the *b* loci, to generate productive maize infection in which teliospores (chlamydospores) were generated, and the currently used designations for these mating type loci was established. In this same study, it was determined that amphisexual progeny were occasionally generated that had both *a* specificities such that they could productively be paired with any strain possessing a different *b* allele. However, these amphisexual strains were not solopathogenic while diploids heterozygous at both *a* and *b* were solopathogens. This indicated to the authors that *a* compatibility factor were clearly not a primary pathogenicity characteristic, a fact further corroborated by Banuett and Herskowitz (1989). Rowell clearly demonstrated that the functions of *a* and *b* were for fusion and dikaryon vigor and stability, respectively (Rowell, 1955). He also noted that alleles of both factors had to be different in the mating partners to generate the virulent pathogen. Adding much confusion in the literature on basidiomycetes to the uninitiated, molecular data now indicates that by chance it turned out that these designations are opposite in function to those used for the higher basidiomycetes. The functions are: *a*=B and *b*=A for *Ustilago* and mushrooms respectively (Casselton & Olesnicky, 1998; Casselton, 2002).

Further work by Trueheart and Herskowitz generated a cytoduction assay in which cell fusion was strictly controlled by possession of unequal alleles at the *a* locus while the *b* locus played no role whatsoever (Trueheart & Herskowitz, 1992). Later Laity *et al.* (1995) complemented this work showing that heterozygosity at the *b* locus within a strain inhibits further mating. Thus, in summary *a1* strains will fuse with *a2* strains regardless of the condition at the *b* locus, except that once *b* becomes heterozygous the cell will fuse no further with any other strain. Additionally, *a1a2b*$_n$ strains fuse promiscuously with any normal haploid mating type. By addition of charcoal to solid medium, the development of the functional dikaryon can easily be monitored in culture (Day & Anagnostakis, 1971; Holliday, 1974). White filamentous growth is observed on this medium only when both *a* and *b* differ in cospotted compatible strain pairs (Fig. 7.3).

The *b* locus as noted by Rowell and Devay (1954) is multiallelic. Puhalla found only 2 *a* alleles but 18 different *b* mating type alleles in 62 lines from 33 different isolations from the US and Canada. He predicted that there should be no more than 25 distinct alleles of *b* in the population (Puhalla, 1970).

The master control genes of mating and pathogenicity, the *a* and *b* mating type genes, have now been cloned and characterized (Kronstad & Leong, 1989, 1990; Schulz *et al.*, 1990; Froeliger & Leong, 1991; Bolker *et al.*, 1992). Holliday had noted that the *pan1* gene was tightly linked at about 2.5 map units from the *a* mating type locus (Holliday, 1961a, 1974). Froeliger and Leong (1991), using this information, cloned the *a2* mating type determinant by complementation of a *pan1* mutant *a1* strain with a cosmid from a prototrophic *a2* strain. The fact that *a2* was present was confirmed by mating *a1b1/a2* transformant strains with an *a1b2* strain to generate a filamentous and pathogenic dikaryon. The initial cloning of the *a* locus indicated that the *a1* and *a2* allelic sequences are idiomorphs, meaning that they lacked sequence homology. Homologous flanks were then employed to isolate the *a2* mating type idiomorph and similar methods were used to confirm its function (Froeliger & Leong, 1991). With this backdrop the *a* mating type genes were sequenced and Bolker *et al.* (1992) demonstrated that the mating type specificity in each idiomorph is determined by two genes. One gene encodes a lipopeptide mating

Figure 7.3 *2b* or not *2b*, that is the question. The upper portion of the photo shows an *a1b1* strain overlaid with an *a2b2* strain yielding the white filamentous dikaryotic growth of the compatible mating reaction. The bottom portion shows an *a1b1* strain overlaid with an *a2b1* strain yielding yeast growth. These results are indicative of the feature that two different *b* alleles are essential for the formation of the filamentous (and pathogenic) dikaryon.

factor, and the other a pheromone receptor. Thus, the *a* locus possesses *mfa* and *pra*, two tightly linked genes that encode secreted pheromone and membrane spanning pheromone receptors respectively (Bolker *et al.*, 1992). Genetic and biochemical data indicate that the interaction between the *U. maydis* pheromones and receptors is quite similar to the events in the *Saccharomyces cerevisiae* paradigm (Banuett, 1998). The pheromone encoded by the *mfa* gene is thought to interact directly with the pheromone receptor product encoded by the *pra* gene of the opposite *a* mating specificity (Spellig *et al.*, 1994). Work to further characterize the important residues of the pheromones has been carried out using synthetic peptides (Szabo *et al.*, 2002). Both pheromones were found to be more affected by C-terminal than N-terminal truncations. Replacement of each amino acid in the Mfa1 pheromone encoded by the *a1* idiomorph revealed that four positions are important for function. A model was derived that suggests structural roles in receptor specificity. Synthetic pheromone has also been shown to cause cell cycle arrest in the G2 phase (Garcia-Muse *et al.*, 2003). This is in contrast with the situation described in ascomycete yeasts such as *S. cerevisiae* and *Schizosaccharomyces pombe*, where pheromone induces cell cycle arrest at G1. The function of the genes at the *a* locus helps explain the fact noted above that a≠ (possession of two different allelic specificities of *a*) is required for a diploid heterozygous at *b* to become filamentous on charcoal mating media (Banuett & Herskowitz, 1989). In addition to its function as a mating attraction system, dikaryon heterozygosity at the *a* locus (in addition to heterozygosity at *b*) also contributes to the *in vitro* production of the postmating dikaryotic filamentous form through an autocrine response, in which the pheromones and receptors of opposite allelic specificity are present within the same cell and, therefore, may continually interact (Banuett & Herskowitz, 1989; Spellig *et al.*, 1994).

Downstream events generating the final response to pheromone appear to involve components similar to those encountered in *S. cerevisiae*. In this budding yeast, signal transduction from the pheromone–receptor interaction to the final cellular responses involves a trimeric G-protein and an MAP kinase cascade with the final phosphorylation and activation of two critical proteins. These proteins are the Ste12p transcription factor, which when activated regulates transcription of target genes, and Far1p which causes cell cycle arrest by inhibition of the kinase activity of the G1 cyclin complex Cdc28-Cln (Valdivieso *et al.*, 1993; Banuett, 1998). In *U. maydis*, none of the four cloned Gα subunits of the trimeric G-proteins appear to be directly involved in transmission of the pheromone signal (Regenfelder *et al.*, 1997; Kruger *et al.*, 1998). As is the case in *S. pombe* (Sipiczki, 1988), a *ras* gene (*ras2*) functions to stimulate filamentous growth through the pheromone responsive MAP kinase cascade (Lee & Kronstad, 2002). Additional work suggested that the cdc25 homolog Sql1 may function as an activator of Ras2 (Muller *et al.*, 2003). An additional finding in this work was that activated Ras1, the product of a second *ras* gene, increased pheromone gene expression. The three members of the pheromone responsive MAP kinase cascade have been identified. These are *ubc4* encoding the ste11p MAPKK kinase homolog (Mayorga & Gold, 1998; Andrews *et al.*, 2000), *fuz7/ubc5* encoding the Ste7p MAPK kinase homolog (Banuett & Herskowitz, 1994; Mayorga & Gold, 1998; Andrews *et al.*, 2000), and the Fus3p and Kss1p MAP Kinase homolog of *ubc3/kpp2* (Mayorga & Gold, 1998, 1999; Muller *et al.*, 1999). A putative adaptor protein Ubc2 may link the MAP kinase cascade with the upstream components of signaling through Ras proteins (Mayorga & Gold, 2001). A gene designated *prf1* encodes an HMG family transcription factor that links the pheromone response pathway to the expression of the *b* locus and thus to pathogenicity (Hartmann *et al.*, 1996). The *prf1* protein has potential phosphorylation sites for both a MAP kinase (presumably *ubc3/kpp2*) and the cyclic AMP-dependent protein kinase (Kahmann *et al.*, 1999; Muller *et al.*, 1999). The putative MAP kinase phosphorylation sites appear important for the biological function of the protein (Muller *et al.*, 1999). The *prf1* gene is required for pathogenicity due to its essential function in the regulation of the *b* mating type genes. This was shown clearly by the fact that constitutive expression of the *b* genes restores pathogenicity in *prf1* mutants (Hartmann *et al.*, 1996). Additional transcription factors are likely involved in transmitting the pheromone responsive MAP kinase and/or cAMP pathway signals besides Prf1. As noted by Lee and Kronstad (2002), epistasis experiments indicated that Ras2 may regulate filamentation via the pheromone responsive MAP kinase cascade including Ubc3, but not through the activation of Prf1. Additionally, work from our laboratory indicates that the MAP kinase cascade is required for acid-induced filamentation while *prf1* is not (Martinez-Espinoza *et al.*, submitted).

Recognition mediated by the products of the *b* genes has been a topic of research focus for understanding self/nonself recognition, because there are at least 25 different specificities at the *b* locus, any of which can function properly with any nonself allele (Puhalla, 1970; Silva, 1972). The *b* mating type locus was first cloned by Kronstad and Leong (1989) and sequenced by the same authors in 1990 (Kronstad & Leong, 1989, 1990). Schulz *et al.* (1990) reported the sequence of four alleles of the *b* locus (Schulz *et al.*, 1990). It was then shown that there were actually two individual genes at the *b* locus (Gillissen *et al.*, 1992). The two *b* locus-encoded products have been designated *bWest* (*bW*) and *bEast* (*bE*) (the original gene isolated by Kronstad and Leong). The bW and bE proteins contain prototypical members of two distinct subgroups of homeodomain proteins found at the mating type loci of basidiomycetes, called HD1 and HD2 respectively (Casselton & Olesnicky, 1998; Casselton, 2002) which interact with each other. Gillissen *et al.* (1992) provided the first genetic evidence of this interaction. Kamper *et al.* (1995) then showed that the bE and bW proteins from one allele could not physically interact, while those derived from dissimilar alleles could. For example, *bW1* and *bE2* gene products form a functional heterodimer while those of *bW1* and *bE1* do not (Kamper *et al.*, 1995). Thus this heterodimer is the functional component of the *b* mating type in *U. maydis*. The alignment of the predicted amino acid sequences of several alleles of the *bE* and *bW* genes revealed that each contains a variable N-terminal region, a central homeodomain-like motif, and a conserved C-terminal region (Kronstad & Leong 1990; Schulz *et al.*, 1990; Gillissen *et al.*, 1992). Specific variable regions of these proteins dictate whether or not there will be a successful interaction and nonself recognition (Yee & Kronstad, 1993, 1998; Kamper et al., 1995). Specificity is mediated by a 40 amino acid region in the variable N-terminal region of *bE*, as demonstrated by the construction of chimeric alleles of *bE1* and *bE2* and chimeric alleles of *bW1* and *bW2* (Yee & Kronstad, 1993), that can be further divided into two subdomains, heterozygosity at either of which is sufficient for compatibility (Yee & Kronstad, 1998). Point and insertional mutation analysis led to a model suggesting cohesive (hydrophobic or polar) interactions of the amino acid R-groups in determination of the ability of interallelic interaction of the *bE* and *bW* polypeptides (Kamper et al., 1995). The lack of such contacts yields an ineffective combination. A similar mechanism for the interaction of b type homeo-domain proteins (referred to as A in these organisms) occurs in holobasidiomycete mating systems (Kues et al., 1992; Asante-Owusu et al., 1996; Casselton & Olesnicky, 1998).

In the bipolar smuts such as *Ustilago hordei*, mating is controlled by a single genetic locus (*MAT*) with two alleles, *MAT*-1 and *MAT*-2. Genes equivalent to those of both *a* and *b* loci of *U. maydis* are in the *MAT* locus (Bakkeren & Kronstad, 1994). Additionally, the *a* and *b* genes are genetically tightly linked (Bakkeren & Kronstad, 1994). This genetic linkage and consequent lack of recombination is the cause for the bipolar mating system in these fungi. The mating type locus spans a 500 kb region showing suppression of recombination located on the largest chromosome of *U. hordei* (Lee et al., 1999).

The *b* locus controls events after cell fusion, necessary for establishment of the infectious, filamentous dikaryon. *lga2*, a gene of unknown function (Urban et al., 1996) located within the *a2* idiomorph, is directly and positively regulated by the *b*-heterodimer (Romeis et al., 2000). Employing inducible promoters to replace those native to *b*, both positively and negatively *b*-transcriptionally regulated genes, are being identified (Brachmann et al., 2001). However, deletion of a number of these genes did not produce any discernible effect on morphology or pathogenicity, indicating that the ones characterized so far do not play a major role in pathogenesis and development. Additionally, genes that when mutated induced expression in haploid cells of the *b* genes, as well as other dikaryon-specific genes, have been identified using another reporter system (Quadbeck-Seeger et al., 2000; Reichmann et al., 2002). Deletion of these genes affects teliosporogenesis and will be discussed in the corresponding section.

7.4 Penetration

7.4.1 Infection structures in U. maydis

Although mating and dikaryon formation can be induced *in vitro*, Snetselaar and Mims (1992) observed that this process occurs more consistently and rapidly on maize leaves. In the resulting hyphae, infection structures that penetrate plant cells directly through the cell walls are produced. All meristematic plant tissues above ground are susceptible to infection by *U. maydis*. In early work concerning the mode of entrance of *U. maydis* into maize, Walter (1934) observed that sporidial germ tubes displayed characteristic curling and swelling prior to penetrating leaf tissue directly through the cell walls. He noted that depressions in the leaf surface are the most common points of penetration. In such areas, hyphae bulge, flatten, and press close to the epidermis, however, unlike typical appressoria, in many cases these structures constrict again, grow farther, coil, and only then produce a penetration peg.

Snetselaar and Mims (1992) published a detailed ultrastructural study showing mating of compatible strains on maize leaf surfaces, formation of infection structures, and penetration of plant epidermis. With a combination of light and electron microscopy techniques, they observed that hyphae formed after mating of compatible haploid strains on plant surfaces produced swollen appressorium-like structures from which penetrating hyphae emerged (Fig. 7.4). These penetrated the cuticle and entered the underlying epidermal cell directly, but remained separated from the host cytoplasm by the invaginated host plasma membrane. Banuett and Herskowitz (1996) reported extensive branching of infection hyphae before penetration, and argued that branching may play a role in this process since each hyphal tip provides an opportunity for penetration. Hyphal entry through stomata has also been reported (Mills & Kotze, 1981; Banuett & Herskowitz, 1996), but the role of this in host colonization is still in doubt since subsequent penetration of plant cells was not confirmed.

Although all above-ground organs of maize can be infected by *U. maydis*, in the field the most prominent galls are commonly found on maize ears. Because in a given ear smutted and healthy kernels were found, it was assumed that kernels can be infected through stigmas (silks) and that this could be a frequent infection

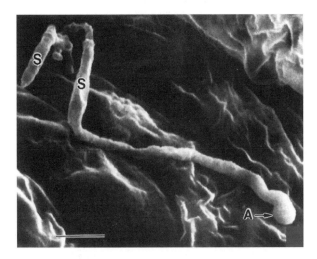

Figure 7.4 Scanning electron micrograph of fixed *Ustilago maydis* appressorium, 12–14 hours after inoculation. Bar = 10 μm. Reproduced from Snetselaar & Mims (1993) with permission of APS press.

pathway under natural conditions. To assess this hypothesis, Snetselaar and Mims (1993) inoculated stigmas with mixtures of compatible haploid strains or a solo-pathogenic diploid strain. Observations of inoculated material using light and electron microscopy techniques confirmed penetration of stigmas. In most penetration sites on the silks, the same appressorium-like structures previously described on leaf surfaces were encountered. Beneath these structures, penetrating hyphae often entered the stigma between epidermal cells and then turned sharply to enter one of the cells. Cross sections showed that the depressions between the two lobes of maize stigmas were the most frequent points of penetration.

Several general conclusions can be drawn regarding early infection events in the *U. maydis*–maize pathosystem. On both leaves and silks, infection structures are formed almost exclusively on epidermal cells that are immature and therefore present little mechanical resistance (Snetselaar & Mims, 1993). Infection structures produced by combinations of compatible haploid strains and diploid solopathogenic strains are indistinguishable. Although the fungus can proliferate near damaged host cells at the points of inoculation, no invasion of such cells has been observed (Snetselaar & Mims, 1992). Maize response to *U. maydis* penetration is not obvious; necrosis or other dramatic symptoms of plant response are not observed (Snetselaar & Mims, 1992). However, anthocyanin production that has been associated with a stress response to fungal penetration is a characteristic symptom of inoculated maize seedlings (Hana, 1929; Banuett & Herskowitz, 1996).

7.4.2   Can haploid strains penetrate plant cells?

According to Walter's and other early reports (Rowell & Devay, 1954; Fischer & Holton, 1957), single haploid sporidia can penetrate plant cells, although the ability of such strains to subsequently colonize plant tissue is very limited. Infection with certain haploid sporidia was reported to cause curl reactions in plant tissue that resembled the hyperplastic phase of normal infection (Rowell & Devay, 1954). This distortion of plant tissue was accentuated by co-inoculation of a single haploid line plus cell-free filtrate of a second compatible haploid strain. In control inoculations in which haploid sporidia plus sterile medium was added, no increase in symptoms was observed. Infection was demonstrated cytologically and by re-isolation of the original strains. We have observed that inoculation with a single haploid strain produces chlorosis indistinguishable from the early symptoms induced by inoculation with compatible strains. The fungus is filamentous in these lesions, however, the plant appears to recover from these lesions a few days after inoculation and further symptoms do not develop (unpublished observations, MDGP). Thus, haploid cells appear to be weakly pathogenic in the absence of a mating partner, being able to penetrate epidermal cells but unable to successfully colonize the plant. In contrast with these observations, hyphae formed by haploid sporidia seem unable to penetrate host cells when immature silks were inoculated (Snetselaar & Mims, 1993), suggesting that only some plant organs are susceptible to penetration by haploid sporidia. It could also be possible that this ability to penetrate plant tissue in the absence of a mating partner is a characteristic inherent only to certain strains.

7.4.3   Regulation of appressorium formation and plant penetration

Appressorium-like structures are observed only on plant surfaces, suggesting that either contact with the plant surface or signals coming from the host induce the formation of these structures in *U. maydis*. To date, little is known about the genetic regulation of formation of infection structures in this fungus. Tight regulation of the cAMP pathway is needed for normal disease development. Mutants with low PKA activity as a result of inactivation of genes in this pathway do not produce any symptoms in inoculated plants. That is the case when inoculating with compatible strains lacking adenylate cyclase activity (*uac1⁻*), the α subunit of the G-protein Gpa3 (*gpa3*) or the catalytic subunit of PKA (*adr1⁻*) (Barrett *et al.*, 1993; Gold *et al.*, 1994; Regenfelder *et al.*, 1997; Durrenberger, 2001). These results taken together show that PKA activity is required for the initial steps of infection. Interestingly, inactivation of any of these genes in a wild-type haploid background leads to filamentous growth showing that filamentous morphology is associated with, but not sufficient for, pathogenic behavior. On the other hand, levels of PKA activity above normal do not appear to affect the penetration process as shown by inoculation with strains with different degrees of PKA activation. For example, *ubc1* (regulatory subunit of PKA)-inactivated mutant strains can infect plants and proliferate within plant tissue, although they are unable to produce galls (Gold *et al.*, 1997; Kruger *et al.*, 2000). Also, dikaryons formed through mating of compatible, activated $gpa3_{Q206L}$ strains, which exhibit elevated PKA activity, could form appressoria and grow through the host epidermal layer normally, with progression of disease being affected only at later stages (Kruger *et al.*, 2000).

Formation of swollen appressorium-like structures and their production of invading hyphae which penetrate epidermal cells appear to be distinct steps in the infection process. Thus, a mutant strain defective in Kpp6 activity, a *b*-regulated MAP kinase, has recently been identified, which is able to produce appresoria but unable to penetrate plant cells (Brachmann *et al.*, 2003). Microscopic observations of plant surfaces after inoculation with compatible strains, both carrying an inactivated mutant allele, $kpp6^{T355A,Y357F}$, showed appressorium formation. However, from the majority of these appressoria only short filaments that failed to penetrate plant cells emerged.

7.4.4   Potential role of lytic enzymes in penetration

Snetselaar and Mims (1993) have suggested that the deformation of the plant cell wall around hyphae growing from cell to cell indicates mechanical penetration. Nevertheless, appressorium-like structures visualized in *U. maydis* are morphologically undifferentiated relative to appressoria formed by other pathogenic fungi and they are not melanized. It is unclear that these structures could penetrate the plant cuticle by mechanical forces alone and thus, other mechanisms such as production of lytic enzymes are potentially involved. However, little is known regarding the expression of lytic enzymes in *U. maydis* and their potential role in the infection process. Schauwecker *et al.* (1995) identified a gene, *egl1*, coding for an endoglucanase, which is not expressed in haploid cells but highly induced in the *b*-dependent filamentous form. However, mutants deleted for this gene were not affected in disease development. Cano-Canchola *et al.* (2000) measured pectate lyase, polygalacturonase, cellulase, and xylanase activities in both haploid and solopathogenic strains, grown on different carbon sources including plant tissues. They also investigated the induction of these lytic enzyme activities over time in inoculated plants. Pectate lyase activity was rapidly induced reaching a peak of activity approximately two days after inoculation, suggesting a potential role in the early steps of infection. All other enzyme activities were induced only at later stages of disease development. As discussed above, there is evidence that haploid strains are able to penetrate host cells although they do not progress further in the disease cycle. Interestingly, pectate lyase activity was induced in inoculations with both haploid and solopathogenic strains (but not in mock infections with water) while all other enzyme activities were induced only during infection with solopathogenic strains. Further studies to establish the role of lytic enzymes in *U. maydis* pathogenicity are necessary.

7.5   Colonization of maize tissue

7.5.1   *Proliferation of hyphae*

An approximate time frame of colonization in laboratory inoculations has been established. Chlorotic spots can be observed as soon as one day after inoculation (Banuett & Herskowitz, 1996). Chlorosis becomes extensive within two to three days following inoculation (Snetselaar & Mims, 1992, 1994; Banuett & Herskowitz, 1996), and anthocyanin production may occur away from the infection site (Banuett & Herskowitz, 1996). Cytological analyses of chlorotic areas indicate that *U. maydis* dikaryotic hyphae can grow inter- as well as intracellularly within host tissue (Christensen, 1963; Callow & Ling, 1973; Luttrell, 1987; Snetselaar & Mims, 1994; Banuett & Herskowitz, 1996). There seems to be no preference for hyphal invasion of particular cell types as hyphae grow within epidermal cells, parenchyma cells and vascular bundle cells within the first four days following inoculation (Snetselaar & Mims, 1994). The deformed rupture site between host cell walls suggests that *U. maydis* may rely on, at least partially, a mechanical means to break cell walls and proliferate from cell to cell (Snetselaar & Mims, 1994). Infected plant cells appear normal with the exception that chloroplasts are sometimes enlarged, containing many starch granules (Snetselaar &

Mims, 1994). Branch primordia that resemble the clamp connections of other basidiomycetes are observed in *U. maydis* hyphae during colonization (Fig. 7.5D) (Christensen, 1963; Snetselaar & Mims, 1994; Banuett & Herskowitz, 1996). However, these structures do not appear to serve the same function that true clamp connections do; they do not fuse with the adjacent hyphal cells (Banuett & Herskowitz, 1996) and nuclear migration into them has not been observed (Snetselaar & Mims, 1994). Additionally, there are some conflicting reports regarding the moment of appearance of these structures. While Banuett and Herskowitz (1996) observed them as soon as one day after inoculation, Snetselaar and Mims (1994) stated that clamp-like structures appear only in binucleate hyphae at later stages of colonization.

During colonization hyphae appear vacuolated and collapsed in some areas (Snetselaar & Mims, 1993; Banuett & Herskowitz, 1996). In contrast to branching that is observed only within plant tissue, migration of cytoplasm to the tip during hyphal extension, leaving behind areas of empty cells, is also a characteristic of the dikaryon formed *in vitro*. Identification of a protein, Kin2, involved in vacuole formation and cytoplasmic migration and the microscopic analysis of the dikaryon formed by fusion of compatible Δ*kin2* strains provided information on the process of plant colonization. *kin2* was identified in a candidate gene approach using degenerate primers designed to amplify a highly conserved region within the motor domain of kinesins (Lehmler *et al.*, 1997). This gene has been shown to encode the heavy chain of conventional kinesin. As previously noted (Day & Anagnostakis, 1971), observations of dikaryons formed *in vitro* by fusion of wild-type cells showed rapid tip growth with all the cytoplasm being moved to the tip, leaving

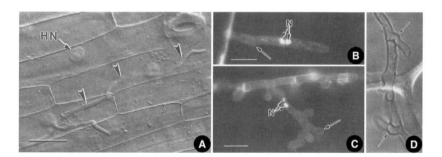

Figure 7.5 Light microscopy of *Ustilago maydis* hyphae in maize leaves. (A) DIC (differential interference contrast) micrograph of a typical unbranched hypha (arrowheads) in leaf epidermal cells, 24 hours after inoculation. HN, host nucleus. Bar = 50 μm. (B) An intracellular dikaryotic (N) hyphal tip, from a squash mount, had a swelling around the septum (arrow) which is reminiscent of clamp connection. Bar = 10 μm. (C) Squash mount; two nuclei (N) were in a lobed hyphal branch (arrow) inside a host cell. Bar = 10 μm. (D) In this hypha in a cleared leaf, the branches and clamps (arrows) seem to have grown in opposite directions. Bar = 10 μm. A and D are DIC micrographs while B and C are epifluorescence micrographs. Reproduced from Snetselaar & Mims (1994) with permission of Cambridge University Press.

behind empty septate cells. These observations were comparable to the situation described during colonization of the plant. In contrast, in dikaryons formed by fusion of Δ*kin2* mutants, hyphal structures remained short and filled with cytoplasm (Lehmler *et al.*, 1997). Further investigation of dikaryotic hyphae formed by Δ*kin2* mutants (Steinberg *et al.*, 1998) showed that they lack the large basal vacuole present in wild-type dikaryons. Instead they contain significantly more 200–400-nm vesicles scattered within the hyphae. Steinberg *et al.* (1998) postulated that Kin2 is involved in vacuole formation and that the accumulation of these vacuoles at the basal end of the tip plays a critical role in supporting cytoplasmic migration. Δ*kin2* strains showed a severe reduction in pathogenicity (Lehmler *et al.*, 1997) suggesting that the process of vacuolization and cytoplasmic migration plays an important role during normal development of the dikaryon *in planta*.

The extent of hyphal ramification may depend on the location of the plant infection. In stigma infections, hyphae grew rapidly without branching; in contrast, more hyphal branching was observed in ovaries and leaves (Snetselaar & Mims, 1993). Hyphae were generally observed to be oriented longitudinally, parallel to the vascular bundles in stigma infections (Snetselaar *et al.*, 2001). Yet they do not appear to grow preferentially in pollen tracts that surround vascular bundles. Interestingly, Snetselaar *et al.* (2001) have shown that the number of smutted kernels is significantly reduced if flowers are pollinated prior to infecting the silks with *U. maydis*. In microscopic observations of silks attached to ovaries that were pollinated before inoculation, the fungus still produced appressoria, penetrated plant cells, and could be seen growing in the silk toward the ovary. Hyphae proliferated normally above the abscission zone formed at the base of the silk to prevent multiple pollen tubes from reaching the ovary, however, very few were found growing beyond that area and into the ovary. The reduction of infected silk observed was postulated to be the result of the inability of *U. maydis* to grow through an area consisting of collapsed cells and disorganized tissue.

7.5.2   Interspecies signaling and colonization-specific fungal gene expression

Although mating can be induced *in vitro*, the resulting dikaryon cannot be maintained in culture. Additionally, structures characteristic of the development of the dikaryon during colonization, such as branching and clamp-like structures, are never observed *in vitro*. This strongly suggests that signals from the plant are important for maintenance of the dikaryon and hence successful colonization. A further piece of evidence to support the hypothesis that signals from the plant induce developmental processes in *U. maydis* is the fact that diploid solopathogenic strains homozygous at the *a* locus are unable to filament *in vitro* but they readily grow as filamentous structures in the infected plant (Banuett & Herskowitz, 1996). Furthermore, mutant strains lacking *ubc1* (regulatory subunit of PKA) are unable to filament *in vitro* but they can colonize the plant and grow filamentous in infected tissue (Gold *et al.*, 1997), suggesting that signals from the plant can provide a factor enhancing hyphal growth in these strains. These results are consistent with the observation that pathogenicity is always associated with filamentous growth in the plant. Developmental events in the proliferating hyphae may be triggered on cue in response to multiple plant signals. However, little is known about such signals and the genes involved in inducing developmental responses in the fungus.

The identification of fungal genes specifically expressed *in planta* is being pursued. Using differential display a maize-induced gene (*mig1*), which encodes a secreted cysteine-containing protein, was identified. *In planta* observations of *mig1* expression using GFP as a reporter system revealed its strong upregulation following penetration until the development of sporogenic hyphae (Basse *et al.*, 2000). Fluorescence was no longer detectable in pigmented teliospores. However, *mig1* appears not to play an essential role during colonization as its deletion did not compromise pathogenicity. The authors suggested that Mig1 could be an enzyme involved in nutrient uptake or acquisition. Alternatively, it may function as an elicitor (Basse *et al.*, 2000). Five additional *mig* genes similar to *mig1* have now been identified (Basse *et al.*, 2002). These genes (*mig2-1*, *mig2-2*, *mig2-3*, *mig2-4* and *mig2-5*) are arranged as direct repeats in a 7.1-kb cluster. Deletion of the entire cluster did not have an effect on pathogenicity. Features of the *mig* genes such as secretion, plant-inducible expression and an even number of cysteines are reminiscent of fungal *avr* genes (reviewed in Basse *et al.*, 2002). Thus far, a gene-for-gene interaction has not been observed in the *U. maydis*–maize interaction. Further characterization of the *mig* genes and other *mig*-like genes will undoubtedly provide key insight into the developmental program of *U. maydis* during pathogenesis.

7.5.3   The biotrophic interface

Little is known of the means by which *U. maydis* acquires nutrients during development inside the host. Analyses of electron micrographs of *U. maydis* invading host tissues have revealed the presence of an intracellular structure (Fig. 7.5C) that somewhat resembles haustoria described in other fungi. However, there is no clear demarcation of an interface analogous to the haustoria of rusts and other biotrophic fungi (Hahn & Mendgen, 2001). Furthermore, these structures are not consistently observed, and when present they are very irregularly branched (Luttrell, 1987; Snetselaar & Mims, 1994). Conceivably, this irregularly branched structure may correspond to the multilobed, sporogenous hyphae that appear intracellularly just prior to spore formation (Banuett & Herskowitz, 1996), rather than to structures formed to obtain nutrients. It is interesting to note that the observations of disease symptoms, such as chlorosis, well in advance of the hyphae of *U. maydis* are not uncommon (Callow & Ling, 1973), suggesting release of toxins and/or degradative enzymes by the fungus. It has also been established that when infected maize leaves are provided with the radioactive $[^{14}C]O_2$, ^{14}C assimilates are increasingly imported into infected tissue, even prior to gall formation (Billett & Burnett, 1978). Nutrient acquisition by the pathogen *in planta* is another area that requires research attention.

7.6 Gall formation and teliosporogenesis

7.6.1 Developmental stages during gall formation and teliosporogenesis

The most remarkable symptom induced in maize infected by *U. maydis* is hypertrophy of plant cells, which is observed macroscopically in the form of tumors. Within these tumors a massive proliferation of fungal sporogenous hyphae takes place that differentiate into diploid teliospores. An interesting aspect of this process is that *U. maydis* appears to be able to induce hypertrophy of plant cells even in the absence of extensive proliferation of fungal hyphae in the galls. Thus, Kruger *et al.* (2000) observed that inoculations with compatible $gpa3_{Q206L}$ mutant strains, in which cAMP signaling is enhanced through an activated allele of *gpa3*, induced tumors similar in size to those induced by wild-type inoculation. However, these tumors contained a much smaller amount of fungal material than normal galls induced by wild-type inoculation.

Fungal development in galls leading toward the formation of mature diploid teliospores is an ordered process with distinctive stages that have been studied in detail, especially in galls formed in leaves and stems. As the fungus proliferates within the plant tumors, an increase in hyphal branching is observed. Increasingly shorter branches are formed, especially at the hyphal tips. These changes appear to signal the switch from vegetative to sporogenous hyphae (Banuett & Herskowitz, 1996). At this stage, the fungus is embedded in a mucilaginous material presumably derived from hyphal walls (Christensen, 1932; Snetselaar & Mims, 1994). Banuett and Herskowitz (1996) observed that hyphae embedded in this matrix tend to stick together even when squeezed out of the plant cell. The tips of the hyphae become lobed followed by hyphal fragmentation that produces segments of one to several cells. There are contradicting reports on whether mycelia fragmentation occurs intra- or intercellularly. Several authors have reported that this process takes place mostly intercellularly (Mills & Kotze, 1981; Snetselaar & Mims, 1994). However, Banuett and Herskowitz (1996) showed in non-fixed samples that proliferation and fragmentation of hyphae occurred within tumor cells (Fig. 7.6, left panel). A possible explanation for these different observations could be artifacts in fixation techniques, but it is also possible that plant cells rupture due to the pressure exerted by fungal hyphae and mucilaginous material. At this stage, karyogamy has presumably taken place as DAPI staining shows a single nucleus per cell (Snetselaar & Mims, 1994; Banuett & Herskowitz, 1996). It has been assumed that for smut fungi meiosis directly follows nuclear fusion. However, in *U. maydis* karyogamy appears to take place early in the sporogenous hyphae and it has been suggested that after nuclear fusion, diploid, nuclei divide mitotically

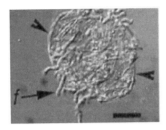 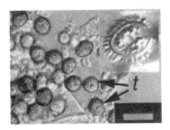

Figure 7.6 Hyphal, *worm-like* stage of teliospore formation and teliospores. (Left) A single plant tumor cell (arrowheads point to cell wall) full of hyphal fragments (f) gives the appearance of a sac full of worms, hence the designation worm-like stage. (Right) Mature teliospores (t) with echinate cell wall are shown here. The inset shows a higher magnification of a single mature teliospore. Scale bars = 20 μm. Reproduced from Banuett & Herskowitz (1996) with permission of The Company of Biologists LTD.

giving rise to masses of diploid, uninucleate cells (Snetselaar & Mims, 1994). Diploid nuclei are clearly capable of mitotic division because diploid-budding sporidia are easily isolated from immature galls (Holliday, 1961b; Snetselaar & Mims, 1992, 1994). Following fragmentation, hyphal cells become rounded and deposition of the secondary cell wall takes place. Cell walls are yellow-brown at first and become dark brown later, and show a characteristic echination in mature teliospores (Fig. 7.6, right panel). As the number of mature teliospores increases, tumors become dark. These processes are not synchronized in different tumors, and even in different parts of the same tumor teliospore at different stages of development are encountered (Banuett & Herskowitz, 1996). When silks are infected, tumors develop in the form of enlarged, hollow ovaries with galled ovary walls (Snetselaar *et al.*, 2001).

7.6.2 Genetic regulation of gall formation and teliosporogenesis

Heterozygosity at the *a* locus is not required for gall formation. Diploid strains of *U. maydis* have been isolated from nature (Christensen, 1963) and can also be forced in the laboratory by selection for prototrophy by complementing auxo-trophic markers (Holliday, 1961a, 1974; Puhalla, 1969). Pure cultures of diploids, heterozygous at the *a* and *b* loci (*a≠ b≠*), which induce tumors and produce mature teliospores, are referred to as diploid solopathogenic strains. Interestingly, diploids heterozygous at the *b* locus but homozygous at the *a* locus (*a= b≠*) are also solo-pathogenic. This result indicates that although heterozygosity at the *a* locus is required for mating, once mating has occurred the *a* locus is dispensable for patho-genicity. That is, the presence of *b* products encoded by alleles of different specifi-cities in a single cell is sufficient for completion of the disease cycle. This conclusion has been confirmed by artificial generation of heterozygosity at *b* in a haploid background by transformation with a second *b* allele (Kronstad & Leong, 1989; Bolker *et al.*, 1995a; Giasson & Kronstad, 1995). Strains constructed in this way are able to induce gall formation and produce mature teliospores. It is worthwhile to note here that galls produced by haploid or diploid solopathogenic strains are often fewer and smaller than those produced by inoculation with compatible haploid strains (Christensen, 1963; Holliday, 1974; Kronstad & Leong, 1989; Giasson & Kronstad, 1995), even when both types of hyphae contain the same genetic information (SEG, unpublished observations). These observations suggest that the presence of two separate nuclei during colonization by the dikaryon some-how enhances disease severity, presumably by a signaling process that takes place between the two haploid nuclei.

Interestingly, in a screen for mutants deficient in filament formation during mating, Banuett (1991) identified a mutation that bypasses the need for different *b* alleles to induce tumors. Inoculation with compatible *rtf1⁻* (repressor of tumor formation) haploids carrying the same *b* allele resulted in gall formation. However, the hypothesis that if the *rtf* mutation bypasses the need for different *b* alleles, then haploid *rtf1⁻* strains should be pathogenic was only partially fulfilled; only one of the *rtf1⁻* haploid strains isolated was able to induce galls and only in 20% of the inoculated plants. This result suggests that mating itself or different *a* alleles are required to induce full pathogenicity in an *rtf1⁻* background. The *rtf* mutation is recessive and appears to affect a gene near to, but distinct from, *b*.

To add complexity to this area of study, there is a report that suggests that external signals may provide the stimuli necessary to induce full pathogenicity in *U. maydis* in the absence of a *b*-heterodimer. Co-inoculations of certain combinations of haploid strains with lines of another smut fungus, *Sphacelotheca reiliana*, were reported to induce gall formation in maize (albeit smaller than normal) (Rowell & Devay, 1954). Analysis of these galls revealed occasional presence of structures that resembled mature teliospores and were able to germinate and produce sporidia. Interestingly, the progeny were always of the same mating type as the original haploid *U. maydis* strain used for the inoculations. It was concluded that gall formation occurred in the absence of fusion between the two organisms.

Gall induction, teliosporogenesis and cAMP signaling are intertwined proc-esses. The cAMP signaling pathway appears to play a critical role in gall formation especially during the induction of tumors, hyphal fragmentation, and the formation of pre-spores (Lee *et al.*, 2003). The fact that mutant strains with different levels of perturbation of cAMP signaling pathway arrest fungal development at different stages shows that changes in PKA activity play an important role in regulating these transitions. For example, mutant strains with constitutive PKA activity achieved by disruption of *ubc1*, the regulatory subunit of PKA, are able to infect and colonize plant tissue but unable to induce gall formation (Gold *et al.*, 1997). Kruger *et al.* (2000) showed that even moderate activation of the cAMP signaling pathway has an effect on tumor morphology and the amount of fungal material in the tumors formed. They identified mutant alleles in *gpa3* ($gpa3_{Q206L}$) and *ubc1* ($ubc1_{R321Q}$), both of which presumably result in activation of the cAMP pathway. However, in contrast to $\Delta ubc1$ mutant, infection with two compatible $gpa3_{Q206L}$ or $ubc1_{R321Q}$ strains led to 50 and 12% of infected plants developing tumors respect-ively. In addition to this decrease in disease severity in comparison with wild-type infections, tumors induced by these mutant strains were altered in morphology and did not produce teliospores. Detailed microscopic observations showed arrest of fungal development after hyphae had formed lobed tips. Kruger and co-workers concluded that these mutant strains represent different levels of cAMP pathway activation, with the activation due to $ubc1_{R321Q}$ allele intermediate between the $gpa3_{Q206}$ and $\Delta ubc1$ strains. These differences in cAMP pathway deregulation would account for the differences observed in disease severity among the mutants, with the lowest severity found in a mutant with the highest level of cAMP pathway activation. In contrast to the pathogenic behavior of mutants with an activated cAMP pathway, strains with low PKA activity due to inactivating mutations in genes of the cAMP pathway, such as adenylate cyclase (*uac1*) or the catalytic subunit of PKA (*adr1*), do not induce any symptoms in the host plant. Taken together, these results suggest that the level of cAMP and PKA activity is high during penetration and initial colonization of plant tissue, while a drop in PKA activity signals the induction of gall formation.

Mutations in genes encoding proteins that are putative targets for PKA have also been shown to affect gall and teliospore formation. A search for suppressors of the filamentous phenotype of the *adr1* mutant identified a gene, *hgl1*, whose product is a potential target for phosphorylation by PKA (Durrenberger, 2001). Inoculation of plants with compatible combinations of *hgl1* mutant strains produced galls, generally larger than those produced by wild-type strains but lacking mature, darkly pigmented teliospores. Close examination of these tumors showed that the fungus is unable to progress from the stage of hyphal fragmentation to teliospore development. Interestingly, haploid cells harboring this mutation become pigmented when cultured *in vitro*.

In addition to PKA, other protein kinases have been shown to be important for disease development and specifically for gall formation. Thus, mutants in *ukb1*, a gene that encodes a putative serine/threonine protein kinase (Abramovitch *et al.*, 2002), are able to colonize plant tissue but incapable of inducing gall formation. Plants inoculated with compatible haploid strains, both deleted for *ukc1*, another protein kinase, or diploid strains harboring the same mutation, did not produce galls (Durrenberger & Kronstad, 1999).

A number of additional genes involved in gall formation and teliosporogenesis have been identified. As previously discussed, the *b* genes encode homeodomain proteins that when derived from different alleles form a heterodimeric complex after cell fusion (Gillissen *et al.*, 1992; Kamper *et al.*, 1995). This complex controls the switch to the filamentous dikaryon and subsequent pathogenic development (Bolker *et al.*, 1995b). Thus, a subset of the genes that are either directly or indirectly under the regulation of the *b*-heterodimer may play a role in pathogenicity. Based on this rationale, Quadbeck-Seeger *et al.* (2000) searched for components of the *b*-dependent regulatory cascade, using a gene encoding an endoglucanase, *egl1*, specifically expressed in the dikaryon (Schauwecker *et al.*, 1995) as a reporter gene. Identification of mutations that lead to expression of *egl1* in haploid cells and complementation of one such mutant identified a gene coding for a protein, Rum1, which exhibited similarity to the human retinoblastoma binding protein 2. In a similar screen, Reichmann *et al.* (2002) identified a gene encoding Hda1, a protein with histone deacetylase activity. Deletion of either gene led to expression of several genes known to be *b*-regulated, in haploid cells as well as induction of the *bE* and *bW* genes themselves. These mutations had a similar effect on disease development; both blocked teliospore formation after karyogamy at the stage of hyphal fragmentation. However, detailed microscopic observation revealed that tumors induced by Δ*hda1* or Δ*rum1* strains have small and large areas of fragmented hyphae, respectively. Thus, the block in teliospore development appears to occur earlier in Δ*hda1* mutants than in Δ*rum1* mutants. As discussed above, it has been suggested that after karyogamy *U. maydis* proliferates within the galled tissue by mitotic division of the diploid cells before mature teliospores are formed. Reichmann *et al.* (2002) postulate that while in Δ*hda1* mutants teliospore development appears to be blocked right after karyogamy and before mitotic division of the diploid nuclei, Δ*rum1* mutants are able to divide mitotically after nuclear fusion but they are unable, at this stage, to reprogram their development in the direction of spore maturation. For both mutants, the lack of mature teliospores was postulated to be the result of the deregulation of a set of genes whose temporal or spatial misexpression prevents the completion of the disease cycle. It was also hypothesized that Hda1 functions in a complex with Rum1 and also independently from it (Reichmann *et al.*, 2002).

Other genes implicated in teliospore maturation include *fuz1*, a gene of unreported function. A *fuz1⁻* strain was identified in a screen for mutations that affected filamentous growth (Banuett & Herskowitz, 1996). Inoculation with compatible *fuz1⁻* strains produced very small galls lacking teliospores. Microscopic analysis of these galls showed that teliospore formation is arrested at the stage where the fungus becomes embedded in a mucilaginous matrix and starts to fragment. The mucilaginous matrix was not produced and the hyphae at this stage showed abnormal morphology (Banuett & Herskowitz, 1996).

7.6.3 Completion of sexual cycle in vitro

U. maydis is a semiobligate parasite; although it does not absolutely require its host for growth, it can complete its sexual cycle only within the plant. As indicated above, diploids can be forced *in vitro*. However, production of sexual spores capable of meiosis is achieved only within the galled tissue, suggesting the need for signaling from the plant. However, Ruiz-Herrera *et al.* (1999) have reported the completion of *U. maydis* cell cycle *in vitro* with the formation of teliospore-like structures by incubating compatible strains or diploids of *U. maydis* on maize callus. When plant callus and fungus were separated by a porous membrane, numerous round and double-walled structures somewhat resembling teliospores were observed in the fungus. Although these authors reported genetic recombination between the compatible haploid strains used to inoculate maize calli, segregation frequencies were very unlike those expected. Neither teliospore-like structures were observed nor were recombinants isolated, when haploid strains of the same mating type were used. It was also noted that the callus plant cells suffered morphological alterations only when compatible haploid strains or solopathogenic diploids were grown on them. Teliospore-like structures resembled teliospores only in the presence of a darkly pigmented double wall but they lacked the characteristic sculpted walls of mature teliospores called echination. Similar structures have been reported in culture due to stress (Durrenberger & Kronstad, 1999). Differences in the architecture of teliospore-like and mature teliospores induced in the plant, together with the unusual segregation frequencies reported, show the need for more work to confirm the completion of the sexual cycle of the fungus *in vitro*.

7.6.4 *Role of phytohormones in gall formation*

Infection of maize by *U. maydis* leads to obvious morphological alterations in the host plant. Because of this very specialized interaction it has been hypothesized that phytohormones, either produced by the fungus or induced in the plant, likely play a role in tumor formation. Moulton (1942) found that galled tissue contained an auxin in much higher amounts than uninfected tissue. This auxin was later identified to be indole-3-acetic acid (IAA) (Wolf, 1952; Turian & Hamilton, 1960). In more recent years, attempts have been made to identify mutant *U. maydis* strains impaired in their ability to produce IAA in culture, as a first step toward elucidating the putative role of this auxin in pathogenicity. Guevara-Lara *et al.* (2000) have characterized a number of mutants with reduced IAA production *in vitro*. Unfortunately, the only strain null for IAA production *in vitro* available to them, named *udi-1* (Sosa-Morales *et al.*, 1997), was auxotrophic for L-methionine. Because L-methionine auxotrophic strains are compromised for pathogenicity (Holliday, 1961a; Fischer *et al.*, 2001) and show reduced *in vitro* mating reactions (Fischer *et al.*, 2001), the effect of *udi-1* mutation on disease severity was confounded in this IAA⁻ strain. Attempts to obtain IAA⁻/meth⁺ segregants generated strains with only partial loss of IAA production *in vitro*. Thus, although inoculation with these strains suggests a role for IAA production in gall formation, these results were not conclusive as even wild-type strains exhibit variability in IAA production *in vitro*.

Another approach to study the role of IAA in gall formation has been to identify genes predicted to be involved in IAA production, and generate mutant strains deleted for such genes. Because *U. maydis* is able to produce IAA *in vitro* from tryptophan, the role of an indole-3-acetaldehyde dehydrogenase, Iad1, which was thought to be involved in this pathway, was investigated (Basse *et al.*, 1996). However, *iad1⁻* mutant strains were fully pathogenic and in this mutant background IAA was still produced *in vitro* from tryptophan, suggesting a different pathway for IAA synthesis.

7.7 Survival

Teliospores of *U. maydis* are believed to overwinter in the soil and plant residues (Christensen, 1963). Fischer (1936) showed that teliospores can be stored for two years without loss of viability. However, little is known about the survival of

U. maydis teliospores in soil under different environmental conditions. Unsterilized soils can be detrimental to teliospore germination (Christensen, 1963) and therefore, over-wintering in galls and decaying plant material may protect the spores from destructive compounds or microbes. Galls are water-repellent and therefore protect the teliospores from moisture (Christensen, 1963), which may reduce the chance of unwanted germination. Another survival strategy employed by *U. maydis* is the production of large numbers of spores. For example, a single large gall may contain over 200 billion teliospores (Christensen, 1963). Moreover, sporidia bud off and increase in numbers as well. Indeed, these secondary sporidia are believed to be the primary infectious agents (Alexopoulus *et al.*, 1996). Such large quantities of inoculum may also increase the likelihood of mating occurring near an appropriate infection court. Additionally, both teliospores and sporidia can be transported by wind over hundreds of miles, thus increasing the range over which *U. maydis* can survive. Sporidia of *U. maydis* can also be desiccated for one to five months without loss of viability, but alternate periods of freezing and thawing are harmful to sporidia (Christensen, 1963).

7.8   Genetic diversity in *U. maydis*

Current work with this fungus concentrates mainly on a small number of laboratory strains. However, early work showed the existence of a high degree of variability in the severity of infection caused by different strains. Stakman and Christensen (1927) reported high levels of variability in the number and size of galls produced by teliospore collections from different locations in the US when tested against ten maize lines. More recent studies have examined genetic variability in *U. maydis* using molecular techniques. Kinscherf and Leong (1988) reported considerable chromosome length polymorphisms among various laboratory and field strains of *U. maydis*. Among field and laboratory isolates, none shared exactly the same karyotype, even though the majority followed a similar general pattern. These authors found that laboratory strains present more chromosome length polymorphisms than field strains, probably due to chromosomal rearrangements. However, this appears not to affect their pathogenicity since these strains retained their ability to mate and produce galls in the plant. A high degree of polymorphism was also found at the chromosome ends when telomeric sequences were used as probes (Sanchez-Alonso *et al.*, 1996). Differences between a standard laboratory strain (FB2) and a wild isolated strain were evident when telomere-associated sequences were analyzed in detail (Sanchez-Alonso & Guzman, 1998). Using RFLP with a set of 23 different probes, Valverde *et al.* (2000) also reported a high level of diversity among isolates from five different locations in Mexico. Additionally, a high degree of variability among strains has also been found at the *b* locus. The N-terminal regions of the *bE*- and *bW*-encoded proteins are highly variable. At this locus, the requirement of heterozygosity for mating and pathogenicity is hypothesized to act as a strong selective force to maintain variability. To test this hypothesis, Zambino *et al.* (1997) designed a method to quickly assay *b* variations. The method uses RFLP of polymerase chain reaction (PCR) products amplified from the hypervariable regions of the *b* locus. In field isolates from four locations of Minnesota, 18 different *b* mating types were found to be widespread and occur at similar frequencies. The authors concluded that a high level of variability might be maintained at this locus even in local populations. In spite of these reports on variability, no pathogenic races have been designated.

7.9 Host resistance

The variation in susceptibility to *U. maydis* among maize hybrids is very high. The mechanisms of resistance to smut are not well understood, but various components of resistance have been postulated to be associated with physiological age and morphological characteristics of the maize variety (Christensen, 1963 and references contained therein). Sweet corn varieties are slightly more susceptible than field corn varieties because of low resistance in the germplasm from which modern sweet corn hybrids are derived (du Toit & Pataky, 1999a). Resistance to smut in field corn varieties has been increased by eliminating lines with susceptibility to *U. maydis* from breeding programs (du Toit & Pataky, 1999b). Field corn breeding has led to a reversal of the situation prevailing in mid-1930s where sweet corn hybrids were highly resistant while field corn varieties that were both high yielding and resistant were difficult to produce (Walter, 1934). Traditional germplasm-screening methods rely mostly on natural infection to assess resistance to smut. In such trials, environmental conditions, growth stage at which infection takes place, and the amount of inoculum are not controlled. Nonetheless, these breeding methods have produced resistant varieties and have apparently been successful in preventing large-scale, damaging epidemics.

New methods of inoculating maize with *U. maydis* in more controlled trials, such as the silk channel method, have been developed (Vincelli & Nesmith, 1994; du Toit & Pataky, 1999b). This standardization of the method of inoculation should improve the efficiency of breeding for common smut resistant varieties of sweet corn (Fig. 7.7). In addition, further advances can still be made using techniques such as: quantitative trait loci (QTL) approaches to determine the genetic basis of

Figure 7.7 Inoculating maize with *Ustilago maydis* using the silk channel method.

resistance and susceptibility of corn to *U. maydis*; determination and modeling of gene flow via monitoring population structure and dynamics of *U. maydis*; examination of population selection pressure and species divergence in the evolutionary relationship of *U. maydis* with the other members of the family Ustilaginae; and the comparison of relationships between populations of smut fungi to determine the rates of inbreeding and migration between them (Zambino *et al.*, 1997).

In recent years, the use of molecular techniques to induce resistance to smut has open new possibilities. A set of potential candidate genes to be used in this approach are the so-called *killer toxins* of *U. maydis* (Puhalla, 1968; Martinez-Espinoza *et al.*, 2002). The killer phenotype is based on the secretion of a toxin which kills sensitive *U. maydis* cells without being in direct contact with them (Schmitt & Breinig, 2002). The genetic bases for the three killer phenotypes in *U. maydis* are the cytoplasmic- ally inherited dsRNA viruses (Koltin & Day, 1976a; Park *et al.*, 1994). The viruses UmV-P1 (Park *et al.*, 1996), UmV-P4 (Park *et al.*, 1994), and UmV-P6 (Tao *et al.*, 1990) produce toxins KP1, KP4, and KP6 respectively. KP1 and KP6 are preprotox- ins that are precursors of the final α/β- heterodimeric protein toxin. Preprotoxins are processed via posttranslational modification in the endoplasmic reticulum and golgi apparatus and finally secreted as the functional heterodimer. KP4 is a monomeric protein toxin which requires no glycosylation or processing other than the signal peptide cleavage necessary for secretion (Park *et al.*, 1994). The three killer toxins have different specificities, such that a strain resistant to KP1 is susceptible to KP4 and KP6 (Al-Aidroos & Bussey, 1978). Resistance is due to recessive alleles of these genes and may be related to modification of membrane receptors for these toxins (Koltin & Day, 1975, 1976b; Finkler *et al.*, 1992). No strain has yet been found which is resistant to all the three toxins (Koltin & Day, 1975). Both KP4 and KP6 have been cloned, sequenced, and expressed in both heterologous and homologous systems (Kinal *et al.*, 1991, 1995; Park *et al.*, 1996). In tobacco, the KP4 toxin was found to be expressed at a much higher level than the KP6 toxin and functioned by blocking calcium-regulated signal transduction in *U. maydis* (Gage *et al.*, 2001). A cDNA, using the ubiquitin promoter to drive expression of the KP4 sequence from *U. maydis*, was transferred by biolistic transformation into several wheat varieties known to be susceptible to stinking smut (*Tilletia tritici*) infection (Clausen *et al.*, 2000). Using transgenic seeds from the varieties tested, it was shown that both inhib- ition of fungal growth and varying degrees of stable inheritance occurred depending on the variety. *In planta*, these investigators showed that the improvement of resistance was statistically significant after inoculation with stinking smut. Killer toxin gene expression in transgenic maize has not been reported but it will be inter- esting to see if these killer toxins can function as antifungal agents in the field.

7.10   Conclusions

Ustilago maydis provides an excellent model for the study of fungal development and host–pathogen interactions. There are many powerful tools that have been generated to assist researchers in their studies on *Ustilago* species. There are several research groups focused on various aspects of the biology of this fungus which have contributed significantly to our understanding. As described in this chapter, mating, morphogenesis, and pathogenicity have been the major foci of research attention to date. Much of the effort has indicated the intertwined nature of signaling pathways that crosstalk to influence all of these processes.

The genomic era promises more rapid and comprehensive analysis of the biology of *U. maydis* in all areas of interest. Recently, a cohesive international effort facilitated and managed by the Whitehead Institute for Genome Research has carried out the publicly funded genomic sequencing of *U. maydis*. Both Lion Bioscience AG/ Bayer CropScience and Exelixis Inc. had generated genomic sequences of this fungus and they donated this information to the Whitehead effort. With wide access to the *U. maydis* genomic sequence (http://www.genome.wi.mit.edu/annotation/ fungi/ustilago_maydis/index.html), methods for understanding the biology of this fascinating plant pathogenic fungus will become even more powerful. Carrying forth this momentum further on an international level is evident by the first international *Ustilago* conference (Kronstad, 2003) organized by Regine Kahmann and Flora Banuett. These same organizers are actively planning a second meeting in 2004. Discussions at the 2003 22nd Fungal Genetics Conference, Asilomar, were very encouraging regarding initiation of international cooperation on use of the genomic sequence. This cooperation is expected to initially involve genome annotation, production of full-length cDNAs, and functional genomics including expression analysis and a genome-wide gene deletion set.

A thus far understudied area with advancing tools is the plant side of the interaction. With over \$130 million invested by the US National Science Foundation's Plant Genome Research Grants on maize-genome-related projects since 1998, this situation is likely to change soon. The promise of tools such as large microarray chips for analysis of plant gene expression concurrently with that of the infecting fungus offers an obvious avenue of exploration.

Acknowledgements

The authors thank Dr Henry Ngugi for his excellent help in critical review of this chapter.

References

Abramovitch, R., Yang, G. & Kronstad, J. (2002) The *ukb1* gene encodes a putative protein kinase required for bud site selection and pathogenicity in *Ustilago maydis*. *Fungal Genet. Biol.*, **37**, 98–108.

Agrios, G.N. (1997) *Plant Pathology*, Academic Press, San Diego.

Al-Aidroos, K. & Bussey, H. (1978) Chromosomal mutants of *Saccharomyces cerevisiae* affecting the cell wall binding site for killer factor. *Can. J. Microbiol.*, **24**, 228–237.

Alexopoulus, C.J., Mims, C.W. & Blackwell, M. (1996) *Introductory Mycology*, Wiley, New York.

Andrews, D.L., Egan, J.D., Mayorga, M.E. & Gold, S.E. (2000) The *Ustilago maydis ubc4* and *ubc5* genes encode members of a MAP kinase cascade required for filamentous growth. *Mol. Plant Microbe Interact.*, **13**, 781–786.

Asante-Owusu, R.N., Banham, A.H., Bohnert, H.U., Mellor, E.J. & Casselton, L.A. (1996) Heterodimerization between two classes of homeodomain proteins in the mushroom *Coprinus cinereus* brings together potential DNA-binding and activation domains. *Gene*, **172**, 25–31.

Bakkeren, G. & Kronstad, J.W. (1994) Linkage of mating-type loci distinguishes bipolar from tetrapolar mating in basidiomycetous smut fungi. *Proc. Natl. Acad. Sci. USA*, **91**, 7085–7089.

Banuett, F. (1991) Identification of genes governing filamentous growth and tumor induction by the plant pathogen *Ustilago maydis*. *Proc. Natl. Acad. Sci. USA*, **88**, 3922–3926.

Banuett, F. (1998) Signalling in the yeasts: an informational cascade with links to the filamentous fungi. *Microbiol. Mol. Biol. Rev.*, **62**, 249–274.

Banuett, F. & Herskowitz, I. (1989) Different *a*-alleles of *Ustilago maydis* are necessary for maintenance of filamentous growth but not for meiosis. *Proc. Natl. Acad. Sci. USA*, **86**, 5878–5882.

Banuett, F. & Herskowitz, I. (1994) Identification of *fuz7*, a *Ustilago maydis* MEK/MAPKK homolog required for *a*-locus-dependent and -independent steps in the fungal life cycle. *Genes Dev.*, **8**, 1367–1378.

Banuett, F. & Herskowitz, I. (1996) Discrete developmental stages during teliospore formation in the corn smut fungus, *Ustilago maydis*. *Development*, **122**, 2965–2976.

Barrett, K.J., Gold, S.E. & Kronstad, J.W. (1993) Identification and complementation of a mutation to constitutive filamentous growth in *Ustilago maydis*. *Mol. Plant Microbe Interact.*, **6**, 274–283.

Basse, C.W., Lottspeich, F., Steglich, W. & Kahmann, R. (1996) Two potential indole-3-acetalde-hyde dehydrogenases in the phytopathogenic fungus *Ustilago maydis*. *Eur. J. Biochem.*, **242**, 648–656.

Basse, C.W., Stumpferl, S. & Kahmann, R. (2000) Characterization of a *Ustilago maydis* gene specifically induced during the biotrophic phase: evidence for negative as well as positive regulation. *Mol. Cell. Biol.*, **20**, 329–339.

Basse, C.W., Kolb, S. & Kahmann, R. (2002) A maize-specifically expressed gene cluster in *Ustilago maydis*. *Mol. Microbiol.*, **43**, 75–93.

Billett, E.E. & Burnett, J.H. (1978) The host–parasite physiology of the maize smut fungus, *Ustilago maydis*. II. Translocation of ^{14}C-labelled assimilates in smutted maize plants. *Physiol. Plant Pathol.*, **12**, 103–112.

Bolker, M., Urban, M. & Kahmann, R. (1992) The *a* mating type locus of *Ustilago maydis* specifies cell signaling components. *Cell*, **68**, 441–450.

Bolker, M., Bohnert, H.U., Braun, K.H., Gorl, J. & Kahmann, R. (1995a) Tagging pathogenicity genes in *Ustilago maydis* by restriction enzyme-mediated integration (REMI). *Mol. Gen. Genet.*, **248**, 547–552.

Bolker, M., Genin, S., Lehmler, C. & Kahmann, R. (1995b) Genetic regulation of mating and dimorphism in *Ustilag maydis*. *Can. J. Bot.*, **37**, S320–S325.

Brachmann, A., Weinzierl, G., Kamper, J. & Kahmann, R. (2001) Identification of genes in the *bW/bE* regulatory cascade in *Ustilago maydis*. *Mol. Microbiol.*, **42**, 1047–1063.

Brachmann, A., Schirawski, J., Muller, P. & Kahmann, R. (2003) An unusual MAP kinase is required for efficient penetration of the plant surface by *Ustilago maydis*. *EMBO J.*, **22**, 2199–2210.

Callow, J.A. & Ling, I.T. (1973) Histology of neoplasms and chlorotic lesions in maize seedlings following injection of sporidia of *Ustilago maydis* (Dc) Corda. *Physiol. Plant Pathol.*, **3**, 489.

Cano-Canchola, C., Acevedo, L., Ponce-Noyola, P., Flores-Martinez, A., Flores-Carreon, A. & Leal-Morales, C.A. (2000) Induction of lytic enzymes by the interaction of *Ustilago maydis* with *Zea mays* tissues. *Fungal Genet. Biol.*, **29**, 145–151.

Casselton, L.A. (2002) Mate recognition in fungi. *Heredity*, **88**, 142–147.

Casselton, L.A. & Olesnicky, N.S. (1998) Molecular genetics of mating recognition in basidiomycete fungi. *Microbiol. Mol. Biol. Rev.*, **62**, 55–70.

Christensen, J.J. (1932) Studies on the genetics of *Ustilago zeae. Phytopathologische Zeitschrift*, **4**, 129–188.

Christensen, J.J. (1963) Corn Smut caused by *Ustilago maydis. Monograph No. 2. Am. Phytopathol. Soc.*, Saint Paul.

Clausen, M., Krauter, R., Schachermayr, G., Potrykus, I. & Sautter, C. (2000) Antifungal activity of a virally encoded gene in transgenic wheat. *Nat. Biotechnol.*, **18**, 446–449.

Day, P.R. & Anagnostakis, S.L. (1971) Corn smut dikaryon in culture. *Nat. New Biol.*, **231**, 19–20.

du Toit, L.J. & Pataky, J.K. (1999a) Effects of silk maturity and pollination on infection of maize ears by *Ustilago maydis. Plant Dis.*, **83**, 621–626.

du Toit, L.J. & Pataky, J.K. (1999b) Variation associated with silk channel inoculation for common smut of sweet corn. *Plant Dis.*, **83**, 727–732.

Durrenberger, F. & Kronstad, J. (1999) The *ukc1* gene encodes a protein kinase involved in morphogenesis, pathogenicity and pigment formation in *Ustilago maydis. Mol. Gen. Genet.*, **261**, 281–289.

Durrenberger, F., Wong, K. & Kronstad, J.W. (1998) Identification of a cAMP-dependent protein kinase catalytic subunit required for virulence and morphogenesis in *Ustilago maydis. Proc. Natl. Acad. Sci. USA*, **95**, 5684–5689.

Durrenberger, F., Laidlaw R.D. & Kronstad, J.W. (2001) The *hgl1* gene is required for dimorphism and teliospore formation in the fungal pathogen *Ustilago maydis. Mol. Microbiol.*, **41**, 337–348.

Finkler, A., Peery, T., Tao, J., Bruenn, J. & Koltin, I. (1992) Immunity and resistance to the Kp6 toxin of *Ustilago maydis. Mol. Gen. Genet.*, **233**, 395–403.

Fischer, G.W. (1936) The longevity of smut spores in herbarium specimens. *Phytopathology*, **26**, 1118–1127.

Fischer, G.W. & Holton, C.S. (1957) *Biology and Control of the Smut Fungi*, Ronald Press, New York.

Fischer, J.A., McCann, M.P. & Snetselaar, K.M. (2001) Methylation is involved in the *Ustilago maydis* mating response. *Fungal Genet. Biol.*, **34**, 21–35.

Froeliger, E.H. & Leong, S.A. (1991) The *a* mating-type alleles of *Ustilago maydis* are idiomorphs. *Gene*, **100**, 113–122.

Gage, M.J., Bruenn, J., Fischer, M., Sanders, D. & Smith, T.J. (2001) KP4 fungal toxin inhibits growth in *Ustilago maydis* by blocking calcium uptake. *Mol. Microbiol.*, **41**, 775–785.

Garcia-Muse, T., Steinberg, G. & Perez-Martin, J. (2003) Pheromone-induced G(2) arrest in the phytopathogenic fungus *Ustilago maydis. Eukaryot. Cell*, **2**, 494–500.

Garrido, E. & Perez-Martin, J. (2003) The *crk1* gene encodes an Ime2-related protein that is required for morphogenesis in the plant pathogen *Ustilago maydis. Mol. Microbiol.*, **47**, 729–743.

Giasson, L. & Kronstad, J.W. (1995) Mutations in the *myp1* gene of *Ustilago maydis* attenuate mycelial growth and virulence. *Genetics*, **141**, 491–501.

Gillissen, B., Bergemann, J., Sandmann, C., Schroeer, B., Bolker, M. & Kahmann, R. (1992) A two-component regulatory system for self/non-self recognition in *Ustilago maydis. Cell*, **68**, 647–657.

Gold, S., Duncan, G., Barrett, K. & Kronstad, J. (1994) cAMP regulates morphogenesis in the fungal pathogen *Ustilago maydis. Genes Dev.*, **8**, 2805–2816.

Gold, S.E., Brogdon, S.M., Mayorga, M.E. & Kronstad, J.W. (1997) The *Ustilago maydis* regulatory subunit of a cAMP-dependent protein kinase is required for gall formation in maize. *Plant Cell*, **9**, 1585–1594.

Guevara-Olvera, L., Xoconostle-Cazares, B. & Ruiz-Herrera, J. (1997) Cloning and disruption of the ornithine decarboxylase gene of *Ustilago maydis*: evidence for a role of polyamines in its dimorphic transition. *Microbiology*, **143**, 2237–2245.

Guevara-Lara, F., Valverde, M.E. & Paredes-Lopez, O. (2000) Is pathogenicity of *Ustilago maydis* (huitlacoche) strains on maize related to *in vitro* production of indole-3-acetic acid? *World J. Microbiol. Biotechnol.*, **16**, 481–490.

Hahn, M. & Mendgen, K. (2001) Signal and nutrient exchange at biotrophic plant–fungus interfaces. *Curr. Opin. Plant Biol.*, **4**, 322–327.

Hana, W.F. (1929) Studies in the physiology and cytology of *Ustilago zeae* and *Sorosporium reilianum. Phytopathology*, **19**, 415–442.

Hartmann, H.A., Kahmann, R. & Bolker, M. (1996) The pheromone response factor coordinates filamentous growth and pathogenicity in *Ustilago maydis. EMBO J.*, **15**, 1632–1641.

Holliday, R. (1961a) The genetics of *Ustilago Maydis. Genet. Res.*, **2**, 204–230.

Holliday, R. (1961b) Induced mitotic crossing-over in *Ustilago maydis. Genet. Res.*, **2**, 231–248.

Holliday, R. (1974) *Ustilago maydis.* In: *Handbook of Genetics* (ed. R.C. King), pp. 575–595, Plenum, New York.

Huber, S., Lottspeich, F. & Kamper, J. (2002) A gene that encodes a product with similarity to dioxygenases is highly expressed in teliospores of *Ustilago maydis. Mol. Genet. Genomics*, **267**, 757–771.

Kahmann, R., Basse, C. & Feldbrugge, M. (1999) Fungal–plant signalling in the *Ustilago maydis*–maize pathosystem. *Curr. Opin. Microbiol.*, **2**, 647–650.

Kamper, J., Reichmann, M., Romeis, T., Bolker, M. & Kahmann, R. (1995) Multiallelic recognition: nonself-dependent dimerization of the bE and bW homeodomain proteins in *Ustilago maydis. Cell*, **81**, 73–83.

Keon, J.P.R., Jewitt, S. & Hargreaves, J.A. (1995) A gene encoding gamma-adaptin is required for apical extension growth in *Ustilago maydis. Gene*, **162**, 141–145.

Kinal, H., Tao, J. & Bruenn, J.A. (1991) An expression vector for the phytopathogenic fungus, *Gene*, **98**, 129–134.

Kinal, H., Park, C.M., Berry, J.O., Koltin, Y. & Bruenn, J.A. (1995) Processing and secretion of a virally encoded antifungal toxin in transgenic tobacco plants. Evidence for a Kex2p pathway in plants. *Plant Cell*, **7**, 677–688.

Kinscherf, T.G. & Leong, S.A. (1988) Molecular analysis of the karyotype of *Ustilago maydis. Chromosoma*, **96**, 427–433.

Koltin, Y. & Day, P.R. (1975) Specificity of *Ustilago maydis* killer proteins. *Appl. Microbiol.*, **30**, 694–696.

Koltin, Y. & Day, P.R. (1976a) Inheritance of killer phenotypes and double-stranded RNA in *Ustilago maydis. Proc. Natl. Acad. Sci. USA*, **73**, 594–598.

Koltin, Y. & Day, P.R. (1976b) Suppression of killer phenotype in *Ustilago maydis. Genetics*, **82**, 629–637.

Kronstad, J.W. (2003) Castles and cuitlacoche: the first international *Ustilago* conference. *Fungal Genet. Biol.*, **38**, 265–271.

Kronstad, J.W. & Leong, S.A. (1989) Isolation of two alleles of the *b* locus of *Ustilago maydis. Proc. Natl. Acad. Sci. USA*, **86**, 978–982.

Kronstad, J.W. & Leong, S.A. (1990) The *b* mating-type locus of *Ustilago maydis* contains variable and constant regions. *Genes Dev.*, **4**, 1384–1395.

Kruger, J., Loubradou, G., Regenfelder, E., Hartmann, A. & Kahmann, R. (1998) Crosstalk between cAMP and pheromone signalling pathways in *Ustilago maydis. Mol. Gen. Genet.*, **260**, 193–198.

Kruger, J., Loubradou, G., Wanner, G., Regenfelder, E., Feldbrugge, M. & Kahmann, R. (2000) Activation of the cAMP pathway in *Ustilago maydis* reduces fungal proliferation and teliospore formation in plant tumors. *Mol. Plant Microbe Interact.*, **13**, 1034–1040.

Kues, U., Richardson, W.V., Tymon, A.M., Mutasa, E.S., Gottgens, B., Gaubatz, S., Gregoriades, A. & Casselton, L.A. (1992) The combination of dissimilar alleles of the A alpha and A beta gene complexes, whose proteins contain homeodomain motifs, determines sexual development in the mushroom *Coprinus cinereus. Genes Dev.*, **6**, 568–577.

Laity, C., Giasson, L., Campbell, R. & Kronstad, J. (1995) Heterozygosity at the *b* mating-type locus attenuates fusion in *Ustilago maydis. Curr. Genet.*, **27**, 451–459.

Lee, N. & Kronstad, J.W. (2002) *ras2* controls morphogenesis, pheromone response, and pathogenicity in the fungal pathogen *Ustilago maydis. Eukaryot. Cell*, **1**, 954–966.

Lee, N., Bakkeren, G., Wong, K., Sherwood, J.E. & Kronstad, J.W. (1999) The mating-type and pathogenicity locus of the fungus *Ustilago hordei* spans a 500-kb region. *Proc. Natl. Acad. Sci. USA*, **96**, 15026–15031.

Lee, N., De'Souza, C.A. & Kronstad, J.W. (2003) Of smuts, blasts, mildews and blights: cAMP signaling in phytopathogenic fungi. *Annu. Rev. Phytopathol.*, **41**, 399–427.

Lehmler, C., Steinberg, G., Snetselaar, K.M., Schliwa, M., Kahmann, R. & Bolker, M. (1997) Identification of a motor protein required for filamentous growth in *Ustilago maydis. EMBO J.*, **16**, 3464–3473.

Loubradou, G., Brachmann, A., Feldbrugge, M. & Kahmann, R. (2001) A homologue of the transcriptional repressor Ssn6p antagonizes cAMP signalling in *Ustilago maydis*. *Mol. Microbiol.*, **40**, 719–730.

Luttrell, E.S. (1987) Relations of hyphae to host-cells in smut galls caused by species of *Tilletia, Tolyposporium*, and *Ustilago*. *Can. J. Bot.*, **65**, 2581–2591.

Martinez-Espinoza, A.D., Garcia-Pedrajas, M.D. & Gold, S.E. (2002) The *Ustilaginales* as plant pests and model systems. *Fungal Genet. Biol.*, **35**, 1–20.

Mayorga, M.E. & Gold, S.E. (1998) Characterization and molecular genetic complementation of mutants affecting dimorphism in the fungus *Ustilago maydis*. *Fungal Genet. Biol.*, **24**, 364–376.

Mayorga, M.E. & Gold, S.E. (1999) A MAP kinase encoded by the *ubc3* gene of *Ustilago maydis* is required for filamentous growth and full virulence. *Mol. Microbiol.*, **34**, 485–497.

Mayorga, M.E. & Gold, S.E. (2001) The *ubc2* gene of *Ustilago maydis* encodes a putative novel adaptor protein required for filamentous growth, pheromone response and virulence. *Mol. Microbiol.*, **41**, 1365–1379.

Mills, L.J. & Kotze, J.M. (1981) Scanning electron-microscopy of the germination, growth and infection of *Ustilago maydis* on maize. *Phytopathology*, **102**, 21–27.

Moulton, J.E. (1942) Extraction of auxin from maize, from smut tumors of maize and from *Ustilago zeae*. *Bot. Gazete*, **103**, 725–729.

Mueller, P., Feldbruegge, M. & Kahmann, R. (2001) A MAP kinase module regulating filamentous growth, mating and pathogenic development in *Ustilago maydis*. *Fungal Genetics Newsletter. XXI Fungal Genetics Conference*, Suppl. 48, 98.

Muller, P., Aichinger, C., Feldbrugge, M. & Kahmann, R. (1999) The MAP kinase *kpp2* regulates mating and pathogenic development in *Ustilago maydis*. *Mol. Microbiol.*, **34**, 1007–1017.

Muller, P., Katzenberger, J.D., Loubradou, G. & Kahmann, R. (2003) Guanyl nucleotide exchange factor Sql2 and Ras2 regulate filamentous growth in *Ustilago maydis*. *Eukaryot. Cell*, **2**, 609–617.

O'Donnell, K.L. & McLaughlin, D.J. (1984) Ultrastructure of meiosis in *Ustilago maydis*. *Mycologia*, **76**, 468–485.

Orth, A.B., Rzhetskaya, M., Pell, E.J. & Tien, M. (1995) A serine (threonine) protein-kinase confers fungicide resistance in the phytopathogenic fungus *Ustilago maydis*. *Appl. Environ. Microbiol.*, **61**, 2341–2345.

Park, C.M., Bruenn, J.A., Ganesa, C., Flurkey, W.F., Bozarth, R.F. & Koltin, Y. (1994) Structure and heterologous expression of the *Ustilago maydis* viral toxin Kp4. *Mol. Microbiol.*, **11**, 155–164.

Park, C.M., Berry, J.O. & Bruenn, J.A. (1996) High-level secretion of a virally encoded anti-fungal toxin in transgenic tobacco plants. *Plant Mol. Biol.*, **30**, 359–366.

Puhalla, J.E. (1968) Compatibility reactions on solid medium and interstrain inhibition in *Ustilago maydis*. *Genetics*, **60**, 461–474.

Puhalla, J.E. (1969) Formation of diploids of *Ustilago maydis* on agar medium. *Phytopathology*, **59**, 1771–1772.

Puhalla, J.E. (1970) Genetic studies on the *b* incompatibility locus of *Ustilago maydis*. *Genet. Res.*, **16**, 229–232.

Quadbeck-Seeger, C., Wanner, G., Huber, S., Kahmann, R. & Kamper, J. (2000) A protein with similarity to the human retinoblastoma binding protein 2 acts specifically as a repressor for genes regulated by the *b* mating type locus in *Ustilago maydis*. *Mol. Microbiol.*, **38**, 154–166.

Regenfelder, E., Spellig, T., Hartmann, A., Lauenstein, S., Bolker, M. & Kahmann, R. (1997) G proteins in *Ustilago maydis*: transmission of multiple signals? *Embo J.*, **16**, 1934–1942.

Reichmann, M., Jamnischek, A., Weinzierl, G., Ladendorf, O., Huber, S., Kahmann, R. & Kamper, J. (2002) The histone deacetylase Hda1 from *Ustilago maydis* is essential for teliospore development. *Mol. Microbiol.*, **46**, 1169–1182.

Romeis, T., Brachmann, A., Kahmann, R. & Kamper, J. (2000) Identification of a target gene for the bE–bW homeodomain protein complex in *Ustilago maydis*. *Mol. Microbiol.*, **37**, 54–66.

Rowell, J.B. (1955) Functional role of compatibility factors and an *in vitro* test for sexual compatibility of haploid lines of *Ustilago zea*. *Phytopathology*, **45**, 370–374.

Rowell, J.B. & Devay, J.E. (1954) Genetics of *Ustilago zeae* in relation to basic problems of its pathogenicity. *Phytopathology*, **44**, 356–362.

Ruiz-Herrera, J., Leon-Ramirez, C., Cabrera-Ponce, J.L., Martinez-Espinoza, A.D. & Herrera-Estrella, L. (1999) Completion of the sexual cycle and demonstration of genetic recombination in *Ustilago maydis in vitro*. *Mol. Gen. Genet.*, **262**, 468–472.

Sanchez-Alonso, P. & Guzman, P. (1998) Organization of chromosome ends in *Ustilago maydis*. RecQ-like helicase motifs at telomeric regions. *Genetics*, **148**, 1043–1054.

Sanchez-Alonso, P., Valverde, M.E., Paredes-Lopez, O. & Guzman, P. (1996) Detection of genetic variation in *Ustilago maydis* strains by probes derived from telomeric sequences. *Microbiology-UK*, **142**, 2931–2936.

Schauwecker, F., Wanner, G. & Kahmann, R. (1995) Filament-specific expression of a cellulase gene in the dimorphic fungus *Ustilago maydis*. *Biol. Chem.*, **376**, 617–625.

Schmitt, M.J. & Breinig, F. (2002) The viral killer system in yeast: from molecular biology to application. *FEMS Microbiol. Rev.*, **26**, 257–276.

Schulz, B., Banuett, F., Dahl, M., Schlesinger, R., Schafer, W., Martin, T., Herskowitz, I. & Kahmann, R. (1990) The *b* alleles of *Ustilago maydis*, whose combinations program pathogenic development, code for polypeptides containing a homeodomain-related motif. *Cell*, **60**, 295–306.

Silva, J. (1972) Alleles at the *b* incompatibility locus in Polish and North American populations of *Ustilago maydis* (DC) Corda. *Physiol. Plant Pathol.*, **2**, 333–337.

Sipiczki, M. (1988) The role of sterility genes (*ste* and *aff*) in the initiation of sexual development in *Schizosaccharomyces pombe*. *Mol. Gen. Genet.*, **213**, 529–534.

Snetselaar, K.M. & McCann, M.P. (1997) Using microdensitometry to correlate cell morphology with the nuclear cycle in *Ustilago maydis*. *Mycologia*, **89**, 689–697.

Snetselaar, K.M. & Mims, C.W. (1992) Sporidial fusion and infection of maize seedlings by the smut fungus *Ustilago maydis*. *Mycologia*, **84**, 193–203.

Snetselaar, K.M. & Mims, C.W. (1993) Infection of maize stigmas by *Ustilago maydis* – light and electron-microscopy. *Phytopathology*, **83**, 843–850.

Snetselaar, K.M. & Mims, C.W. (1994) Light and electron-microscopy of *Ustilago maydis* hyphae in maize. *Mycol. Res.*, **98**, 347–355.

Snetselaar, K.M., Carfioli, M.A. & Cordisco, K.M. (2001) Pollination can protect maize ovaries from infection by *Ustilago maydis*, the corn smut fungus. *Can. J. Bot.* **79**, 1390–1399.

Sosa-Morales, M.E., Guevara-Lara, F., Martinez-Juarez, V.M. & Paredes-Lopez, O. (1997) Production of indole-3-acetic acid by mutant strains of *Ustilago maydis* (maize smut huitlacoche). *Appl. Microbiol. Biotechnol.*, **48**, 726–729.

Spellig, T., Bolker, M., Lottspeich, F., Frank, R.W. & Kahmann, R. (1994) Pheromones trigger filamentous growth in *Ustilago maydis*. *EMBO J.*, **13**, 1620–1627.

Stakman, E.C. & Christensen, J.J. (1927) Heterothallism in *Ustilago zeae*. *Phytopathology*, **17**, 827–834.

Steinberg, G., Schliwa, M., Lehmler, C., Bolker, M., Kahmann, R. & McIntosh, J.R. (1998) Kinesin from the plant pathogenic fungus *Ustilago maydis* is involved in vacuole formation and cytoplasmic migration. *J. Cell Sci.*, **111**, 2235–2246.

Stevenson, J.A. & Johnson, A.G. (1944) The nomenclature of the cereal smut fungi. *Plant Dis. Rep.*, **28**.

Szabo, Z., Tonnis, M., Kessler, H. & Feldbrugge, M. (2002) Structure–function analysis of lipopeptide pheromones from the plant pathogen *Ustilago maydis*. *Mol. Genet. Genomics*, **268**, 362–370.

Tao, J.S., Ginsberg, I., Banerjee, N., Held, W., Koltin, Y. & Bruenn, J.A. (1990) *Ustilago maydis* Kp6 killer toxin – structure, expression in *Saccharomyces cerevisiae*, and relationship to other cellular toxins. *Mol. Cell. Biol.*, **10**, 1373–1381.

Tillet, M. (1766) Observation sur la maladie du Mais ou ble de Turquie. *Mem. Acad. Sci. Paris*, **1760**.

Trueheart, J. & Herskowitz, I. (1992) The *a* locus governs cytoduction in *Ustilago maydis*. *J. Bacteriol.*, **174**, 7831–7833.

Turian, G. & Hamilton, R.H. (1960) Chemical detection of 3-indolylacetic acid in *Ustilago zeae* tumors. *Biochimica Biophysica Acta*, **41**, 148–150.

Urban, M., Kahmann, R. & Bolker, M. (1996) The biallelic *a* mating type locus of *Ustilago maydis*: remnants of an additional pheromone gene indicate evolution from a multiallelic ancestor. *Mol. Gen. Genet.*, **250**, 414–420.

Valdivieso, M.H., Sugimoto, K., Jahng, K.Y., Fernandes, P.M. & Wittenberg, C. (1993) FAR1 is required for posttranscriptional regulation of CLN2 gene expression in response to mating pheromone. *Mol. Cell Biol.*, **13**, 1013–1022.

Valverde, M.H., Vandemark, G.J., Martinez, O., Paredes-Lopez, O. (2000) Genetic diversity of *Uslilago maydis* strains. *World J. Microbiol. Biotechnol.*, **16,** 49–55.

Vincelli, P. & Nesmith, W.C. (1994) *Reactions of sweetcorn hybrids to four diseases*, Cooperative Extension Service, University of Kentucky, Lexington, KY.

Walter, J.M. (1934) The mode of entrance of *Ustilago zeae* into corn. *Phytopathology*, **24**, 1012–1020.

Weinzierl, G., Leveleki, L., Hassel, A., Kost, G., Wanner, G. & Bolker, M. (2002) Regulation of cell separation in the dimorphic fungus *Ustilago maydis*. *Mol. Microbiol.*, **45**, 219–231.

Wolf, F.T. (1952) The production of indole acetic acid by *Ustilago zeae*, and its possible significance in tumor formation. *Proc. Natl. Acad. Sci. USA*, **38**, 106–111.

Wosten, H.A., Bohlmann, R., Eckerskorn, C., Lottspeich, F., Bolker, M. & Kahmann, R. (1996) A novel class of small amphipathic peptides affect aerial hyphal growth and surface hydrophobicity in *Ustilago maydis*. *EMBO J.*, **15**, 4274–4281.

Yee, A.R. & Kronstad, J.W. (1993) Construction of chimeric alleles with altered specificity at the *b* incompatibility locus of *Ustilago maydis*. *Proc. Natl. Acad. Sci. USA*, **90**, 664–668.

Yee, A.R. & Kronstad, J.W. (1998) Dual sets of chimeric alleles identify specificity sequences for the bE and bW mating and pathogenicity genes of Ustilago maydis. *Mol. Cell Biol.*, **18**, 221–232.

Zambino, P., Groth, J.V., Lukens, L., Garton, J.R. & May, G. (1997) Variation at the *b* mating type locus of *Ustilago maydis*. *Phytopathology*, **87**, 1233–1239.

8 *Blumeria graminis* f. sp. *hordei*, an obligate pathogen of barley

Maike Both and Pietro D. Spanu

8.1 Introduction

Barley powdery mildew is caused by *Blumeria graminis* f. sp. *hordei*, a member of the Erysiphaceae family of the ascomycete fungi. The taxonomy of the powdery mildews has been debated. One recent proposal has seen the Erysiphaceae divided into a number of tribes, one of which – the Blumeriae – includes the genus *Blumeria* (formerly grouped with *Erysiphe*) (Braun *et al.*, 2002). *B. graminis* is further subdivided into *formae speciales* which are able to infect particular species of grass; of these, *B. graminis* f. sp. *hordei* and *B. graminis* f. sp. *tritici* infect barley and wheat respectively and are of great agronomic importance. *B. graminis* f. sp. *hordei* is probably the best-studied member of this group and it is assuming the role of a model organism for investigating the biology of the powdery mildews.

The most evident symptoms of the disease are white powdery pustules on the plant surface (Figs 8.1A and 8.1B). Like all powdery mildews, *B. graminis* f. sp. *hordei* is an obligate biotrophic pathogen in that it requires a host for growth, development and completion of its life cycle. The fungus absorbs nutrients but does not kill the cells it penetrates. In heavy infections, this determines a significant sink for photosynthetically fixed carbon that ultimately reduces growth and crop yields. For this reason, in certain agricultural systems, the resources spent in controlling powdery mildews of cereals are significant (Hewitt, 1998).

The obligate nature of these biotrophic interactions results from an exquisite integration of the physiology of the pathogen and of the plant and an exceptionally high level of compatibility between the two organisms. This also means that it is not possible to grow *B. graminis* f. sp. *hordei* in axenic cultures, which limits the studies that are possible using biochemical and molecular biological approaches. Nonetheless, there has been significant progress in understanding the underlying mechanisms of growth, development and infection. A comprehensive treatise on the powdery mildews has been recently compiled and includes an in-depth analysis of many aspects of the biology of these fungi (Bélanger *et al.*, 2002). In this chapter, we review some of the recent advances in this field by describing briefly the life cycle of *B. graminis* f. sp. *hordei* and then focusing our discussion on individual stages of infection.

Figure 8.1 (A) Sporulating colonies of *B. graminis* f. sp. *hordei* infecting barley. (B) The typical powdery appearance of the pustule is due to the production of abundant conidia that are dispersed by air currents.

8.2   Life cycle – an overview

B. graminis f. sp. *hordei* can reproduce sexually and asexually. The sexual cycle results in the formation of cleistothecia that are able to survive an inclement environment (e.g. over winter), and produce ascospores when conditions are favourable. In addition to their role in survival, the sexual structures enable the fungus to undergo recombination and evolve rapidly. This is evident in the ability of *B. graminis* f. sp. *hordei* to adapt swiftly to strong selection forces such as those generated by large areas of host monoculture containing specific resistance genes against powdery mildews, and the extensive use of fungicides.

In contrast, the asexual cycle (Fig. 8.2) consists of very rapid succession of host infection structures and production of extremely abundant air-borne conidia whose dissemination results in very effective spread of the disease within a short time. The asexual life cycle has received most attention, possibly because of its importance for the spread of powdery mildew disease, but also because it proceeds in a very ordered and synchronous fashion, which facilitates biochemical and molecular biological studies of the mechanisms underlying development. For this reason, we will limit our review to a discussion of the asexual cycle.

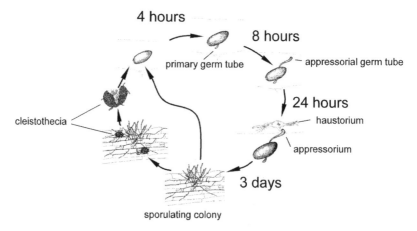

Figure 8.2 Diagrammatic representation of the life cycle of *B. graminis* f. sp. *hordei*.

The asexual life cycle begins when wind-dispersed conidia (Fig. 8.3A) land on the surface of a barley leaf. Within seconds of surface contact, the conidia become attached to the leaf by releasing an extracellular matrix (ECM) (Carver *et al.*, 1999; Wright *et al.*, 2002a). Two hours after that the conidia form primary germ tubes (Fig. 8.3B), which touch the leaf surface but normally do not elongate more than a few micrometers. The primary germ tube is thought to play a role in the uptake of water from the host and in perception of leaf-derived signals, possibly through a cuticular peg (Edwards, 2002). A second hypha, the appressorial germ tube, then emerges from the opposite end of the conidium. The appressorial germ tube elongates and four hours after inoculation it begins to swell to form an appressorium; eight to twelve hours after the initial contact, fully developed hooked appressoria are visible (Fig. 8.3C).

A peg grows underneath the appressorium (Fig. 8.3D) to penetrate directly through the cuticle and plant cell wall, by a combination of force and enzymatic breakdown of cutins and cellulose (Francis *et al.*, 1996; Pryce-Jones & Gurr, 1999). The hypha that develops inside the epidermal cell differentiates into a multidigitate haustorium but does not rupture the host's plasma membrane. The haustorium is a complex structure that consists of a central, elliptical, swollen body from the end of which a small number of finger-like processes emerge (Fig. 8.3E). The haustorium is responsible for the absorption of the majority of nutrients from the host plant. The extrahaustorial membrane, which surrounds the haustorium, is continuous with the plasma membrane of the epidermal cell. The extrahaustorial matrix lies between the extrahaustorial membrane and the fungal haustorial cell wall. When the haustorium is established, secondary hyphae grow on the plant surface and develop an extensive external mycelium from which further penetration pegs enter into epidermal cells to form other haustoria. Four days after inoculation, the external mycelium differentiates erect, aerial conidiophores that in turn produce large numbers of conidia (Figs 8.1B and 8.3F). The sheer quantity

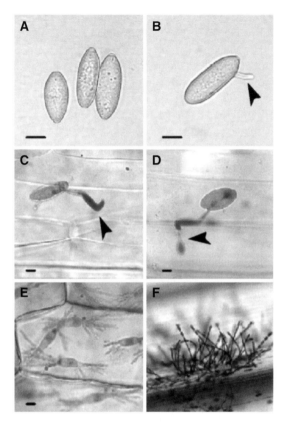

Figure 8.3 The main stages of development of the asexual cycle of *B. graminis* f. sp. *hordei.* (A) Asexual conidia; (B) Conidium germination on an artificial surface to produce a primary germ tube (arrow); (C) Conidium germinating on a barley leaf showing a primary and a secondary, appressorial germ tube (arrow) which is differentiating into a mature appressorium; (D) A hooked appressorium develops a penetration peg (arrow) from which a haustorium primordium is formed; (E) haustoria in the cells of infected barley epidermis; note that there can be more than one haustorium per cell. All bars = 10 μm. (F) Mature, sporulating colony showing erect conidiophores producing chains of conidia.

of wind-dispersed conidia – a pustule can produce up to 200 000 conidia (Hall *et al.*, 2000) – enables the disease to spread very rapidly through fields and over large areas.

8.3   Conidia and germination

The air-borne conidia produced by the external mycelia do not germinate until they come into contact with a solid substrate. Conidia appear not to remain viable for very long periods after their formation when they come in contact with a solid surface.

However, they can probably remain alive for long periods when suspended in air – as is demonstrated by their ability to be dispersed by wind over significant distances. In any case, they have the capacity to grow and produce fully formed infection structures. It is, therefore, assumed that conidia possess sufficient nutrients to sustain the early developmental stages prior to the formation of haustoria. It is not known exactly which nutrients are stored, but cytochemical techniques show that the cytoplasm of *B. graminis* f. sp. *hordei* conidia contains glycoproteins (Roberts *et al.*, 1996). Ungerminated conidia also contain a large number of RNA transcripts that encode genes necessary for metabolic processes involved in germination and early development. Thomas and co-workers first described these genes from this stage by characterising *expressed sequence tags* (ESTs) (Thomas *et al.*, 2001). These studies identified ESTs of genes that encode enzymes that function in all aspects of metabolism; the same authors then quantified the transcript levels by Serial Analysis of Gene Expression (SAGE) (Thomas *et al.*, 2002). There are large numbers of ribosomal proteins in ungerminated and germinated conidia. It is intriguing that genes involved in protein degradation are well represented too. This indicates that during germination there is active protein turnover involving synthesis and degradation that might allow amino acid recycling. There is also evidence that glycogen breakdown is involved in this process: a gene encoding a glycogen-debranching enzyme was detected, and its transcripts accumulated during germination. Glycogen is a storage compound in *Magnaporthe grisea* (Thines *et al.*, 2000) and it might be used as a storage compound in *B. graminis* f. sp. *hordei* as well. Germination also sees an increase in the abundance of genes encoding glycolytic enzymes; these then diminish during formation of the appressoria. Other genes whose transcripts were detected include chitinases, which may be involved in the degradation of the conidial cell wall during germ tube emergence and cell wall formation, and enzymes that catalyse the elongation of 1,3-β-glucans. The conidia of *B. graminis* f. sp. *tritici* contain fatty acids. Intriguingly, the fatty acid composition changes in time: fatty acids showed significant elongation and a drastic decrease in their total amount as the conidia aged (Muchembled *et al.*, 2000). It is unclear what the biological significance of these changes might be.

8.4   Surface perception and early events following contact with the surface

The conidium perceives its position relative to the solid surface: this is demonstrated by the observation that the point of emergence of the primary germ tube is determined within the first minute of contact. Primary germ tube emergence occurs close to the host leaf surface. The area of contact required to induce primary germ tube emergence is that of a needle tip or a spider's thread (Wright *et al.*, 2000). After surface perception, the conidium immediately releases an ECM. This is a fast response: if the surface is hydrophobic, the ECM is exuded 20 seconds after the first contact. On hydrophilic surfaces (glass or cellulose), this is less rapid (Carver *et al.*, 1999). The ECM is produced by the entire surface of the conidium, which becomes more hydrophilic, and then flows via capillary forces towards the substrate (Kunoh *et al.*, 1988; Carver *et al.*, 1999). ECM is also produced by the germ tubes and the appressoria.

One of the simplest functions of the ECM might be to attach the conidium to the leaf prior to infection. The experimental evidence to support this is a correlation between the quantity of ECM deposited and the force of attachment to a substratum (Wright *et al.*, 2002a). On glass and silanized glass, the ECM forms a pad under the conidium. This pad is larger on hydrophobic, silanized glass than on untreated hydrophilic glass, and the planar nature of the artificial surface causes the liquid to coalesce. On the host plant, however, only small ECM deposits appear where projections of the conidium touch the plant surface (Wright *et al.*, 2002b).

The chemical composition of the ECM depends on which organ it is produced by: the ECM formed by germ tubes and appressoria contains higher concentrations of protein than that formed on the surfaces by the conidium (Carver *et al.*, 1999). This result might explain the finding that ECM exuded from conidia can be washed off, whereas germ tube exudates are insoluble (Carver *et al.*, 1995; Nicholson & Kunoh, 1995). There is evidence that the ECM is required for the further development of the germ tubes (Nicholson *et al.*, 1993). Biochemical analysis indicates that esterases are present in the ECM, and that they are discharged in two stages during the first 15 minutes of development. The ECM deposited by conidia eroded the barley cuticle (Kunoh *et al.*, 1990). It is probable that this effect is due to the presence of cutinase activity in the ECM (Pascholati *et al.*, 1992). Given the small volumes of ECM deposited on the host surface, a complete degradation of the cuticle for penetration purposes seems unlikely. Instead, the ECM and the enzymes in it might serve as a means of detecting the surface by enzymatic degradation of some of its components. Degradation of cutins could be involved in recognition of the host: the cutinase can release cutin monomers which then determine the point of germ tube emergence (Carver & Ingerson, 1987) and induce appropriate development of appressoria (Pascholati *et al.*, 1992; Francis *et al.*, 1996).

Once the conidium is firmly attached to the substratum, a peg emerges from the primary germ tube, breaks the host cuticle and touches the plant cell wall. This is thought to be the means by which *B. graminis* f. sp. *hordei* takes up water and low molecular weight compounds from the plant surface (Edwards, 2002). These might also act as host recognition factors and cause directed growth of the germ tubes (Nielsen *et al.*, 2000). Full development that leads to the formation of functional haustoria has only been observed on the surfaces of host plants. On artificial surfaces, *B. graminis* f. sp. *hordei* conidia germinate and can form some structures that resemble those of a normal infection but development is arrested; in some cases, abnormal structures such as multiple primary germ tubes or elongated appressorial germ tubes are produced. The nature of the surface determines which of these are formed and how far the germling develops before growth stops. For example, on cellulose and host plant epidermis, appressorium formation can normally be observed (Carver & Ingerson, 1987) – these are termed *inductive surfaces*. On other media, e.g. glass and agar, development is arrested at an early stage after formation of the appressorial germ tube: these surfaces are usually referred to as *non-inductive*. The mechanisms that sense the surface appear to have a very refined capacity of discrimination. For example, two types of cellulose membrane support different degrees of development (Kinane *et al.*, 2000). One type of cellulose (Cellulose 1) allows similar rates of formation of the appressorial germ tube and appressorium as that observed on the plant surface, even if development is slower than that on the plant. On the other type (Cellulose 2) *B. graminis* f. sp. *hordei* forms smaller numbers of appressorial germ tube and appressoria: Cellulose 2 is, therefore, *semi-inductive*.

There is evidence that the chemical composition of the surface is important: several compounds of the host plant surface are linked to the successful development of infection. For example, cutins elicit appressorial germ tube formation and penetration. When glass slides are coated with cutin monomers, the proportion of conidia that formed appressorial germ tubes increases. Moreover, ebelactones, inhibitors of esterases and *B. graminis* f. sp. *hordei* cutinases, inhibit appressorial germ tube formation on cutin-coated slides. The ebelactones also reduce the number of appressorial germ tubes formed, and infection of barley. Thus, *B. graminis* f. sp. *hordei* degrades the plant cuticle with cutinase, and utilises the cutin monomers as signals necessary for further development of infection structures (Francis *et al.*, 1996). Tsuba and co-workers found evidence of an elicitor from the barley cuticle that induces appressorium formation in *B. graminis* f. sp. *hordei* (Tsuba *et al.*, 2002). By examining the appressorium-inducing capacity of several fractions of barley cuticle waxes, an aldehyde-containing fraction was identified as the most inductive. The major component of this fraction is a C_{26} aldehyde. Non-host plants, such as cabbage, have similar aldehydes but they are of different chain lengths (C_{28} and C_{30}) and do not induce appressorium formation to the same degree. Another clue to the importance of the contact surface chemistry was found: film-forming agents applied to barley leaves reduce both conidial germination and appressorium formation (Sutherland & Walters, 2002).

B. graminis f. sp. *hordei* does not only perceive soluble signals on the host surface, but can discriminate surface hydrophobicity. Conidia develop further on a hydrophobic surface compared to a hydrophilic one (Carver *et al.*, 1996). This might be linked to how tightly the conidia adhere to the substratum. Binding of conidia to hydrophobic surfaces (e.g. dry barley coleoptile) is stronger and results in a higher proportion of germlings that develop appressoria compared to those that adhere to a hydrophilic substrate (e.g. pre-wetted barley coleoptile). A polycation (poly-L-ornithine) that increases the attachment of conidia to coleoptiles in wet conditions also increases the frequency of appressorium formation (Yamaoka & Takeuchi, 1999).

8.5 Signal transduction during early development of *B. graminis* f. sp. *hordei*

In the preceding sections of this chapter, we have discussed how *B. graminis* f. sp. *hordei* perceives the surface on which it germinates and produces infection structures. Only if the germling colonises an appropriate host can it continue its full differentiation programme that ultimately lead to the establishment of infection and the completion of the asexual life cycle. It is evident that *B. graminis* f. sp. *hordei* needs to transduce the primary signals detected from the surface of the barley leaf into the expression of functions that leads to the formation of a fully functional, compatible interaction. In recent years, some attention has been devoted to understanding these signal transduction pathways. The experimental strategies used include: measurement of the levels of known messengers during development, use of inhibitors and activators of known signal transduction enzymes, cloning and analysis of the expression of genes that encode these signal transduction elements, and using these genes to complement signal transduction mutants in other fungi. Only one developmental programme, which necessarily passes through infection of barley, leads to the completion of the life cycle. The consequence of this is that mutational and genetic analysis adopted in other fungi is not possible in *B. graminis* f. sp. *hordei*. In the following sections, we review the evidence that some of the known developmental signalling pathways are active and functional in *B. graminis* f. sp. *hordei*.

8.5.1 *cAMP and PKA*

The signalling pathway involving cyclic AMP (cAMP) and its downstream target protein kinase A (PKA) are well established and are highly conserved elements in the transduction of signals in fungi and animals. In a general model, an extracellular signal is perceived from the cell by a specific receptor, coupled to a stimulatory G-protein. This activates an adenylate cyclase to synthesise cAMP from ATP. cAMP can, in turn, activate PKA by binding to its two regulatory subunits which leads to the dissociation of the tetrameric PKA, release and activation of the two catalytic subunits. The catalytic subunits are then able to phosphorylate specific substrate molecules (Alberts *et al.*, 2002).

The cAMP and PKA signalling pathway is connected with infection and pathogenicity in various plant pathogens (Kronstad, 1997). For example, the corn smut pathogen *Ustilago maydis* requires the cAMP pathway for successful infection of its host and for the switch from budding growth to a filamentous, infectious dikaryon (Gold *et al.*, 1994; Kronstad, 1997). In the rice blast fungus *M. grisea*, cAMP and PKA are essential to the development and function of appressoria (Mitchell & Dean, 1995; Xu *et al.*, 1997). cAMP is also considered a pathogenicity factor in the chestnut blight fungus *Cryphonectria parasitica*, where the levels of cAMP influence the virulence of the fungus (Chen *et al.*, 1996). These facts and the finding that signalling pathways are highly conserved in eukaryotes have led to the assumption that cAMP could be a universal requirement for infection and virulence. The following studies have found evidence that the cAMP/PKA-dependent pathway is active in *B. graminis* f. sp. *hordei*. A catalytic subunit of *B. graminis* f. sp. *hordei* cAMP-dependent PKA (cPKA) was identified and characterised. The gene transcript is present in ungerminated conidia and in conidia germinating on barley. PKA enzyme is also active in ungerminated conidia (Hall *et al.*, 1999). The *cpka* gene of *B. graminis* f. sp. *hordei* restores pathogenicity in a mutant of the rice blast fungus *M. grisea* that lacks a *cpkA* (Bindslev *et al.*, 2001). This is an important finding, because, apart from demonstrating that the *B. graminis* f. sp. *hordei* cPKA is biologically functional, it also indicates that signalling components regulating pathogenicity-related events are conserved in different plant pathogenic fungi.

The levels of cAMP and PKA increase just before germination and formation of the primary germ tube. A second peak of PKA activity preceded emergence of the appressorial germ tube. Activators of adenylate cyclase such as forskolin or cholera toxin induce higher levels of cAMP in the germling. These activators also increase the rate of primary germ tube formation. Thus, the first peak of activity is functionally correlated with germination and development of the primary germ tube. However, if the activators are added immediately after inoculation, development of appressorial germ tube is inhibited. This inhibition is reversed (and appressorial germ tube formation is stimulated) if the activators are added after one hour from inoculation. Thus the presence of two peaks of cAMP and PKA and an intervening period of lower activity are necessary for appropriate differentiation of the appressorial germ tube. There is no evidence, however, that cAMP and PKA are involved in subsequent development and differentiation of appressoria (Kinane *et al.*, 2000). These results were in part corroborated by the measurement of *B. graminis* f. sp. *hordei* cPKA mRNA using RT-PCR (Zhang *et al.*, 2001). The highest transcript levels are found when primary germ tubes are formed, but they are lower in ungerminated conidia and in the appressorial germ tube and appressorium stages. Although higher levels of mRNA do not accompany the second peak of cPKA activity found by Kinane & Oliver (2000), it is possible that the levels of activity are the result of enzyme activation (by cAMP) rather than the result of an increase in mRNA accumulation.

8.5.2   *MAP kinases*

Another well-established signal transduction pathway involves the mitogen-activated protein kinase (MAPK). MAPK, like PKA, is a member of the serine/threonine kinase family. In general, a signal is perceived by a receptor, which activates the first component of the MAPK cascade: a MAP kinase kinase kinase (MAPKKK). This passes the signal via phosphorylation to the MAP kinase kinase (MAPKK), which in turn activates the MAPK. The target of the MAPK can be a protein or a transcription factor. This three-component MAPK cascade is highly conserved in eukaryotic organisms (Banuett, 1998). MAP kinases have been identified in several fungal pathogens, and are functionally linked to events such as appressorium formation and penetration (*M. grisea*) and virulence (*U. maydis*) (Xu, 2000). The importance of the MAPK signalling pathway to other fungi as well, such as *Botryotinia fuckeliana*, has led to the belief that MAP kinases might also be involved in *B. graminis* f. sp. *hordei* development.

Using degenerate primers, Zhang and Gurr (Zhang & Gurr, 2001) cloned and characterised two MAPK genes (*mpk1* and *mpk2*) from *B. graminis* f. sp. *hordei*. A semi-quantitative estimate of MAPK mRNA based on RT-PCR showed that expression levels of *mpk1* increase during germination of *B. graminis* f. sp. *hordei* conidia; *mpk2* transcripts were not detected in ungerminated conidia, but appear at the primary germ tube and the appressorial germ tube and appressorium stages. Thus, MAP kinases might be involved in controlling germination and differentiation of *B. graminis* f. sp. *hordei*. Kinane and Oliver (Kinane & Oliver, 2000) measured endogenous MAPK activity and observed the influence of inhibitors of the MAPK pathway on *B. graminis* f. sp. *hordei* conidia placed to germinate on inductive Cellulose 1. They also determined the influence of several known activators of the MAPK pathway on conidia germinating on the semi-inductive Cellulose 2. MAPK activity in conidial extracts increases immediately after surface contact and remains at high levels for eight hours, after which the activity drops by about 50%. The MAPK activators sphingosine and PAF-16 stimulate appressorial germ tube formation and appressoria development on Cellulose 2 without affecting germination. Conversely, the MAPK inhibitor PD 98059 reduces the frequency of appressorial germ tube formation and appressoria development. In this case, there is also no effect on germination.

There is a connection between the MAPK and the cAMP/PKA pathways: activators of the cAMP pathway (cholera toxin, pertussis toxin and forskolin) increase significantly the MAPK activity in germinating spores, whereas the activators of MAPK do not affect cAMP levels. The simplest explanation for this is that the cAMP/PKA element is upstream of a MAPK cascade. In addition, cAMP and MAPK activators, when applied together, increase the rates of appressorial germ tube and appressorium differentiation on semi-inductive Cellulose 2 (Kinane & Oliver, 2003).

8.5.3   PKC

The protein kinase C (PKC) was originally so named because it was found to be a Ca^{2+}-dependent kinase. However, PKC now includes a large family of serine/ threonine kinases with different properties: the cPKC or classical Ca^{2+}-dependent PKC; the nPKC or novel PKC which act independently of Ca^{2+}; and the aPKC or atypical PKC which show a certain sequence homology to cPKC and nPKC. PKC genes similar to the mammalian nPKCs have been isolated from several fungi, amongst which are also the plant pathogens *M. grisea* and *B. fuckeliana*. In *M. grisea*, PKC functions in appressorium formation (Thines *et al.*, 1997a,b).

Two PKC genes, *pkc1* and *pkc*-like, have been identified in *B. graminis* f. sp. *hordei* (Zhang *et al.*, 2001). Semi-quantitative RT-PCR was used to assay the transcript abundance during germination and early development. *pkc1* mRNA is present at all stages tested and is slightly elevated during formation of the primary germ tube. The *pkc*-like mRNA follows a similar pattern of expression as that seen for PKA: it is detected only during primary germ tube formation but not at other stages (Zhang *et al.*, 2001; Wheeler *et al.*, 2003). PKC enzyme activity is detectable in both ungerminated conidia and germlings with a primary germ tube. An activator of PKC, the phorbol ester 12-myristate 13-acetate, enhances PKC activity *in vitro* and increases the proportion of conidia forming appressorial germ tubes. Inhibitors of mammalian PKC do not have any effect on the *B. graminis* f. sp. *hordei* PKC activity *in vitro* or on development (Zhang *et al.*, 2001). These results are a clue that *B. graminis* f. sp. *hordei* PKC plays a role in differentiation of the appressorial germ tube, but further research is needed to confirm and extend these findings.

8.5.4   G-proteins

Ras proteins belong to the family of monomeric GTPases. They transmit signals from the surface of the cell. Ras is activated by guanine nucleotide exchange factors, which stimulate the dissociation of GDP and the uptake of GTP. In the GTP-bound form, Ras is activated and can relay the signal to other signalling proteins. GTPase-activating proteins (GAPs) inactivate Ras by stimulating the hydrolysis of bound GTP back to GDP. Thus, Ras functions as a switch by cycling between these two stages (Alberts *et al.*, 2002).

Research on *B. graminis* f. sp. *hordei* strains that are resistant to the fungicide quinoxyfen shows that differentiation in *B. graminis* f. sp. *hordei*, GTP-binding proteins and Ras-type signalling are related (Wheeler *et al.*, 2003). Quinoxyfen is a fungicide that is active particularly against powdery mildews because it reduces germination, induces formation of abnormally long and thin appressorial germ tubes, and decreases conidia production. Quinoxyfen-resistant *B. graminis* f. sp. *hordei* mutants have arisen in the field. Differential Display RT-PCR identified a gene that is down-regulated in one of the resistant mutants. The gene encodes a protein with homology to GAPs. Normally, *B. graminis* f. sp. *hordei* GAP mRNA is present in the conidia and then decreases after germination and is no longer detected one day after inoculation. At later stages of colony development, when the fungus is sporu-lating abundantly, GAP mRNA is again detectable. Quinoxyfen appears to prevent this decrease in GAP mRNA in the wild-type *B. graminis* f. sp. *hordei*, whereas quinoxyfen-resistant mutants behave as if the fungicide were not present. The authors suggest that decrease in GAP mRNA causes an activation of Ras function and a transmission of stimuli that lead to normal differentiation (Wheeler *et al.*, 2003).

8.6   Penetration

B. graminis f. sp. *hordei* invades the tissue of the host plant by direct penetration of epidermal cells. The hyphae that enter the cells develop from surface appressoria.

The plant cell wall is ruptured and a penetration peg grows through it but does not breach the plants' plasma membrane. There has been some debate as to whether the mechanism *B. graminis* f. sp. *hordei* employs to breach the epidermal cell wall relies on chemical degradation and loosening of cell wall components or whether the penetration peg uses physical force to do so. Over thirty years ago, ultrastructural data suggested that *B. graminis* f. sp. *hordei* uses both chemical degradation of the cell wall components and mechanical pushing apart of the fibrils evident in the papilla (Edwards & Allen, 1970). More recently, this view has received further support: *B. graminis* f. sp. *hordei* has enzymes capable of degrading cell walls (Kunoh *et al.*, 1990; Francis *et al.*, 1996) and it expresses genes homologous to pectinases (Suzuki *et al.*, 1998). Indirect evidence that cellulases are produced by *B. graminis* f. sp. *hordei* was obtained by immunolocalisation using monoclonal antibodies that recognise cellobiohydrolases from *Trichoderma* (Pryce-Jones & Gurr, 1999); although it is possible that the antibodies used might cross-react with epitopes on molecules other than cellulolytic enzymes, the data suggest that *B. graminis* f. sp. *hordei* does indeed secrete limited amounts of hydrolases at the precise point of penetration. Pryce-Jones and collaborators also demonstrated that the *B. graminis* f. sp. *hordei* appressoria are capable of generating significant turgor pressure on the surface of the plants (Pryce-Jones & Gurr, 1999). Such pressure (estimated to be up to 4 MPa) is likely to be used to generate sufficient force to break through the cell wall and is certainly high enough to counteract the turgor pressure of the epidermal cells (<1 MPa).

The processes required for successful penetration are complex. For example, removal of cuticle components and artificially changing the composition of the plant surface results in failure of the fungus to penetrate successfully and leads to collapse of the appressoria (Iwamoto *et al.*, 2002). Furthermore, it appears that the fungus induces susceptibility at the site of penetration and to the cells immediately surrounding it. This phenomenon is local and, although its molecular and biochemical basis is not understood, it might be mediated by alterations in the levels of the phenolic compounds produced by the barley cells (Lyngkjaer *et al.*, 2001).

8.7   The haustorium

Once inside the epidermal cell, *B. graminis* f. sp. *hordei* develops a haustorium. The haustorium is a nucleate cell that is devoted to the uptake of nutrient from the host and is thus central to the establishment of pathogenicity. First, this appears as an ellipsoid, which then grows finger-like processes at either end. Haustoria have a degree of autonomy and totipotence. Once they are established, it is possible to remove all fungal structures from the surface of the plant and they can grow and regenerate a viable colony. The host's plasma membrane surrounds the whole structure at all times and is sometimes called the extrahaustorial membrane. Compared to the plasma membrane the extrahaustorial membrane is thicker and has different biochemical characteristics (Spencer-Phillips & Gay, 1981; Gay *etal.*, 1987). The compartment between the extrahaustorial membrane and the haustorial cell wall is the extrahaustorial matrix and is sealed at the haustorial neck with a neckband. Electron microscopy coupled with immunogold labelling was used to localise threonine-hydroxyproline-rich glycoproteins in the *B. graminis* f. sp. *hordei* pathosystem. The most pronounced accumulation of these proteins occurs in the extrahaustorial matrix. Thus, the host plant establishes a modified barrier between itself and the pathogen (Hippesanwald *etal.*, 1994). *B. graminis* haustoria have proven recalcitrant to isolation unlike those of the related pea powdery mildew *Erysiphe pisi* and those of the rust fungus *Uromyces fabae*, and this has posed significant limitations to our understanding of how they function. Thus, we will review here briefly what is known of the appressoria from other obligate biotrophic pathogens including *E. pisi* and *U. fabae*.

Cytochemical studies show that there are clear differences between the fungal haustorial plasma membrane and the plant's extrahaustorial membrane: the *E. pisi* haustorial plasma membrane has a high phosphatase activity, and it also contains specific glycoproteins (Mackie *etal.*, 1991, 1993). There is a high histochemical ATPase activity in *E. pisi* haustoria but the extrahaustorial membranes lose their ATPase activity (Spencer-Phillips & Gay, 1981). ATPase activity in extrahaustorial membranes of *B. graminis* f. sp. *hordei* is also missing. Conversely, ATPase activity can be detected indirectly in haustorial plasma membranes of *B. graminis* f. sp. *hordei*. The loss of ATPase activity probably causes a loss of solute retention capacity of the host cell, enabling the fungus to take up nutrients. (Gay *etal.*, 1987).

Mackie and co-workers have characterised monoclonal antibodies that bind specifically to glycoproteins of the haustorial plasma membranes, but not to the mycelial plasma membranes of *E. pisi*. These glycoproteins are produced in the early stages of haustorial development (Mackie *etal.*, 1991, 1993). There are also specific glycoproteins located in the extrahaustorial plasma membrane. Two monoclonal antibodies (UB9 and UB11) recognised epitopes in the pea plasma membrane (Roberts *etal.*, 1993); UB9 recognises a glycoprotein in the plant plasma membrane. It does not label the extrahaustorial plasma membrane of infected leaves in early stages of haustorium development. UB11 shows specificity to the extrahaustorial matrix in early stages of haustorium development, but does not bind to plant plasma membranes. These results confirm that there is specific molecular differentiation in the haustorial complex. It is not clear by which mechanisms *B. graminis* f. sp. *hordei* acquires nutrients from the plant but sugar and amino acid transporters are involved in nutrient uptake through haustoria in other biotrophic fungi (Hahn *etal.*, 1997; Hahn & Mendgen, 1997; Voegele *etal.*, 2001). In *B. graminis* f. sp. *tritici*, glucose is taken up from the plant into the mycelium, possibly by an H^+/glucose symport system for carbohydrate uptake (Sutton *etal.*, 1999).

In *E. pisi*, the haustorium takes up principally glucose, not sucrose, and also utilises glutamine (Clark & Hall, 1998). The glucose uptake is an active transport, as application of various inhibitors of sugar uptake interfere with the process. Also, invertase activity is induced in the plant apoplast by infection. This suggests that sucrose is broken down in the apoplast to fructose and glucose and the fungus then absorbs glucose.

In *U. fabae*, a hexose transporter is predominantly expressed in haustoria and is localised to the haustorial plasma membrane. This transporter shows a substrate specificity for D-glucose and D-fructose. An H^+-symport mechanism operates here too (Voegele *et al.*, 2001). The H^+/ATPase activity of fungal structures is highest in haustoria, and is developmentally regulated (Struck *et al.*, 1996, 1998). H^+/ATPase is involved in nutrient uptake: the proton gradient across the membrane can drive the active transport of nutrients from the extrahaustorial matrix. In a haustorial library of *U. fabae*, a fungal amino acid transporter was characterised; this transporter is restricted to haustoria and is localised to the haustoria plasma membrane itself (Hahn *et al.*, 1997).

8.8   Vegetative growth and sporulation

Once the haustoria have developed and become fully functional, the infection is established and the fungus taps into the host's metabolic resources to provide energy and material to support further growth and development. Hyphae on the outer surface of the leaves grow and produce more haustoria in the epidermal cells underneath the colony. Single epidermal cells can host multiple haustoria and, for a while at least, continue to live and supply the fungus. Three to four days after initial contact, conidiophores differentiate on the external hyphae and produce abundant conidia for further dissemination. At this stage, it is evident that the fungus manipulates the plant's metabolism to maintain the nutrient supply. Measurements of the physiology of infected leaves show that respiration of barley is stimulated within 24 hours after inoculation (McAinsh *et al.*, 1989). *B. graminis* f. sp. *tritici* alters the source–sink relationships of the plant: the fungus triggers a rise of acid invertase activity 48 hours after infection, which then remains high for several days. This results in an accumulation of soluble carbohydrates and consequent increase in the partitioning of photosynthate into starch. Interestingly, these alterations are not merely the result of an increased sink of nutrients: the imbalance remains even if the fungus is killed with fungicides. Thus, there appears to be a signal that actively resets the host's metabolism (Wright *et al.*, 1995a). Apart from creating a novel sink for fixed carbon and other nutrients, powdery mildew infections also interfere with overall photosynthesis. Thus, rubisco and the activity of the Calvin cycle enzymes are reduced in infected leaves (Scholes *et al.*, 1994; Wright *et al.*, 1995b): this means that *B. graminis* f. sp. *hordei* causes a gradual decline in the rate of photosynthesis and premature loss of chlorophyll in infected parts of the host, while carbohydrates and amino acids are made available to the fungus in the process. It is not only the fungus that affects the metabolism of the host, but the converse is also true: the metabolic status of the plant has an effect on the physiology of the fungus. For example, if ambient CO_2 increases from 350 ppm to 700 ppm, carbohydrates accumulate in the leaves and powdery mildew colonies grow faster (Hibberd, 1996).

The alterations just described take place in infected leaves of growing plants. If the leaves are placed under conditions that favour rapid senescence (e.g. after detachment), the *green islands* phenomenon is observed: areas of the leaf around the infection site remain green for a long time while the rest of the tissue dies (Schulze-Lefert & Vogel, 2000). This is probably due to interference in the senescence. It would be interesting to determine how the fungus modulates senescence in infected tissues. Towards the end of the season and after multiple cycles of infection mediated by the dispersal of asexual conidia, the powdery mildews enter into the sexual phase and cleistothecia are formed. At these later developmental stages, the generative mycelium can store nutrients and progressively becomes more independent of nutrition supply by the host, providing its own metabolites for the developing cleistothecia. It is interesting to note that when cleistothecia are formed, the haustoria become predominantly encased and colonies seemed to take up fewer nutrients: thus the fungus is once more nutritionally autonomous (Gotz & Boyle, 1998).

8.9   Outlook

Although *B. graminis* f. sp. *hordei* is an obligate biotroph and has limited the experimental analysis of the interaction, much has been achieved in understanding how the fungus attacks its host and achieves the exquisite compatibility necessary for its survival and reproduction. In this chapter, we have seen how a combination of different experimental approaches, ranging from microscopy to physiology, biochemistry and molecular biology, have contributed to this field. Further developments will depend on the continuing use of these techniques, but might also benefit from a genetic analysis of the fungus. This has indeed already been successfully exploited by several researchers in the study and identification of avirulence determinants in gene-for-gene type interactions (Brown & Simpson, 1994; Jensen *et al.*, 1995; Pedersen *et al.*, 2002). The application of reverse genetic techniques to powdery mildews relies on the ability to transform the fungus. Transient (Christiansen *et al.*, 1995) and stable (Chaure *et al.*, 2000) transformations have been reported but have so far proven to be difficult to reproduce effectively. Until this is achieved, expression of powdery mildew genes in heterologous systems such as *M. grisea* (Bindslev *et al.*, 2001) will have to be relied on. In the future, development of microarrays for the global analysis of gene expression in *Blumeria*, as well as the complete sequence of the *Blumeria* genome, might provide further knowledge of the biology of infection by powdery mildews. This, in turn, will contribute significantly to the development and deployment of durable and effective disease control in crops of great commercial value.

References

Alberts, B., Johnson, A., Lewis, J., Raff, M., Roberts, K. & Walter, P. (eds) (2002) *Molecular Biology of the Cell,* Garland Science, New York.

Banuett, F. (1998) *Microbiol. Mol. Bio. Rev.,* **62**, 249–274.

Bélanger, R.R., Bushnell, W.R., Dik, A.J. & Carver, T.L.W. (2002) *The Powdery Mildews: A Comprehensive Treatise,* APS Press, St Paul, Minnesota.

Bindslev, L., Kershaw, M.J., Talbot, N.J. & Oliver, R.P. (2001) *Mol. Plant Microbe Interact.,* **14**, 1368–1375.

Braun, U., Cook, R.T.A., Inman, A.J. & Shin, H.-D. (2002) In: *The Powdery Mildews: A Comprehensive Treatise* (eds R.R. Bélanger, W.R. Bushnell, A.J. Dik & T.L.W. Carver), pp. 13–55, APS Press, St Paul, Minnesota.

Brown, J.K. & Simpson, C.G. (1994) *Curr. Genet.,* **26**, 172–178.

Carver, T.L.W. & Ingerson, S.M. (1987) *Physiol. Mol. Plant Pathol.,* **30**, 359–372.

Carver, T.L.W., Thomas, B.J. & Ingersonmorris, S.M. (1995) *Can. J. Bot.,* **73**, 272–287.

Carver, T.L.W., Ingerson, S.M. & Thomas, B.J. (1996) *Influences of Host Surface Features on Development of* Erysiphe graminis *and* Erysiphe pisi, BIOS Scientific Publishers, Oxford.

Carver, T.L.W., Kunoh, H., Thomas, B.J. & Nicholson, R.L. (1999) *Mycol. Res.,* **103**, 547–560.

Chaure, P., Gurr, S.J. & Spanu, P. (2000) *Nat. Biotechnol.,* **18**, 205–207.

Chen, B.S., Gao, S.J., Choi, G.H. & Nuss, D.L. (1996) *Proc. Natl. Acad. Sci. USA,* **93**, 7996–8000.

Christiansen, S.K., Knudsen, S. & Giese, H. (1995) *Curr. Genet.,* **29**, 100–102.

Clark, J.I.M. & Hall, J.L. (1998) *New Phytologist,* **140**, 261–269.

Edwards, H.H. (2002) *Can. J. Bot.,* **80**, 1121–1125.

Edwards, H.H. & Allen, P.J. (1970) *Phytopathology,* **60**, 1504–1509.

Francis, S.A.D., Dewey, F.M. & Gurr, S.J. (1996) *Physiol. Mol. Plant Pathol.,* **49**, 201–211.

Gay, J.L., Salzberg, A. & Woods, A.M. (1987) *New Phytologist,* **107**, 541–548.

Gold, S., Duncan, G., Barrett, K. & Kronstad, J. (1994) *Genes Dev.,* **8**, 2805–2816.

Gotz, M. & Boyle, C. (1998) *Plant Dis.,* **82**, 507–511.

Hahn, M. & Mendgen, K. (1997) *Mol. Plant Microbe Interact.,* **10**, 427–437.

Hahn, M.N.U., Struck, C., Göttfert, M. & Mendgen, K. (1997) *Mol. Plant Microbe Interact.,* **10**, 438–445.

Hall, A.A., Bindslev, L., Rouster, J., Rasmussen, S.W., Oliver, R.P. & Gurr, S.J. (1999) *Mol. Plant Microbe Interact.,* **12**, 960–968.

Hall, A.A., Carver, T.L.W., Zhang, Z. & Gurr, S.J. (2000). In: *Encyclopedia of Microbiology,* Vol. 3 (ed., B.a.H. Alexander), pp. 269–276, Academic Press.

Hewitt, H. (1998) *Fungicides in Crop Protection,* CAB International.

Hibberd, J. (1996) *New Phytologist,* **134**, 551–551.

Hippesanwald, S., Marticke, K.H., Kieliszewski, M.J. & Somerville, S.C. (1994) *Protoplasma,* **178**, 138–155.

Iwamoto, M., Takeuchi, Y., Takada, Y. & Yamaoka, N. (2002) *Physiol. Mol. Plant Pathol.,* **60**, 31–38.

Jensen, J., Jensen, H.P. & Jorgensen, J.H. (1995) *Hereditas,* **122**, 197–209.

Kinane, J. & Oliver, R.P. (2000) *Fungal Genet. Biol.,* **39**, 94–102.

Kinane, J. & Oliver, R.P. (2003) Evidence that the appressorial development in barley powdery mildew is controlled by MAP kinase activity in conjunction with the cAMP activity. *Fungal Genet. Biol.,* **39**, 94–102.

Kinane, J., Dalvin, S., Bindslev, L., Hall, A., Gurr, S. & Oliver, R. (2000) *Mol. Plant Microbe Interact.,* **13**, 494–502.

Kronstad, J.W. (1997) *Trends Plant Sci.,* **2**, 193–199.

Kunoh, H., Yamaoka, N., Yoshioka, H. & Nicholson, R.L. (1988) *Exp. Mycol.,* **12**, 325–335.

Kunoh, H., Nicholson, R.L., Yosioka, H., Yamaoka, N. & Kobayashi, I. (1990) *Physiol. Mol. Plant Pathol.,* **36**, 397–407.

Lyngkjaer, M.F., Carver, T.L.W. & Zeyen, R.J. (2001) *Physiol. Mol. Plant Pathol.,* **59**, 243–256.

Mackie, A.J., Roberts, A.M., Callow, J.A. & Green, J.R. (1991) *Planta,* **183**, 399–408.

Mackie, A.J., Roberts, A.M., Green, J.R. & Callow, J.A. (1993) *Physiol. Mol. Plant Pathol.*, **43**, 135–146.

McAinsh, M.R., Ayres, P.G. & Hetherington, A.M. (1989) *Plant Sci.*, **64**, 221–230.

Mitchell, T.K. & Dean, R.A. (1995) *Plant Cell*, **7**, 1869–1878.

Muchembled, J., Sahraoui, A.L., Grandmougin-Ferjani, A. & Sancholle, M. (2000) *Biochem. Soc. Trans.*, **28**, 875–877.

Nicholson, R.L. & Kunoh, H. (1995) *Can. J. Bot.*, **73**, S609–S615.

Nicholson, R.L., Kunoh, H., Shiraishi, T. & Yamada, T. (1993) *Physiol. Mol. Plant Pathol.*, **43**, 307–318.

Nielsen, K.A., Nicholson, R.L., Carver, T.L.W., Kunoh, H. & Oliver, R.P. (2000) *Physiol. Mol. Plant Pathol.*, **56**, 63–70.

Pascholati, S.F., Yoshioka, H., Kunoh, H. & Nicholson, R.L. (1992) *Physiol. Mol. Plant Pathol.*, **41**, 53–59.

Pedersen, C., Rasmussen, S.W. & Giese, H. (2002) *Fungal Genet. Biol.*, **35**, 235–246.

Pryce-Jones, E.C.T. & Gurr, S.J. (1999) *Physiol. Mol. Plant Pathol.*, **55**, 175–182.

Roberts, A.M., Mackie, A.J., Hathaway, V., Callow, J.A. & Green, J.R. (1993) *Physiol. Mol. Plant Pathol.*, **43**, 147–160.

Roberts, D.R., Mims, C.W. & Fuller, M.S. (1996) *Can. J. Bot.*, **74**, 231–237.

Scholes, J.D., Lee, P.J., Horton, P. & Lewis, D.H. (1994) *New Phytologist*, **126**, 213–222.

Schulze-Lefert, P. & Vogel, J. (2000) *Trends Plant Sci.*, **5**, 343–348.

Spencer-Phillips, P.T.N. & Gay, J.L. (1981) *New Phytologist*, **89**, 393–398.

Struck, C., Hahn, M. & Mendgen, K. (1996) *Fungal Genet. Biol.*, **20**, 30–35.

Struck, C., Siebels, C., Rommel, O., Wernitz, M. & Hahn, M. (1998) *Mol. Plant Microbe Interact.*, **11**, 458–465.

Sutherland, F. & Walters, D.R. (2002) *Eur. J. Plant Pathol.*, **108**, 385–389.

Sutton, P.N., Henry, M.J. & Hall, J.L. (1999) *Planta*, **208**, 426–430.

Suzuki, S., Komiya, Y., Tsuyumu, S., Kunoh, H., Carver, T.L.W. & Nicholson, R.L. (1998) *Ann. Phytopathol. Soc. Japan*, **64**, 160–166.

Thines, E., Eilbert, F., Sterner, O. & Anke, H. (1997a) *FEMS Microbiol. Lett.*, **151**, 219–224.

Thines, E., Eilbert, F., Sterner, O. & Anke, H. (1997b) *FEMS Microbiol. Lett.*, **156**, 91–94.

Thines, E., Weber, R.W. & Talbot, N.J. (2000) *Plant Cell*, **12**, 1703–1718.

Thomas, S.W., Glaring, M.A., Rasmussen, S.W., Kinane, J.T. & Oliver, R.P. (2002) *Mol. Plant Microbe Interact.*, **15**, 847–856.

Thomas, S.W., Rasmussen, S.W., Glaring, M.A., Rouster, J.A., Christiansen, S.K. & Oliver, R.P. (2001) *Fungal Genet. Biol.*, **33**, 195–211.

Tsuba, M., Katagiri, C., Takeuchi, G., Yakada, Y. & Yamaoka, N. (2002) *Physiol. Mol. Plant Pathol.*, **60**, 51–57.

Voegele, R.T., Struck, C., Hahn, M. & Mendgen, K.W. (2001) *Proc. Natl. Acad. Sci. USA*, **98**, 8133–8138.

Wheeler, I., Hollomon, D.W., Gustavson, G., Mitchell, J.C., Longhurst, C., Zhang, Z. & Gurr, S. (2003) *Mol. Plant Pathol.*, **4**, 177–186.

Wright, A.J., Carver, T.L.W., Thomas, B.J., Fenwick, N.I.D., Kunoh, H. & Nicholson, R.L. (2000) *Physiol. Mol. Plant Pathol.*, **57**, 281–301.

Wright, D.P., Baldwin, B.C., Shephard, M.C. & Scholes, J.D. (1995a) *Physiol. Mol. Plant Pathol.*, **47**, 237–253.

Wright, D.P., Baldwin, B.C., Shephard, M.C. & Scholes, J.D. (1995b) *Physiol. Mol. Plant Pathol.*, **47**, 255–267.

Wright, A.J., Thomas, B.J. & Carver, T.L.W. (2002a) *Physiol. Mol. Plant Pathol.*, **61**, 217–226.

Wright, A.J., Thomas, B.J., Kunoh, H., Nicholson, R.L. & Carver, T.L.W. (2002b) *Physiol. Mol. Plant Pathol.*, **61**, 163–178.

Xu, J.R. (2000) *Fungal Genet. Biol.*, **31**, 137–152.

Xu, J.R., Urban, M., Sweigard, J.A. & Hamer, J.E. (1997) *Mol. Plant Microbe Interact.*, **10**, 187–194.

Yamaoka, N. & Takeuchi, Y. (1999) *Physiol. Mol. Plant Pathol.*, **54**, 145–154.

Zhang, Z.G. & Gurr, S.J. (2001) *Gene*, **266**, 57–65.

Zhang, Z., Priddey, G. & Gurr, S. (2001) *Mol. Plant Pathol.*, **2**, 327–337.

9 The *Phytophthora infestans*–potato interaction

Pieter van West and Vivianne G.A.A. Vleeshouwers

9.1 Introduction

Phytophthora infestans was one of the first organisms identified as the cause of a plant disease. Only more than 150 years after its discovery are we slowly beginning to understand some of the fundamental molecular processes that make it such a devastating pathogen of potato and tomato plants. This chapter highlights recent discoveries in research on *P. infestans*, with specific emphasis on molecular and cellular processes that occur during the disease cycle. We discuss developmental processes of the pre-infection stages of *P. infestans*, from zoosporogenesis through zoospore liberation, encystment, germination, and the formation of appressoria-like structures, as well as the molecular processes taking place during colonisation of the plant, to the formation of sporangia and oospores. In addition, we will discuss the plant response towards *P. infestans* penetration and colonisation.

9.2 History of late blight

Phytophthora infestans (Mont.) de Bary is commonly known as the causal agent of late blight on potato. It is also known as the Irish potato famine pathogen, because it totally destroyed Ireland's potato crop in the mid-1840s. This resulted in widespread poverty, mass starvation, and immense economic and sociological changes, including emigration of large numbers of people to the United States and elsewhere (Bourke, 1991). It is estimated that Ireland lost about two million inhabitants due to starvation and emigration. Probably no other single plant pathogen has ever caused such extensive human suffering (Erwin & Ribeiro, 1996).

In 1842, von Martius was one of the first to suggest that a micro-organism could be the actual cause of a plant disease (von Martius, 1842). The British mycologist, Berkeley, who translated von Martius' paper into English, welcomed this new concept. Their views countered the widespread idea that fungi grew on organisms that previously became putrefied by *dampness* or similarly mysterious factors; many considered fungi and bacteria as the result, but not the cause, of disease (Bourke, 1991).

Shortly after von Martius' paper, the first epidemics of the late blight disease of potato occurred in Europe and in the north-eastern United States (Bourke, 1991; Peterson *et al.*, 1992). Again there was great debate about the cause. Finally in 1876, Anton de Bary determined conclusively that a micro-organism, which he named *Phytophthora* (*plant destroyer*) *infestans*, was the cause of late blight (de Bary, 1876).

More than 150 years after the great famine, late blight has remained a disreputable disease worldwide. The disease has reached epidemic proportions in Europe, North America, and Russia, due to the development of resistance to phenylamide oomycetecides in many populations of the pathogen and due to the widespread occurrence of new genotypes (Deahl *et al.*, 1991; Drenth *et al.*, 1993; Fry & Goodwin, 1997). At present, potatoes cannot be grown without prophylactic application of oomycetecides, in most areas of the world.

9.3   Economic and social impact of *Phytophthora* plant pathogens

P. infestans is a re-emerging pest ever since its discovery. It still causes major epidemics in both potato and tomato crops worldwide. For example, in 2000, 15% of the total potato crop of Russia was destroyed due to late blight. Undoubtedly, such severe epidemics could trigger a new, potentially catastrophic potato famine (Schiermeier, 2001). Worldwide losses in potato production caused by late blight, and measures to control the disease have been estimated at $5 billion annually (Duncan, 1999).

P. infestans is only the tip of the iceberg when it comes down to estimating disease losses caused by the total group of plant pathogenic *Phytophthora* species. There are about 60 species in the genus *Phytophthora* that cause various disease symptoms, including root rot, fruit rot, foliar blight, and stem blight on many economically important plants (Erwin & Ribeiro, 1996). A few examples of root rot pathogens are *Phytophthora sojae* on soybean, *Phytophthora fragariae* on strawberries, *Phytophthora cryptogea* on many plant species including tomato and cucumber, and *Phytophthora cinnamomi* on various woody plant species. Other species cause leaf blight symptoms. *P. infestans* is a good example of such a pathogen along with *Phytophthora porri* on leek. There are also many fruit rot pathogens such as *Phytophthora capsici* on various plants, and *Phytophthora palmivora* and *Phytophthora megakarya* on cocoa pods. The classification of *Phytophthora* species on the basis of which part of the plant it predominantly infects is quite arbitrary because most *Phytophthora* species are able to cause disease symptoms on all plant tissues, above and below ground.

The environmental damage caused by *Phytophthora* diseases in natural ecosystems can be tremendous, due to difficulties in controlling the spread of the disease. An example of a severe ecological tragedy is sudden oak death, a disease caused by *Phytophthora ramorum*, which has emerged recently along the Pacific coast of the United States. *P. ramorum* is destroying oak trees and is probably also spreading to other trees, such as redwoods, and to other regions in North America (Knight, 2002; Rizzo *et al.*, 2002). Likewise, *P. cinnamomi*, which has a very wide host range, infecting over 900 species of plants (Zentmyer, 1980), has caused severe epidemics in the jarrah tree forests in Western Australia (Podger *et al.*, 1965; Podger, 1972) as well as more recent outbreaks across the world.

The Irish potato famine is therefore absolutely not limited to a historic reference. In reality, many *Phytophthora* epidemics are only just being kept under control by the use of prophylactic oomycetecides.

9.4 *Phytophthora infestans* and its taxonomic position

The genus *Phytophthora* is closely related to the genus *Pythium* and both are classified under the family Pythiaceae. This group, commonly referred to as the water moulds, belongs to the class Oomycetes in the order Peronosporales. Historically, oomycetes have been classified under the kingdom Fungi due to some of their fungus-like characteristics (Barr, 1983; Dick, 1995). However, studies of cell wall composition (Bartnicki-Garcia & Wang, 1983), metabolism (Vogel, 1964; Pfyffer *et al.*, 1990), and rRNA sequence (Förster *et al.*, 1990; Illingworth *et al.*, 1991) indicate that oomycetes are better classified with chrysophytes, diatoms, and golden-brown algae in the kingdom Straminopila (Baldauf *et al.*, 2000; Margulis *et al.*, 2000; Kamoun, 2003).

Among the eukaryotic plant pathogens, oomycetes form a unique group that must have gained the ability to infect plants independently of the *true* fungi. Some biological features of oomycetes are relatively uncommon in *true* fungi (Zentmyer, 1983; Griffith *et al.*, 1992). For example, oomycetes are for the major part of their life cycle diploid, whereas the majority of the true fungi are haploid. The cell wall of oomycetes is predominantly composed of cellulose and β-glucans; unlike fungal cell walls, they contain little chitin (Bartnicki-Garcia & Wang, 1983; Leal-Morales *et al.*, 1997; Mort-Bontemps *et al.*, 1997; Werner *et al.*, 2002). Also oomycete species have coenocytic mycelium with no, or few, septa, whereas most true fungi have distinct septa. *Phytophthora* species require an exogenous source of β-hydroxy sterols for sporulation, as they are unable to perform epoxidation of squalene into sterols (Hendrix, 1970; Elliott, 1983). Furthermore, *Phytophthora* species are thiamine auxotrophs and therefore need an exogenous source of thiamine for growth (Erwin & Ribeiro, 1996). A key feature of oomycetes is that they can produce zoospores. These are biflagellated cells with one whiplash and one tinsel flagellum (Hemmes, 1983). The rootlet morphology of the flagella is comparable to that of the heterokont algae (Barr, 1981). Furthermore, members of the Pythiaceae are uniquely resistant to polyene antibiotics such as pimaricin and nystatin (Eckert & Tsao, 1962).

9.5 The disease cycle of *Phytophthora infestans*

P. infestans is generally considered a specialised pathogen causing disease on potato and tomato crops (Fig. 9.1), although natural infection of plants outside the genera *Solanum* and *Lycopersicon* have been reported (Erwin & Ribeiro, 1996).

Figure 9.1 Potato leaf infected with *Phytophthora infestans*. The infected leaf becomes necrotic and turns black, rendering the blighted look of infected potato fields and illustrating the disease name late blight.

The disease cycle of *P. infestans* is well studied (Fig. 9.2). Pathogenesis by *P. infestans* mainly involves asexual growth. Infections typically begin when sporangia land on the surface of a leaf. During wet conditions and at temperatures below 12°C, zoospores are released from the sporangium (Fig. 9.3A). After a motile period, the zoospores encyst and produce a germ tube. Direct germination of the sporangia is also possible and occurs mainly at temperatures above 12°C. The germ tube is able to differentiate into an appressorium, or appressorium-like structure (Fig. 9.3B). A penetration peg is formed that pierces the cuticle and penetrates the underlying plant cell. Subsequently, an infection vesicle is formed (Fig. 9.3C,D) in the epidermal cell and hyphae grow into the mesophyll cell layers both intra- and intercellularly (Fig. 9.3E,F). Occasionally, intracellular haustorial feeding structures are formed (Fig. 9.3F). After three to four days, *P. infestans* starts to grow saprophytically in the necrotised centre of the growing lesion. Hyphae emerge through the stomata, and sporangiophores are formed which produce numerous new sporangia on the underside of the leaf. The sporangia do not desiccate and are relatively short-lived (Judelson, 1997a). Infected foliage first becomes yellow and then water-soaked and eventually turns black. Lesions may occur anywhere on the leaf, petiole, or even on the stem of the potato plant. Tubers become infected later in the season. In the early stages, slightly brown or purple blotches appear on the skin. In damp soils, the disease progresses rapidly, and the tubers decay either before or after harvest.

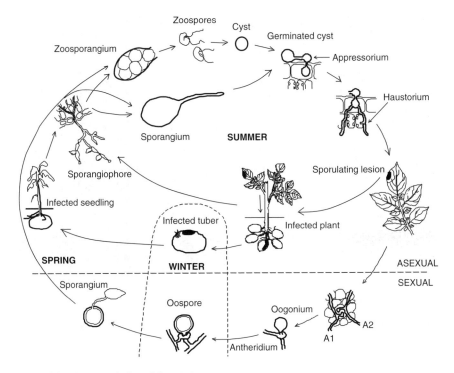

Figure 9.2 Life cycle of *Phytophthora infestans*.

9.5.1   *Molecular and cellular events during the disease cycle of* P. infestans

Various research groups around the world are now exploring the molecular developmental processes of *P. infestans* and other oomycetes quite extensively. Here,
we give an overview of the molecular events in *P. infestans* that take place during
the pre-infection and colonisation processes.

9.5.1.1   *Release of zoospores from sporangia*

The asexual sporangia of *P. infestans* germinate either directly through a germ
tube or indirectly by releasing zoospores first (Fig. 9.3A). The sporangia are multi-
nucleated and remain so until cytokinesis is initiated by a cold shock that eventually
compartmentalises single nuclei within each zoospore. This process is called sporangial cleavage and takes place during wet conditions. Subsequently, the zoospores
are released from the sporangia. Divalent cations, and in particular Ca^{2+}, play
an important role during sporangial cleavage and the release of zoospores. For
example, Judelson and Roberts (2002) found that the calmodulin antagonist trifluoro-
perazine specifically inhibited sporangial cleavage, and that the calcium channel

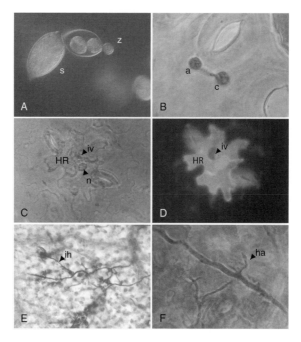

Figure 9.3 (A) Release of *P. infestans* zoospores (z) from a sporangium (s). The picture was taken using Differential Interference Contrast microscopy. Magnification 1000x. (B) Cyst (c) of *P. infestans* has formed a germ tube and an appressorium (a) on the leaf surface of potato. The picture was taken using Phase Contrast microscopy. Magnification 1000x. (C, D) Hypersensitive response (HR) in a resistant wild *Solanum* interspecific hybrid (*Solanum arnezii* × *Solanum hondelmannii*) after penetration of *P. infestans*. An infection vesicle (iv) is formed in the penetrated epidermal cell, and the condensed nucleus (n) of the neighbouring cell has moved to the penetration site. The penetrated cell and the neighbouring cell immediately show the characteristics of the HR: granular cytoplasm (C, Phase Contrast microscopy) and bright autofluorescence when illuminated with UV light (D, Fluorescence microscopy). Also two neighbouring stomatal guard cells display the HR. The pictures were taken 22 hours after inoculation. Magnification 400x. (E) Histochemical localisation of glucuronidase activity in leaves of a susceptible potato cultivar after infection with a *P. infestans ipiO-gus* transformant. Invading hyphae (ih) growing through the potato leaf, four days after inoculation. Magnification 100x. (F) Compatible interaction between *P. infestans* and a susceptible potato cultivar. Hyphae are forming digit-like haustoria (ha), which are penetrating parenchyma cells in the mesophyll. The absence of visible plant cell responses is typical of this biotrophic stage. The picture was taken using Phase Contrast microscopy. Magnification 400x.

blocker verapamil decreased the efficiency of sporangial release significantly. In *P. cinnamomi*, two distinct changes in cytoplasmic, free Ca^{2+} were associated with sporangium differentiation (Jackson & Hardham, 1996). Within the first minute of cold shock, Ca^{2+} concentration rose rapidly, but within ten minutes this had fallen back again to almost the initial resting concentration. A second gradual and more long-lasting increase in Ca^{2+} concentration occurred during the process of cytoplasmic cleavage and was thought to be associated with the regulation of cytokinesis. It is possible that these changes in Ca^{2+} are regulated via a novel kinase gene that was recently identified in *P. infestans* (Judelson & Roberts, 2002). Expression of this kinase was first detected soon after chilling sporangia in water. The transcript persisted in motile zoospores and in germinated cysts but was not detected in other tissues. The predicted protein sequence resembled Ca^{2+}- and calmodulin-regulated protein kinases but lacked an apparent C-terminal regulatory or protein-association domain typically found in such proteins. Gene silencing studies were unsuccessful and presently the function of this kinase remains unknown (Judelson & Roberts, 2002).

Recently, two genes involved in signal transduction pathways of *P. infestans*, a G-protein α subunit (*Pigpa1*) and a G-protein β subunit (*Pigpb1*), were isolated and were found to have major roles in sporangial differentiation (Laxalt *et al.*, 2002; Latijnhouwers, 2003). Both genes are differentially expressed during asexual development with the highest mRNA levels in sporangia. In *Pigpa1*-silenced mutants, zoospore production was affected by as much as 45% and they turned 6–8 times more frequently, causing them to swim relatively short distances compared to wild type zoospores. In addition, attraction to surfaces (negative geotaxis), chemotaxis towards glutamic and aspartic acid, and autoaggregation were all affected, suggesting that the G-alpha protein play a major role in these processes (Latijnhouwers, 2003). Autoaggregation is a process whereby some zoospore species swim spontaneously together in groups of hundreds or thousands. Encystment may follow and further zoospores are then actively attracted to the cyst aggregate (Reid *et al.*, 1995; van West *et al.*, 2003). In root pathogenic oomycete species, autoaggregation is believed to assist in a successful infection by increasing the inoculum pressure. Whether autoaggregation is also relevant to typical leaf pathogens, such as *P. infestans*, is at present unknown.

Osmoregulation and turgor pressure within the sporangium during zoospore release in *Phytophthora* species are probably regulated by cytoplasmic proline (Ambikapathy *et al.*, 2002). High concentrations of proline were found in the cytoplasm of zoospores in the differentiating sporangia of *Phytophthora nicotianae*. It is anticipated that high proline concentrations in the cytoplasm may counterbalance a hyperosmotic, extracellular sporangial lumen. Proline is expelled rapidly after release of the zoospores from the sporangium, thereby preventing osmotic rupture of the zoospore membrane. Once released, the zoospores are unable to build up turgor pressure and instead osmoregulation is maintained by the contractile water expulsion vacuole which pumps water out of the cytoplasm (Ambikapathy *et al.*, 2002).

9.5.1.2   *Encystment, germination, and appressoria formation*

Encystment of zoospores is a rapid process involving detachment of the two flagella, and secretion of glycoproteins and other molecules to form the primary cell wall (Hardham, 2001). Cysts become firmly attached to the plant surface due to the release of these adhesive molecules that are pre-formed in small peripheral vesicles, which are situated near the groove of the ventral surface (Sing & Bartnicki-Garcia, 1975; Hardham & Gubler, 1990). The molecular composition of the adhesion substance is still unknown, but antibodies that react specifically with this material from *P. cinnamomi* recognise a protein of just over 200 kDa (Hardham & Gubler, 1990).

From experimental data, it is clear that calcium-signalling events also play a major role in encystment and cyst germination (Warburton & Deacon, 1998; Connolly *et al.*, 1999). Experiments using the calcium-dependent fluorophore Fura-2 showed that encystment of zoospores of *Phytophthora nicotianae* was correlated with an initial Ca^{2+} influx, followed by a larger, more progressive efflux of Ca^{2+} over a period of 20–30 min (Connolly *et al.*, 1999; Warburton & Deacon, 1998). The latter efflux of intracellular Ca^{2+} correlated with the time of germination of the cysts. This phase was inhibited by TMB-8, which inhibits the release of Ca^{2+} from internal calcium stores, and the effect of this inhibition was remediated by addition of high external Ca^{2+}.

Also G-protein-mediated signalling is hypothesised to play a major role during encystment and germination. This is supported by expression analysis of the G-protein α and β subunit genes of *P. infestans*, which are both expressed during these developmental stages (Laxalt *et al.*, 2002). Latijnhouwers *et al.* (2002) demonstrated that mechanical agitation of zoospores, which is often used to induce synchronised encystment, resulted in increased levels of the phospholipid, phosphatidic acid. Interestingly, treatment of zoospores with mastoparan, which is a G-protein activator, also resulted in an increase of phosphatidic acid via activation of phospholipase-D. Likewise, encystment was also induced by treatment with n- and sec-butanol, which are both able to activate phospholipase-D. Therefore, it can be concluded that a G-protein-regulated signalling mechanism activates a phospholipase-D pathway that is involved in encystment stimulation (Latijnhouwers *et al.*, 2002).

Considerable efforts are being made to identify genes specifically up-regulated during the pre-infection stages of the disease cycle. In one screen for such genes from *P. infestans*, a gene family encoding cyst-germination-specific acidic repeat (*Car*) proteins was identified. *Car* gene expression was found specifically during cyst germination and in the appressorium stage (Gornhardt *et al.*, 2000). *Car* proteins are homologous to mammalian mucins and are localised on the surface of the pre-infection structures. These adhesive proteins may therefore be important components of the mucous layer that protects the germinated cyst from desiccation, physical damage, or adverse effects of the plant defence response (Gornhardt *et al.*, 2000).

Pieterse *et al.* (1993a) employed a differential screening assay to identify *in planta*-induced genes or *ipi*-genes (see also Section 9.5.1.4). These are genes whose expression is specifically induced during colonisation. It is thought that such genes may encode essential pathogenicity factors (Pieterse *et al.*, 1993a). Interestingly, one of the nine identified *ipi*-genes, *ipiB*, turned out to be predominantly expressed in the pre-infection stages (Pieterse *et al.*, 1994a; van West *et al.*, 1998; van West, 2000). The *ipiB* gene constitutes a family of at least three members, *ipiB1*, *ipiB2*, and *ipiB3*, which are clustered in a head-to-tail arrangement. They encode three similar glycine-rich proteins that consist mainly of two repeats with the core sequences, A/V-G-A-G-L-Y-G-R and G-A-G-Y/V-G-G (Pieterse *et al.*, 1994b). *IpiB* expression was only observed in zoospores, cysts, and in germinated cysts. The latter showed the highest mRNA levels, suggesting that the IPI-B proteins may be structural proteins, associated with the cell wall, and that they may be involved in the development of appressoria (van West & Govers, unpublished data; van West *et al.*, 2003).

Attempts to identify genes specifically induced during the cyst germination and appressoria stages are in progress. A proteomics-based approach in combination with a suppression subtractive hybridisation (SSH) method is being followed (Grenville, Avrova, Taylor, Birch, & van West, unpublished data). SSH is a method for generating cDNA libraries highly enriched for differentially expressed genes (Diatchenko *et al.*, 1999; Birch & Kamoun, 2000). It is a highly sensitive technique targeting low-abundance transcripts, which would not be picked up by differential display or cDNA-AFLPs (amplified fragment Length polymorphisms) (Birch & Kamoun, 2000), or in proteomic analyses. The combination of SSH and proteomics thus maximises chances of identifying appressorial stage-specific genes. Initial results have shown up-regulation of several genes in the appressorial stage. Also several newly synthesised proteins were found on 2D gels containing proteins from germinated cysts and also appressoria of *P. infestans*. These are now being analysed in more detail (Grenville, Avrova, Taylor, Birch, & van West, unpublished data).

9.5.1.3 Mating

P. infestans is a heterothallic organism with two known mating types, A1 and A2. These represent compatibility types differing in hormone production and response, rather than dimorphic sexual forms (Judelson, 1997a). In response to hormones, male and female gametangia (antheridia and oogonia) are formed. Haploid nuclei are generated in the gametangia. An A1 and an A2 nuclei fuse and an oospore containing one diploid nucleus is generated (Fig. 9.2). The oospore matures rapidly and a thick wall is formed which enables the oospore to survive in the soil for many years. After germination of the oospore, new progeny, either A1 or A2, are able to infect newly planted tubers or stems and leaves, which come into contact with the soil (Drenth *et al.*, 1995).

A genetic understanding of mating type evolved from studies using both *P. infestans* and *P. parasitica* (Judelson *et al.*, 1995; Judelson, 1996; Fabritius & Judelson, 1997). A1 strains were shown to be heterozygotes at the single mating type locus (A/a) while A2 strains were homozygous (a/a). Therefore, it is expected that the two mating types segregate in a 1:1 ratio. However, the inheritance of mating type reveals several forms of non-Mendelian behaviour in almost every cross. Often the ratios of A1 and A2 types are distorted. For example, in some cases a 10-fold excess of A1 mating type isolates is found (Judelson, 1996; van der Lee *et al.*, 1997). In addition, many isolates appear to contain recessive detrimental loci linked to the mating type, which act as a system of balanced lethals (Judelson *et al.*, 1995). This becomes apparent by the recovery of only two of the four possible combinations of mating type-linked alleles within the progeny of crosses. Occasionally, crosses are generated that yield either sterile or self-fertile progeny (Judelson, 1996, 1997b).

RAPD markers were used to construct a genetic map that contains the *P. infestans* mating type locus (Judelson *et al.*, 1995; Judelson 1996). Subsequently, several of these markers and a bacterial artificial chromosome (BAC) library were employed to generate a physical contig spanning the mating type locus. Analyses of this contig revealed a large region of heteromorphism within the area linked to the mating type, between the *A* and *a* chromosome homologues. The mating type locus was narrowed to a 60–70 kb interval via genetic mapping of candidate genes that were identified from a cDNA library after hybridising DNA fragments of the BAC clone (Randall *et al.*, 2003). Sequence analysis of 35 cDNAs, believed to be closest to the mating type locus, did not result in the identification of an obvious gene that may determine mating type such as putative transcription factors, genes for hormone biosynthesis, or transmembrane receptors (Randall *et al.*, 2003).

Recently, Fabritius *et al.* (2002) identified several genes from *P. infestans* that are all up-regulated during sexual development, via an SSH approach. One protein showed very strong homology to an extracellular glycoprotein elicitor from *P. sojae* (Sacks *et al.*, 1995). Another one showed weak homology to the elicitin gene family (Kamoun *et al.*, 1999b). Three proteins were identified that had similarities to proteins that are known to interact with RNA, namely a ribonuclease activator, the pumilio family of RNA-binding proteins, and RNase H. It is speculated that the RNase-related proteins could negatively regulate protein translation during sexual development, and that the RNA-binding protein may stabilise transcripts needed for germination (Fabritius *et al.*, 2002).

A member of the Puf family of translational regulators was also found to be expressed during sexual and asexual sporulation in *P. infestans* (Cvitanich & Judelson, 2003). Up to 51% amino acid identity was found in the RNA-binding domain of the protein. Detailed expression studies using transformants expressing a promoter fused with the β-glucuronidase (GUS) gene demonstrated that expression was first detected in male and female gametangial initials, and persisted in mature oospores. Expression was also noted in hyphal tips just prior to asexual sporulation, in sporangiophores, in mature sporangia, and in zoospores. However, the signal quickly disappeared once spores made the transition to hyphal growth. Therefore, it may be possible that the Puf protein exhibits a general RNA-binding activity in *P. infestans*, which functions as a protector of transcripts in spores until germination is initiated.

Besides mating, *P. infestans* is probably also able to generate genetically new strains by parasexuality. For example, heterokaryotic strains of *P. sojae* can be formed naturally when hyphae of two different genotypes fuse and the nuclei are mixed (Layton & Kuhn, 1990). Attempts to obtain hyphal fusion during growth of *P. infestans in planta* are so far unsuccessful (Judelson, 1997a; Judelson & Yang, 1998). However, heterokaryotic and polyploid strains were obtained *in vitro*, by mixing zoospores of two different strains (Judelson & Yang, 1998).

9.5.1.4 In planta *gene expression*

At present, little is known about the molecular events taking place in the hyphal cells of *P. infestans* during colonisation of plant tissue. However, it is anticipated that with the progress of modern molecular biology techniques it will be possible to begin to unravel the molecular processes involved in the interaction. Here, we focus on some experimental approaches that several researchers have undertaken to identify *P. infestans* genes expressed during colonisation of the plant.

It seems likely that a successful colonisation requires the up- or down-regulation of particular genes of *P. infestans*. A differential hybridisation approach resulted in the isolation of nine *ipi*-genes (see also Section 9.5.1.2) (Pieterse *et al.*, 1993a). Two of these *ipi*-genes encode homologues of ubiquitin (encoded by *ubi3R*) and calmodulin (*calA*) respectively (Pieterse *et al.*, 1991, 1993b). A third gene, *ipiO*, codes for a protein with unknown function (Pieterse *et al.*, 1994b). The IPI-O protein contains a putative signal peptide and a putative cell attachment sequence consisting of the three amino acids, arginine, glycine, and aspartic acid (RGD). *In vitro* expression studies revealed that expression of *ipiO* also occurs in zoospores, cysts, and germinating cysts of *P. infestans* (Pieterse *et al.*, 1994a; van West *et al.*, 1998). Cytological assays on leaves colonised by *P. infestans* transformants expressing β-GUS, under the control of the *ipiO* promoter, showed that expression of *ipiO* is specifically localised to the subapical and vacuolated areas at the tips of invading hyphae, indicating that *ipiO* is expressed during the biotrophic stage of the infection process. It is hypothesised that IPI-O may have its function at the interface between the invading hyphae and the plant cells (van West *et al.*, 1998).

Beyer *et al.* (2002) employed a SSH approach to identify *P. infestans* genes induced during the infection of potato. Mycelium of *P. infestans* was placed in contact with the host plant to encourage expression of potential *in planta*-induced genes, and then separated from the plant tissue after 4, 8, or 24 hours. A differential cDNA library was generated by SSH, comparing the so-called induced mycelium with mycelium incubated in water. Twenty per cent of about 100 generated cDNA fragments showed increased transcript levels in mycelium within the first 24 hours after exposure to a potato leaf. Some of the identified genes include an amino acid transporter, a D-xylose-proton symporter, a spliceosome-associated factor, an ABC transporter, and a cell division control protein with homology to the cdc6 protein from *Xenopus*. These results confirm that nutrient uptake via transporter proteins must be an important process for *in planta* growth.

Several small- and large-scale sequencing projects of cDNA or expressed sequence tags (ESTs) from *P. infestans* and other oomycetes are in progress or have been completed (Kamoun *et al.*, 1999b; Waugh *et al.*, 2000; Qutob *et al.*, 2000; Tyler, personal communication; van West, unpublished data). About 16 000 ESTs will be deposited in the *Phytophthora* Genome Consortium (PGC) database along with about 40 000 *P. sojae* ESTs (Waugh *et al.*, 2000; Qutob *et al.*, 2000; https://xgi.ncgr.org/pgc/). At present, ESTs from four interaction-cDNA libraries of *P. infestans* infected potato tubers are in the database alongside a cDNA library of *in vitro* grown mycelium. In addition, many *P. sojae* interaction-cDNA clones are present in the database. These so-called *interaction transcriptomes* (Birch & Kamoun, 2000) of *Phytophthora* spp. and their host plants will be very informative to deduct which *Phytophthora* genes are expressed during the interaction. Fortunately, discriminating the *Phytophthora* ESTs from the plant ESTs is not a major problem because *Phytophthora* genes have a much higher GC content (Qutob *et al.*, 2000; Kamoun, 2003). *Phytophthora* ESTs have an average GC content of about 58% and for plant ESTs this is about 46%. This finding, along with the availability of novel computational tools and algorithms, led to the development of PexFinder (Torto *et al.*, 2003), which is an algorithm for automated identification of extracellular proteins from EST data sets (http://www.oardc.ohio-state.edu/phytophthora/pexfinder). The program identified 261 ESTs corresponding to a set of 142 non-redundant *Pex* (*Phytophthora* extracellular proteins) cDNAs out of 2147 ESTs from the public databases. Interestingly, of these, 78 *Pex* cDNAs were novel showing no significant matches to public databases.

To identify which of the *Pex* cDNAs encode effector proteins that are able to manipulate plant processes, high throughput, functional-expression assays in plants were performed on 63 of the identified cDNAs using an *Agrobacterium tumefaciens* binary vector carrying the potato virus X (PVX) genome (Kamoun *et al.*, 1999a, 2002). This led to the discovery of two novel necrosis-inducing cDNAs, *crn1* and *crn2*, that belong to a large and complex protein family in *P. infestans*. Further characterisation of the *crn* genes indicated that they are both expressed in *P. infestans* during colonisation of the host plant tomato and that *crn2* induces defence response genes in tomato (Torto *et al.*, 2003).

Another group of about 11 proteins that encode Kazal-type serine protease inhibitors was identified from *P. infestans* again by data mining from EST libraries (Kamoun, 2003; Tian & Kamoun, 2003). The protease inhibitors contained one to three Kazal-type serine protease inhibitor domains. One of the extracellular protease inhibitors (EPIs), EPI1, was found to inhibit and interact with tomato P69 subtilisin-like proteases, suggesting that a novel type of defence–counterdefence crosstalk between plants and *P. infestans* may exist. Interestingly, EPI1 was found to protect degradation of another secreted protease inhibitor, EPI2, in intercellular fluids of tomato. This finding clearly suggests that inhibition of plant proteases will help keep secreted proteins from *P. infestans* unharmed in the apoplast of the plant (Tian & Kamoun, 2003).

At present, only a limited number of pathogenicity factors have been studied in the *P. infestans*–potato interaction. These include mainly enzymes that degrade cell walls such as endocellulases, 1,3-β-glucanases, β-glucosidases, pectin esterases, galactanases, and polygalacturonases (Friend, 1991), and phytotoxins (Seidel, 1961). Also some EST clones have been identified in *P. infestans* and *P. sojae* that encode degradative enzymes such as cutinases, proteases, endo- and exoglucanases, and chitinases (Kamoun *et al.*, 1999a; Qutob *et al.*, 2000; McLeod *et al.*, 2003). Some of these proteins have also been found on 2D-gels of proteins isolated from culture filtrates (Torto *et al.*, 2003; van West, Li, Taylor & Gow, unpublished).

9.6   The plant response

9.6.1   Compatible interactions

Early infection events of *P. infestans* on potato leaves are similar in both susceptible and resistance interactions (Hohl & Suter, 1976; Coffey & Wilson, 1983). Typically, infection starts when penetration of an epidermal cell occurs. In susceptible plants, branching hyphae with haustoria expand from the infection vesicle to neighbouring cells through the intercellular space. At the tip of growing hyphae, the interaction is fully biotrophic. There is no visible indication of defence responses in surrounding plant cells, and hyphae ramify through the mesophyl cell layers.

9.6.2   Incompatible interactions

Resistance to *P. infestans* may occur at the subspecies or variety level (race-specific resistance) or at the species or genus level (non-host resistance). In addition, resistance may be quantitative (partial resistance) with a partial reduction in disease severity. To achieve durable resistance, an improved understanding of the molecular basis of the various types of disease resistance is essential.

Since *Phytophthora* belongs to the unique group of oomycete plant pathogens, it evolved the ability to infect plants independently of true fungi (Kumar & Rzhetsky, 1996; Paquin *et al.*, 1997; van de Peer & de Wachter, 1997). This suggests that oomycetes may have distinct genetic and biochemical mechanisms for interacting with plants. For example, plant saponins target membrane sterols and are toxic to filamentous fungi (Osbourn, 1996a,b) but not to oomycetes as these contain little sterols in their membranes.

9.6.2.1   Race-specific resistance and the hypersensitive response

As in many other pathosystems, race-specific resistance to *P. infestans* in potato is explained by the gene-for-gene model (Flor, 1971). In this concept, the presence of both a resistance (*R*) gene and a corresponding avirulence (*Avr*) gene results in a hypersensitive response (HR), whereas absence of at least one of the two results in disease. The HR generally occurs as a rapid, localised cell death, and is considered as a form of programmed cell death in plants (Mittler & Lam, 1996; Heath, 1998). As a consequence of this local plant response, invading hyphae of *P. infestans* become isolated between dead cells, and no further biotrophic interaction is possible.

9.6.2.2   Avirulence genes

In recent years, substantial progress has been made in the identification of *Avr* genes from *P. infestans*. After making crosses, the segregation of *Avr* phenotypes can be followed in segregating progenies. A molecular genetic linkage map of *P. infestans* was constructed, and *Avr1*, *Avr2*, *Avr3*, *Avr4*, *Avr10*, and *Avr11* were mapped (van der Lee *et al.*, 1997). *Avr3*, *Avr10*, and *Avr11* occurred as a cluster at the telomeric region of Linkage Group VIII, and deletion of that part of the chromosome correlated with gain of virulence on potatoes carrying *R3*, *R10*, and *R11* (van der Lee *et al.*, 2001). Physical mapping is now enabled by construction of *P. infestans* BAC libraries (Randall & Judelson, 1999; Whisson *et al.*, 2001). *Avr*-linked AFLP markers are used to land on the BACs, and contigs are constructed with the aim of positionally cloning *Avr* genes (Whisson *et al.*, 2001).

Another approach was initiated by sequencing a number of candidate genes identified *in silico* from ESTs (Torto *et al.*, 2003) in order to identify SNPs (single nucleotide polymorphisms) associated with race structure from an extensive set of *P. infestans* isolates. This approach has identified a strong candidate for the *Avr3* gene (Armstrong *et al.*, 2003).

In addition, a proteomic approach is being followed to identify race-specific avirulence factors (van West, Li, Taylor & Gow, unpublished data). Several strains that have a large number of *Avr* genes are compared with strains that have none or few *Avr* genes. Secreted proteins are being purified from culture media and are separated by 2D-gel electrophoresis. The protein profiles of the avirulent strains are compared with the profiles of the virulent strains, to look for protein spots that correlated with the avirulence phenotype.

9.6.2.3 Resistance genes

Genetic resistance to *P. infestans* is amply present in wild *Solanum* species (Hoekstra & Seidewitz, 1987; Colon & Budding, 1988; Wastie, 1991; Ruiz De Galarreta *et al.*, 1998; Micheletto *et al.*, 1999). In the past, 11 *R* genes have been introgressed from *Solanum demissum* into potato (Müller & Black, 1952). These *R* genes provide strong resistance to specific races only (Malcolmson & Black, 1966), thereby illustrating the gene-for-gene model. A consequence of this model is that races of the pathogen with a chromosomal deletion (van der Lee *et al.*, 2001) or a mutation in their *Avr* gene(s) can arise and become virulent on particular plant genotypes. Such processes took place in *P. infestans* field isolates; the pathogen adapted to introgressed *R* genes and became virulent on *R* gene-bearing potatoes again.

To date, two *R* genes targeted against late blight have been cloned, both by a positional cloning approach. The first gene was *R1* (Ballvora *et al.*, 2002). As predicted by the gene-for-gene model, *R1* conferred race-specific resistance to a *P. infestans* race 4 isolate, but became infected by a *P. infestans* race 1 isolate. The *R1* gene contains a leucine zipper (LZ) motif, a putative nucleotide binding site (NBS), and several leucine-rich repeats (LRR). This NBS–LRR structure is typical of *R* genes active against other pathogens (Young, 2000). The specificity of recognition is expected to lay in the LRR domain, a hypervariable region which functions during interactions with other proteins (Jones & Jones, 2001). The NBS domain is involved in the signal transduction pathway, leading to the onset of the cell death response (van der Biezen & Jones, 1998), and the LZ may participate in coiled-coiled (CC) secondary protein structure (Lupas, 1996).

Also the second *R* gene, *Rpi-blb*, belongs to this same class of CC-NBS-LRR genes (van der Vossen *et al.*, 2002). Surprisingly, *Rpi-blb* appears to be the same gene as *Rb*, which was mapped and cloned in a different laboratory (Naess *et al.*,

2000). *Rpi-blb/Rb* originates from *Solanum bulbocastanum*, and confers full resistance to all isolates of *P. infestans* tested so far.

The cloning of some of the *S. demissum*-derived *R* genes is in progress in various laboratories. Although *P. infestans* has already overcome these genes, they will be useful to study the mechanism of *P. infestans* recognition at the molecular level. However, the pathogen *Avr* factors, which these are thought to interact with, have yet to be identified. Fortunately for agriculture, also novel *R* genes are being retrieved from *Solanum* germplasm and are subject to cloning and introgression in various laboratories. Although it cannot be excluded that virulent isolates of *P. infestans* can arise, such genes could prove valuable to obtain a resistant potato crop, at least for a number of years.

9.6.2.4   Non-host resistance

Non-host resistance to *P. infestans* occurs in most plant species. Weeds, like *Arabidopsis* plants, that are able to grow in infected fields are excellent examples of non-host resistance. A particularly interesting example is a relative of potato, the black nightshade (*Solanum nigrum*), which survives in *P. infestans*-infected potato fields. Cytological studies of *Arabidopsis* and *S. nigrum* have indicated that an extremely rapid HR occurs upon inoculation with *P. infestans* (Vleeshouwers *et al.*, 2000).

Recent developments suggest that multiple layers of gene-for-gene interactions form the initial barrier to *P. infestans* infection in non-host plants (Kamoun, 2001). In this scenario, a non-host interaction would occur when multiple elicitors or *Avr* gene products are recognised by matching *R* genes. To isolate such non-host *R* genes by a genetic approach is hampered by sexual incompatibility between hosts and non-hosts and by the absence of variation in plant resistance and pathogen virulence. However, there are various examples of *P. infestans* proteins acting as elicitors in non-host plants.

In *Nicotiana* species, resistance to *P. infestans* is diverse, with the intensity of the HR being correlated with the level of resistance (Kamoun *et al.*, 1998a). Most *Nicotiana* spp. are typical non-hosts for *P. infestans*, but some species can become partially infected by aggressive field isolates. *Phytophthora* species produce 10-kD extracellular proteins, termed elicitins, which induce the HR and other biochemical changes associated with defence responses in *Nicotiana* (Ricci *et al.*, 1989; Kamoun *et al.*, 1993a,b, 1997b). *P. infestans* strains deficient in the elicitin protein INF1 (Kamoun *et al.*, 1997b; van West *et al.*, 1999) induce disease lesions on *Nicotiana benthamiana*, suggesting that INF1 functions as an *Avr* factor in this plant species (Kamoun *et al.*, 1998b). Also other members of the elicitin family, such as INF2a and INF2b, induce various degrees of HR symptoms in *Nicotiana* species (Kamoun *et al.*, 1997a, submitted). Functional analysis of such genes is still underway to further dissect and compare the molecular basis of non-host resistance in this *Nicotiana*–*P. infestans* model system (Kamoun, 2001).

Another well-studied non-host for *Phytophthora* is parsley. Following inoculation with *Phytophthora*, parsley cells exhibit a series of morphological and biochemical defence responses that culminate in HR cell death (Hahlbrock *et al.*, 1995; Naton *et al.*, 1996). An extracellular 42-kDa glycoprotein elicitor from *P. sojae*, or more specifically a 13 amino-acid peptide (Pep-13) derived from this protein, induces changes in plasma membrane permeability, oxidative burst, activation of defence genes, and accumulation of defence compounds (Nürnberger *et al.*, 1994). The 42-kD glycoprotein is highly conserved among ten *Phytophthora* species, including *P. infestans*, and was shown to possess calcium-dependent cell wall *trans*-glutaminase (TGAse) activity (Brunner *et al.*, 2002). Mutational analysis within Pep-13 revealed that the same amino acids were indispensable for both TGAse and elicitor activity in potato and parsley. This suggests that Pep-13 functions as a genus-specific elicitor for activation of defence responses in both host and non-host plants (Brunner *et al.*, 2002).

9.6.2.5   Partial resistance

Cytological examination of a series of partially resistant *Solanum* species inoculated by *P. infestans* revealed HR reactions, and on numerous occasions late, or trailing HR, was observed (Vleeshouwers *et al.*, 2000). Partial resistance was genetically analysed in a cross between diploid potato lines, and quantitative trait loci (QTLs) contributing to resistance to late blight were identified (Leonards-Schippers *et al.*, 1994). Also in *Solanum berthaultii* and *Solanum microdontum*, QTLs for late blight resistance were found (Ewing *et al.*, 2000; Sandbrink *et al.*, 2000). Genetic mapping revealed that QTLs correspond to regions of the genome that contain clusters of *R* genes and *R* gene analogues (RGAs) (Leister *et al.*, 1996), which suggest that qualitative and quantitative resistance may share a similar genetic basis.

Various resistance mechanisms have been reported in addition to the HR. As pathogens attempt to feed on the plant by dismantling the cell walls, plants in turn deposit dense materials, such as callose or lignin, to hamper penetration of the cells (Hijwegen, 1963; Aist, 1976). In the case of *P. infestans*, localised cell wall degradation at haustorial penetration sites is accompanied by accumulation of lignin-like material (Friend, 1973) and callose in papillae or collars (Wilson & Coffey, 1980; Cuypers & Hahlbrock, 1988; Gees & Hohl, 1988).

9.6.3   Durable resistance

All known types of resistance to *Phytophthora* are associated with the HR. Recent developments suggest that *R* gene receptors triggered by pathogen elicitors may also mediate non-host resistance and partial resistance phenotypes (Kamoun *et al.*, 1999c). Even though *R* genes are thought to be ineffective in the field over long periods of time, there are attractive theories that suggest that *R* genes could mediate durable resistance. For example, multiple *R* genes recognising unrelated *Avr* targets would be difficult to overcome, as the pathogen would require various independent mutations to become virulent (Staskawicz *et al.*, 1995; Crute & Pink, 1996). In addition, a durable *R* gene could recognise an *Avr* gene that is essential to the pathogen (Staskawicz *et al.*, 1995; Swords *et al.*, 1996; Laugé *et al.*, 1998). Targeting such essential proteins is expected to lead to durable resistance because mutations in the dual *Avr*-virulence gene would result in a severe fitness penalty for the pathogen. Therefore, current research is focusing on studying elicitors and their interaction with the plant at the molecular level. For example, functional genomic strategies have been applied by mining large EST databases in search of candidate elicitor genes using PexFinder (see Section 9.5.1.4) (Torto *et al.*, 2003). cDNA clones coding for *Phytophthora* extracellular proteins have been tested for recognition in plants using various functional assays, e.g. systems based on *Agrobacterium tumefaciens* (van der Hoorn *et al.*, 2000; Kamoun *et al.*, 2003), or potato virus X (Hammond-Kosack *et al.*, 1995; Laugé *et al.*, 1998; Kamoun *et al.*, 1999a), or a combination of both (Qutob *et al.*, 2002; Torto *et al.*, 2003). Using the binary PVX expression system, high throughput screening of *Nicotiana* and *Solanum* germplasm is being performed and resistant plants recognising specific elicitors have been identified (Torto *et al.*, 2003; Vleeshouwers *et al.*, 2002; Vleeshouwers & Kamoun, unpublished results). Unravelling the nature of oomycete *Avr* and elicitor molecules and their interaction with novel *R* genes will aid in understanding the molecular basis of race evolution and in defining sustainable strategies for engineering durable genetic resistance.

9.7   Future perspectives

Conventional breeding for late blight resistance in potato has been relatively unsuccessful. This is partly because biological knowledge of the oomycete–plant interaction is very limited. Therefore, to be able to develop late blight resistance, there should be greater emphasis on studying the biology of the pathogen itself and the interaction with its host. Progress and the impetus for fundamental molecular research on *P. infestans* have in the past been hampered by a lack of a good molecular tool box, the lack of DNA sequence information, and in particular a lack of a route for functional analysis of genes. However, work in the last few years has significantly altered these aspects of research and it is now possible, if not yet easy, to examine the relationship between host and pathogen at the molecular level (for recent reviews see Kamoun, 2003; van West *et al.*, 2003). Gene silencing is beginning to become a robust and generally applicable technology. Now it is used to explore gene-for-gene interactions as well as key aspects of the fundamental molecular processes in the life cycle of *P. infestans*. In addition, new genetic resources are being explored and innovative approaches have become available, e.g. novel functional genomic techniques such as PexFinder and binary PVX expression systems.

   We envisage that in particular the asexual life cycle from zoosporogenesis through zoospore liberation, encystment, germination, appressorium formation, *in planta* colonisation, and growth are important targets for future work on the plant pathology of *P. infestans* and other oomycetes. It is anticipated that functional genomics and proteomic approaches will ultimately lead to new control strategies and in-depth analysis of the biology of *P. infestans* and other oomycete pathogens.

Acknowledgements

The authors would like to thank The Royal Society for providing funding and a personal fellowship to Pieter van West (Royal Society University Research Fellowship) and Laura J. Grenville for critically reading the chapter.

References

Aist, J. (1976) Papillae and related wound plugs of plant cells. *Annu. Rev. Phytopathol.*, **14**, 145–163.

Ambikapathy, J., Marshall, J.S., Hocart, C.H. & Hardham, A.R. (2002) The role of proline in osmoregulation in *Phytophthora nicotianae. Fungal Genet. Biol.*, **35**, 287–299.

Armstrong, M., Avrova, A., Whisson, S. & Birch, P. (2003) Cloning avirulence genes from *Phytophthora infestans. Fungal Genet. Newsletter*, **50**, Suppl. (399), p. 132.

Baldauf, S.L., Roger, A.J., Wenk-Siefert, I. & Doolittle, W.F. (2000) A kingdom-level phylogeny of eukaryotes based on combined protein data. *Science*, **290**, 972–977.

Ballvora, A., Ercolano, M.R., Weiss, J., Meksem, K., Bormann, C.A., Oberhagemann, P., Salamini, F. & Gebhardt, C. (2002) The R1 gene for potato resistance to late blight (*Phytophthora infestans*) belongs to the leucine zipper/NBS/LRR class of plant resistance genes. *Plant J.*, **30**, 361–371.

Barr, D.J.S. (1981) The phylogenetic and taxonomic implications of flagellar rootlet morphology among zoosporic fungi. *Biosystems*, **14**, 359–370.

Barr, D.J.S. (1983) The zoosporic grouping of plant pathogens, entity or non-entity. In: *Zoosporic Plant Pathogens: A Modern Perspective* (ed. S.T. Buczaki), pp. 43–83. Academic Press.

Bartnicki-Garcia, S. & Wang, M.C. (1983) Biochemical aspects of morphogenesis in *Phytophthora*. In: *Phytophthora: Its Biology, Taxonomy, Ecology, and Pathology* (eds D.C. Erwin, S. Bartnicki-Garcia & P.H. Tsao), pp. 121–138. American Phytopathological Society Press, St Paul, MN.

Beyer, K., Jiménez, S.J., Randall, T.A., Lam, S., Binder, A., Boller, T. & Collinge, M.A. (2002) Characterization of *Phytophthora infestans* genes regulated during the interaction with potato. *Mol. Plant Pathol.*, **3**, 473–485.

Birch, P.R.J. & Kamoun, S. (2000) Studying interaction transcriptomes: coordinated analyses of gene expression during plant–microorganism interactions. *Trends in Plant Science*, Supplement: Life Science Research in the 21st Century: A Trends Guide, pp. 77–82.

Bourke, A. (1991) Potato blight in Europe in 1845: the scientific controversy. In: *Phytophthora* (eds J.A. Lucas, R.C. Shattock, D.S. Shaw & L.R. Cooke), pp. 12–24. Cambridge Univ. Press Cambridge, UK.

Brunner, F., Rosahl, S., Lee, J., Rudd, J.J., Geiler, C., Kauppinen, S., Rasmussen, G., Scheel, D. & Nurnberger, T. (2002) Pep-13, a plant defense-inducing pathogen-associated pattern from Phytophthora transglutaminases. *EMBO J.*, **21**, 6681–6688.

Coffey, M. & Wilson, U. (1983) Histology and cytology of infection and disease caused by *Phytophthora*. In: *Phytophthora: Its Biology, Taxonomy, Ecology and Pathology* (eds D.C. Erwin, S. Bartnicki-Garcia & P. Tsao), pp. 289–310. American Phytopathological Society. St Paul, Minnesota, USA.

Colon, L.T. & Budding, D.J. (1988) Resistance to late blight (*Phytophthora infestans*) in ten wild *Solanum* species. *Euphytica Supplement*, 77–86.

Connolly, M.S., Williams, N., Heckman, C.A. & Morris, P.F. (1999) Soybean isoflavones trigger a calcium influx in *Phytophthora sojae. Fungal Genet. Biol.*, **28**, 6–11.

Crute, I.R. & Pink, D.A.C. (1996) Genetics and utilization of pathogen resistance in plants. *Plant Cell*, **8**, 1747–1755.

Cuypers, B. & Hahlbrock, K. (1988) Immunohistochemical studies of compatible and incompatible interactions of potato leaves with *Phytophthora infestans* and of the nonhost response to *Phytophthora megasperma. Can. J. Bot.*, **66**, 700–705.

Cvitanich, C. & Judelson, H.S. (2003) A gene expressed during sexual and asexual sporulation in *Phytopthora infestans* is a member of the Puf family of translational regulators. *Eukaryot. Cell*, **2**, 465–473.

Deahl, K.L., Groth, R.W., Young, R., Sinden, S.L. & Gallegly, M.E. (1991) Occurrence of the A2 mating type of *Phytophthora infestans* in potato fields in the United States and Canada. *Am. Potato J.*, **68**, 717–726.

de Bary, A. (1876) Researches into the nature of the potato fungus, *Phytophthora infestans*. *J. R. Agric. Soc. Engl.*, Ser. 2., **12**, 239–269.

Diatchenko, L., Lukyanov, S., Lau, Y.F. & Siebert, P.D. (1999) Suppression subtractive hybridization: a versatile method for identifying differentially expressed genes. *Methods Enzymol.*, **303**, 349–380.

Dick, M.W. (1995) The Straminipilous fungi. A new classification for the biflagellate fungi and their uniflagellate relatives with particular reference to Lagenidiaceous fungi. *C.A.B. Internat. Mycol. Pap.* No. 168.

Drenth, A., Goodwin, S., Fry, W. & Davidse, L. (1993) Genotypic diversity of *Phytophthora infestans* in the Netherlands revealed by DNA poly-morphisms. *Phytopathology*, **83**, 1087–1092.

Drenth, A., Janssen, E.M. & Govers, F. (1995) Formation and survival of oospores of *Phytophthora infestans* under natural conditions. *Plant Pathol.*, **44**, 86–94.

Duncan, J. (1999) *Phytophthora* – an abiding threat to our crops. *Microbiol. Today*, **26**, 114–116.

Eckert, J.W. & Tsao, P.H. (1962) A selective antibiotic medium for isolation of *Phytophthora* and *Pythium* from plant roots. *Phytopathology*, **52**, 771–777.

Elliott, C.G. (1983) Physiology of sexual reproduction in *Phytophthora*. In: *Phytophthora: Its Biology, Taxonomy, Ecology, and Pathology* (eds D.C. Erwin, S. Bartnicki-Garcia & P.H. Tsao), pp. 71–80. American Phytopathological Society Press, St Paul, MN.

Erwin, D.C. & Ribeiro, O.K. (1996) Introduction to the genus *Phytophthora*. In: *Phytophthora Diseases Worldwide*, pp. 1–7. American Phytopathological Society. St. Paul, MN.

Ewing, E.E., Simko, I., Smart, C.D., Bonierbale, M.W., Mizubuti, E.S.G., May, G.D. & Fry, W.E. (2000) Genetic mapping from field tests of qualitative and quantitative resistance to *Phytophthora infestans* in a population derived from *Solanum tuberosum* and *Solanum berthaultii*. *Mol. Breeding*, **6**, 25–36.

Fabritius, A.L. & Judelson, H.S. (1997) Mating-type loci segregate aberrantly in *Phytophthora infestans* but normally in *Phytophthora parasitica*: implications for models of mating-type determination. *Curr. Genet.*, **32**, 60–65.

Fabritius, A.L., Cvitanich, C. & Judelson, H.S. (2002) Stage-specific gene expression during sexual development in *Phytophthora infestans*. *Mol. Microbiol.*, **45**, 1057–1066.

Flor, H.H. (1971) Current status of the gene-for-gene concept. *Annu. Rev. Phytopathol.*, **78**, 275–298.

Förster, H., Coffey, M.D., Elwood, H. & Sogin, M.L. (1990) Sequence analysis of the small subunit ribosomal RNAs of three zoosporic fungi and implications for fungal evolution. *Mycologia*, **82**, 306–312.

Friend, J. (1973) Resistance of potato to *Phytophthora infestans*. In: *Fungal Pathogenicity and the Plant's Response: 3rd Long Ashton Symposium 1971* (eds R.J.W. Byrde & C.V. Cutting), conference proceedings, pp. 383–396. Academic Press, London.

Friend, J. (1991) The biochemistry and cell biology of interaction. In: *Phytophthora Infestans, the Cause of Late Blight on Potato. Advances in Plant Pathology 7* (eds D.S. Ingram & P.H. Willams), pp. 85–129. Academic Press, London, UK.

Fry, W. & Goodwin, S.B. (1997) Resurgence of the Irish potato famine fungus. *Bioscience*, **47**, 363–371.

Gees, R. & Hohl, H.R. (1988) Cytological comparison of specific (R3) and general resistance to late blight in potato leaf tissue. *Phytopathology*, **78**, 350–357.

Gornhardt, B., Rouhara, I. & Schmelzer, E. (2000) Cyst germination proteins of the potato pathogen *Phytophthora infestans* share homology with human mucins. *Mol. Plant Microbe Interact.*, **13**, 32–42.

Griffith, J.M., Davis, A.J. & Grant, B.R. (1992) Target sites of fungicides to control oomycetes. In: *Target Sites of Fungicide Action* (ed. W. Köller), pp. 69–100. CRC Press, Boca Raton, Fla.

Hahlbrock, K., Scheel, D., Logemann, E., Nurnberger, T., Parniske, M., Reinold, S., Sacks, W.R. & Schmelzer, E. (1995) Oligopeptide elicitor-mediated defense gene activation in cultured parsley cells. *Proc. Natl. Acad. Sci. USA*, **92**, 4150–4157.

Hammond-Kosack, K.E., Staskawicz, B.J., Jones, J.D.G. & Baulcombe, D.C. (1995) Functional expression of a fungal avirulence gene from a modified potato virus X genome. *Mol. Plant Microbe Interac.*, **8**, 181–185.

Hardham, A.R. (2001) The cell biology behind *Phytophthora* pathogenicity. *Australas. Plant Pathol.*, **30**, 91–98.

Hardham, A.R. & Gubler, F. (1990) Polarity of attachment of zoospores of a root pathogen and pre-alignment of the emerging germ tube. *Cell Biol. Int. Rep.*, **14**, 947–956.

Heath, M.C. (1998) Apoptosis, programmed cell death and the hypersensitive response. *Eur. J. Plant Pathol.*, **104**, 117–124.

Hemmes, D.E. (1983) Cytology of *Phytophthora*. In: *Phytophthora: Its Biology, Taxonomy, Ecology, and Pathology* (eds D.C. Erwin, S. Bartnicki-Garcia & P.H. Tsao), pp. 9–40. American Phytopathological Society Press, St Paul, MN.

Hendrix, J.W. (1970) Sterols in growth and reproduction of fungi. *Annu. Rev. Phytopathol.*, **8**, 111–130.

Hijwegen, T. (1963) Lignification, a possible mechanism of active resistance against pathogens. *Neth. J. Plant Pathol.*, **69**, 314–317.

Hoekstra, R. & Seidewitz, L. (1987) Evaluation data on tuber-bearing *Solanum* species. In: *German-Dutch Curatorium for Plant Genetic Resources*. Brauschweig, Federal Republic of Germany.

Hohl, H.R. & Suter, E. (1976) Host–parasite interfaces in a resistant and a susceptible cultivar of *Solanum tuberosum* inoculated with *Phytophthora infestans*: leaf tissue. *Can. J. Bot.*, **54**, 1956–1970.

Illingworth, C.A., Andrews, J.H., Bibeau, C. & Sogin, M.L. (1991) Phylogenetic placement of *Athelia bombacina*, *Aureobasidium pullalans*, and *Colletrotrichum gloeosporioides* inferred from sequence comparisons of small-subunit ribosomal RNAs. *Exp. Mycol.*, **15**, 65–75.

Jackson, S.L. & Hardham, A.R. (1996) A transient rise in cytoplasmic free calcium is required to induce cytokinesis in zoosporangia of *Phytophthora cinnamomi*. *Eur. J. Cell Biol.*, **69**, 180–188.

Jones, D.A. & Jones, J.D.G. (2001) The roles of leucine rich repeats in plant defences. *Adv. Bot. Res. Adv. Plant Pathol.*, **24**, 90–167.

Judelson, H.S. (1996) Genetic and physical variability at the mating type locus of the oomycete, *Phytophthora infestans*. *Genetics*, **144**, 1005–1013.

Judelson, H.S. (1997a) The genetics and biology of *Phytophthora infestans*: modern approaches to a historical challenge. *Fungal Genet. Biol.*, **22**, 65–76.

Judelson, H.S. (1997b) Expression and inheritance of sexual preference and selfing potential in *Phytophthora infestans*. *Fungal Genet. Biol.*, **21**, 188–197.

Judelson, H.S. & Roberts, S. (2002) Novel protein kinase induced during sporangial cleavage in the oomycete *Phytophthora infestans*. *Eukaryot. Cell*, **1**, 687–695.

Judelson, H.S. & Yang, G. (1998) Recombination pathways in *Phytophthora infestans*: polyploidy resulting from aberrant sexual development and zoospore-mediated heterokaryosis. *Mycol. Res.*, **102**, 1245–1253.

Judelson, H.S., Spielman, L.J. & Shattock, R.C. (1995) Genetic mapping and non-Mendelian segregation of mating type loci in the oomycete, *Phytophthora infestans*. *Genetics*, **141**, 503–512.

Kamoun, S. (2001) Nonhost resistance to *Phytophthora*: novel prospects for a classical problem. *Curr. Opin. Plant Biol.*, **4**, 295–300.

Kamoun, S. (2003) Molecular genetics of pathogenic oomycetes. *Eukaryot. Cell*, **2**, 191–199.

Kamoun, S., Klucher, K.M., Coffey, M.D. & Tyler, B.M. (1993a) A gene encoding a host-specific elicitor protein of *Phytophthora parasitica*. *Mol. Plant Microbe Interact.*, **6**, 573–581.

Kamoun, S., Young, M., Glascock, C.B. & Tyler, B.M. (1993b) Extracellular protein elicitors from *Phytophthora*: host-specificity and induction of resistance to bacterial and fungal phytopathogens. *Mol. Plant Microbe Interact.*, **6**, 15–25.

Kamoun, S., Lindqvist, H. & Govers, F. (1997a) A novel class of elicitin-like genes from *Phytophthora infestans*. *Mol. Plant Microbe Interact.*, **10**, 1028–1030.

Kamoun, S., van West, P., de Jong, A.J., de Groot, K.E., Vleeshouwers, V.G.A.A. & Govers, F. (1997b) A gene encoding a protein elicitor of *Phytophthora infestans* is down-regulated during infection of potato. *Mol. Plant Microbe Interact.*, **10**, 13–20.

Kamoun, S., van West, P. & Govers, F. (1998a) Quantification of late blight resistance of potato using transgenic *Phytophthora infestans* expressing β-glucuronidase. *Eur. J. Plant Pathol.*, **104**, 521–525.

Kamoun, S., van West, P., Vleeshouwers, V.G.A.A., de Groot, K.E. & Govers, F. (1998b) Resistance of *Nicotiana benthamiana* to *Phytophthora infestans* is mediated by the recognition of the elicitor protein INF1. *Plant Cell*, **10**, 1413–1425.

Kamoun, S., Honée, G., Weide, R., Laugé, R., Kooman-Gersmann, M., de Groot, K., Govers, F. & de Wit, P.J.G.M. (1999a) The fungal gene *Avr9* and the oomycete gene *inf1* confer avirulence to potato virus X on tobacco. *Mol. Plant Microbe Interact.*, **12**, 459–462.

Kamoun, S., Hraber, P., Sobral, B. Nuss, B. & Govers, F. (1999b) Initial assessment of gene diversity for the oomycete pathogen *Phytophthora infestans* based on expressed sequences. *Fungal Genet. Biol.*, **28**, 94–106.

Kamoun, S., Huitema, E. & Vleeshouwers, V.G.A.A. (1999c) Resistance to oomycetes: a general role for the hypersensitive response? *Trends Plant Sci.*, **4**, 196–200.

Kamoun, S., Dong, S., Hamada, W., Huitema, E., Kinney, D., Morgan, W.R., Styer, A., Testa, A. & Torto., T. (2002) From sequence to phenotype: functional genomics of *Phytophthora*. *Can. J. Plant Pathol.*, **24**, 6–9.

Kamoun, S., Hamada, W. & Huitema, E. (2003) Agrosuppression: a bioassay for the hypersensitive response suited to high-throughput screening. *Mol. Plant Microbe interact.*, **16**, 7–13.

Knight, J. (2002) Fears mount as oak blight infects redwoods. *Nature*, **415**, 251.

Kumar, S. & Rzhetsky, A. (1996) Evolutionary relationships of eukaryotic kingdoms. *J. Mol. Evol.*, **42**, 183–193.

Latijnhouwers, M. (2003) The role of heterotrimeric G-proteins in development and virulence of *Phytophthora infestans*. PhD thesis, Wageningen Agricultural University and the Graduate School of Experimental Plant Sciences, pp. 1–158.

Latijnhouwers, M., Munnik, T. & Govers, F. (2002) Phospholipase D in *Phytophthora infestans* and its role in zoospore encystment. *Mol. Plant Microbe Interact.*, **15**, 939–946.

Laugé, R., Joosten, M.H.A.J., Haanstra, J.P., Goodwin, P.H., Lindhout, P. & De Wit, P.J.G.M. (1998) Successful search for a resistance gene in tomato targeted against a virulence factor of a fungal pathogen. *Proc. Natl. Acad. Sci. USA*, **95**, 9014–9018.

Laxalt, M.A., Latijnhouwers, M., van Hulten, M. & Govers, F. (2002) Differential expression of G protein α and β subunit genes during development of *Phytophthora infestans*. *Fungal Genet. Biol.*, **36**, 137–146.

Layton, A.C. & Kuhn, D.N. (1990) *In planta* formation of heterokaryons of *Phytophthora megasperma* f. sp. *glycinea*. *Phytopathology*, **80**, 602–606.

Leal-Morales, C.A., Gay, L., Fevre, M. & Bartnicki-Garcia, S. (1997) The properties and localization of *Saprolegnia monoica* chitin synthase differ from those of other fungi. *Microbiology*, **143**, 2473–2483.

Leister, D., Ballvora, A., Salamini, F. & Gebhardt, C. (1996) A PCR-based approach for isolating pathogen resistance genes from potato with potential for wide application in plants. *Nature Genet.*, **14**, 421–429.

Leonards-Schippers, C., Gieffers, W., Schafer Pregl, R., Ritter, E., Knapp, S.J., Salamini, F. & Gebhardt, C. (1994) Quantitative resistance to *Phytophthora infestans* in potato: a case study for QTL mapping in an allogamous plant species. *Genetics*, **137**, 67–77.

Lupas, A. (1996) Coiled coils: new structures and new functions. *Trends Biochem. Sci.*, **21**, 375–382.

Malcolmson, J.F. & Black, W. (1966) New *R* genes in *Solanum demissum* Lindl. and their complementary races of *Phytophthora infestans* (Mont.) de Bary. *Euphytica*, **15**, 199–203.

Margulis, L., Dolan, M.F. & Guerrero, R. (2000) The chimeric eukaryote: origin of the nucleus from the karyomastigont in amitochondriate protists. *Proc. Natl. Acad. Sci. USA*, **97**, 6954–6959.

McLeod, A., Smart, C.D. & Fry, W.E. (2003) Characterization of 1,3-glucanase and 1,3;1,4-glucanase genes from *Phytophthora infestans*. *Fungal Genet. Bio.*, **38**, 250–263

Micheletto, S., Andreoni, M. & Huarte, M. (1999) Vertical resistance to late blight in wild potato species from Argentina. *Euphytica*, **110**, 133–138.

Mittler, R. & Lam, E. (1996) Sacrifice in the face of foes: Pathogen-induced programmed cell death in plants. *Trends Microbiol.*, **4**, 10–15.

Mort-Bontemps, M., Gay, L. & Fevre, M. (1997) CHS2, a chitin synthase gene from the oomycete *Saprolegnia monoica*. *Microbiology*, **143**, 2009–2020.

Müller, K.O. & Black, W. (1952) Potato breeding for resistance to blight and virus diseases during the last hundred years. *Zeitschrift für Pflanzenzüchtung*, **31**, 305–318.

Naess, S.K., Bradeen, J.M., Wielgus, S.M., Haberlach, G.T., McGrath, J.M. & Helgeson, J.P. (2000) Resistance to late blight in *Solanum bulbocastanum* is mapped to chromosome 8. *Theor. Appl. Genet.*, **101**, 697–704.

Naton, B., Hahlbrock, K. & Schmelzer, E. (1996) Correlation of rapid cell death with metabolic changes in fungus-infected, cultured parsley cells. *Plant Physiol.*, **112**, 433–444.

Nürnberger, T., Nennstiel, D., Jabs, T., Sacks, W.R., Hahlbrock, K. & Scheel, D. (1994) High affinity binding of a fungal oligopeptide elicitor to parsley plasma membranes triggers multiple defense responses. *Cell*, **78**, 449–460.

Osbourn, A.E. (1996a) Preformed antimicrobial compounds and plant defense against fungal attack. *Plant Cell*, **8**, 1821–1831.

Osbourn, A.E. (1996b) Saponins and plant defence – a soap story. *Trends Plant Sci.*, **1**, 4–9.

Paquin, B., Laforesst, M.J., Forget, L., Roewer, I., Wang, Z., Longcore, J. & Lang, B.F. (1997) The fungal mitochondrial genome project: evolution of fungal mitochondrial genomes and their gene expression. *Curr. Genet.*, **31**, 380–395.

Peterson, P.D., Jr., Campbell, C.L. & Griffith, C.S. (1992) James E. Teschemacher and the cause and management of potato late blight in the United States. *Plant Dis.*, **76**, 754–756.

Pfyffer, G., Boraschi-Gaia, E., Weber, B., Hoesch, L., Orpin, C.G. & Rast, D. (1990) A further report on the occurrence of acylic sugar alcohols in fungi. *Mycol. Res.*, **92**, 219–222.

Pieterse, C.M.J., Risseeuw, E.P. & Davidse, L.C. (1991) An *in planta* induced gene of *Phytophthora infestans* codes for ubiquitin. *Plant Mol. Biol.*, **17**, 799–811.

Pieterse, C.M.J., Riach, M.R., Bleker, T., van den Berg-Velthuis, G.C.M. & Govers, F. (1993a) Isolation of putative pathogenicity genes of the potato late blight *Phytophthora infestans* by differential screening of a genomic library. *Physiol. Mol. Plant Pathol.*, **43**, 69–79.

Pieterse, C.M.J., Verbakel, H.M., Hoek Spaans, J., Davidse, L.C. & Govers, F. (1993b) Increased expression of the calmodulin gene of the late blight fungus *Phytophthora infestans* during pathogenesis on potato. *Mol. Plant–Microbe Interact.*, **6**, 164–172.

Pieterse, C.M.J., Derksen, A.M.C.E., Folders, J. & Govers, F. (1994a) Expression of the *Phytophthora infestans ipi*B and *ipi*O genes *in planta* and *in vitro*. *Mol. Gen. Genet.*, **244**, 269–277.

Pieterse, C.M.J., van West, P., Verbakel, H.M., Brasse, P.W.H.M., van den Berg-Velthuis, G.C.M. & Govers, F. (1994b) Structure and genomic organization of the *ipiB* and *ipiO* gene clusters of *Phytophthora infestans*. *Gene*, **138**, 67–77.

Podger, F.D. (1972) *Phytophthora cinnamomi*, a cause of lethal disease of indigenous plant communities in Western Australia. *Phytopathology*, **62**, 972–981.

Podger, F.D., Doepel, R.F. & Zentmyer, G.A. (1965) Association of *Phytophthora cinnamomi* with a disease of *Eucalyptus marginata* forest in Western Australia. *Plant Dis. Rep.*, **49**, 943–947.

Qutob, D., Hraber, P.T., Sobral, B.W. & Gijzen, M. (2000) Comparative analysis of expressed sequences in *Phytophthora sojae*. *Plant Physiol.*, **123**, 243–254.

Qutob, D., Kamoun, S. & Gijzen, M. (2002) Expression of a *Phytophthora sojae* necrosis-inducing protein occurs during transition from biotrophy to necrotrophy. *Plant J.*, **32**, 361–373.

Randall, T.A. & Judelson, H.S. (1999) Construction of a bacterial artificial chromosome library of *Phytophthora infestans* and transformation of clones into *P. infestans*. *Fungal Genet. Biol.*, **28**, 160–170.

Randall, T.A., Ah Fong, A. & Judelson, H.S. (2003) Chromosomal heteromorphism and an apparent translocation detected using a BAC contig spanning the mating type locus of *Phytophthora infestans*. *Fungal Genet. Biol.*, **38**, 75–84.

Reid, B., Morris, B.M. & Gow, N.A.R. (1995) Calcium-dependent, genus-specific autoaggregation of zoospores of phytopathogenic fungi. *Exp. Mycol.*, **19**, 202–213.

Ricci, P., Bonnet, P., Huet, J.C., Sallantin, M., Beauvais Cante, F., Bruneteau, M., Billard, V., Michel, G. & Pernollet, J.C. (1989) Structure and activity of proteins from pathogenic fungi *Phytophthora* eliciting necrosis and acquired resistance in tobacco. *Eur. J. Biochem.*, **183**, 555–564.

Rizzo, D.M., Garbelotto, M., Davidson, J.M., Slaughter, G.W. & Koike, S.T. (2002) *Phytophthora ramorum* as the cause of extensive mortality of *Quercus* spp. and *Lithocarpus densiflorus* in California. *Plant Dis.*, **86**, 205–214.

Ruiz De Galarreta, J.I., Carrasco, A., Salazar, A., Barrena, I., Iturritxa, E., Marquinez, R., Legorburu, F.J. & Ritter, E. (1998) Wild *Solanum* species as resistance sources against different pathogens of potato. *Potato Res.*, **41**, 57–68.

Sacks, W., Nuernberger, T., Hahlbrock, K. & Scheel, D. (1995) Molecular characterization of nucleotide sequences encoding the extracellular glycoprotein elicitor from *Phytophthora megasperma*. *Mol. Gen. Genet.*, **246**, 45–55.

Sandbrink, J.M., Colon, L.T., Wolters, P.J.C.C. & Stiekema, W.J. (2000) Two related genotypes of *Solanum microdontum* carry different segregating alleles for field resistance to *Phytophthora infestans*. *Mol. Breeding*, **6**, 215–225.

Schiermeier, Q. (2001) Russia needs help to fend off potato famine, researchers warn. *Nature*, **410**, 1011.

Seidel, H. (1961) Untersuchungen über den Nährstoffbedarf und die toxinbildung des pilzes *Phytophthora infestans* (Mont.) de Bary in vollsynthetischen nährlösungen. *Phytopathologische Zeitschrift*, **41**, 1–26.

Sing, V.O. & Bartnicki-Garcia, S. (1975) Adhesion of *Phytophthora palmivora* zoospores: electron microscopy of cell attachment and cyst wall fibril formation. *J. Cell Sci.*, **18**, 123–132.

Staskawicz, B.J., Ausubel, F.M., Baker, B.J., Ellis, J.G. & Jones, J.D.G. (1995) Molecular genetics of plant disease resistance. *Science*, **268**, 661–667.

Swords, K.M.M., Dahlbeck, D., Kearney, B., Roy, M. & Staskawicz, B.J. (1996) Spontaneous and induced mutations in a single open reading frame alter both virulence and avirulence in *Xanthomonas campestris* pv. *vesicatoria avrBs2*. *J. Bacteriol.*, **178**, 4661–4669.

Tian, M. & Kamoun, S. (2003) An extracellular protease inhibitor from *Phytophthora infestans* targets tomato serine proteases: a counter-defence mechanism? *Fungal Genet. Newsletter*, **50** Suppl. (322), p. 113.

Torto, T.A., Li, S., Styer, A., Huitema, E., Testa, A., Gow, N.A.R., van West, P. & Kamoun, S. (2003) EST mining and functional expression assays identify extra-cellular signal proteins from the plant pathogen *Phytophthora*. *Genome Res.*, **13**, 1675–1685.

van de Peer, Y. & de Wachter, R. (1997) Evolutionary relationships among the eukaryotic crown taxa taking into account site-to-site rate variation in 18S rRNA. *J. Mol. Evol.*, **45**, 619–630.

van der Biezen, E.A. & Jones, J.D. (1998) The NB-ARC domain: a novel signalling motif shared by plant resistance gene products and regulators of cell death in animals. *Curr. Biol.*, **8**, R226–227.

van der Hoorn, R.A.L., Laurent, F., Roth, R. & de Wit, P.J.G.M. (2000) Agroinfiltration is a versatile tool that facilitates comparative analyses of *Avr9/Cf-9*-induced and *Avr4/Cf-4*-induced necrosis. *Mol. Plant–Microbe Interact.*, **13**, 439–446.

van der Lee, T., De Witte, I., Drenth, A., Alfonso, C. & Govers, F. (1997) AFLP linkage map of the oomycete *Phytophthora infestans*. *Fungal Genet. Biol.*, **21**, 278–291.

van der Lee, T., Testa, A., van' t Klooster, J., van den Berg Velthuis, G., & Govers, F. (2001) Chromosomal deletion in isolates of *Phytophthora infestans* correlates with virulence on R3, R10, and R11 potato lines. *Mol. Plant Microbe Interact.*, **14**, 1444–1452.

van der Vossen, E., Sikkema, A., Hekkert, B.T.L., Gros, J., Muskens, M., Stiekema, W.J. & Allefs, S. (2002) Cloning of an *R* gene from *Solanum bulbocastanum* conferring complete resistance to *Phytophthora infestans*. In: *GILB Global Initiative on Late Blight* (eds G. Wenzel & I. Wulfert). Hamburg.

van West, P. (2000) Molecular tools to unravel the role of genes from *Phytophthora infestans*. PhD thesis, Wageningen Agricultural University and the Graduate School of Experimental Plant Sciences, pp. 1–160.

van West, P., de Jong, A.J., Judelson, H.S., Emons, A.M.C. & Govers, F. (1998) The *ipi*O gene of *Phytophthora infestans* is highly expressed in invading hyphae during infection. *Fungal Genet. Biol.*, **23**, 126–138.

van West, P., Kamoun, S., van' t Klooster, J. W. & Govers, F. (1999) Internuclear gene silencing in *Phytophthora infestans. Mol. Cell*, **3**, 339–348.

van West, P., Appiah, A.A. & Gow, N.A.R. (2003) Advances in research on root pathogenic oomycetes. *Physiol. Mol. Plant Pathol.*, **62**, 99–113.

Vleeshouwers, V.G.A.A., van Dooijeweert, W., Govers, F., Kamoun, S. & Colon, L.T. (2000) The hypersensitive response is associated with host and nonhost resistance to *Phytophthora infestans. Planta*, **210**, 853–864.

Vleeshouwers, V.G.A.A., Kamphuis, L., Torto, T., Kamoun, S., Jacobsen, E. & Visser, R.G.F. (2002) PVX-based elicitor screening to identify recognition in *Solanum*; towards novel resistances to *Phytophthora infestans.* In: *10th New Phytologist Symposium: Functional Genomics of Plant Microbe Interactions* (eds F. Martin, M.J. Harrison, N.J. Talbot & J. Ingram) Nancy.

Vogel, H.J. (1964) Distribution of lysine pathways among fungi: evolutionary implications. *Am. Nat.*, **98**, 435–446.

von Martius, C.F.P. (1842) Die Kartoffel-epidemie der letzten Jahre oder die Stockfaule und Räude der Kartoffeln. Munich.

Warburton, A.J. & Deacon, J.W. (1998) Transmembrane Ca^{2+} fluxes associated with zoospore encystment and cyst germination by the phytopathogen *Phytophthora parasitica. Fungal Genet. Biol.*, **25**, 54–62.

Wastie, R. (1991) Breeding for resistance. In: *Phytophthora infestans, the Cause of Late Blight of Potato* (eds D. Ingram & P. Williams), pp. 193–224. London: Academic Press.

Waugh, M., Hraber, P., Weller, J., Wu, Y., Chen, G., Inman, J., Kiphart, D. & Sobral, B. (2000) The *Phytophthora* genome initiative database: informatics and analysis for distributed pathogenomic research. *Nucleic Acids Res.*, **28**, 87–90.

Werner, S., Steiner, U., Becher, R., Kortekamp, A., Zyprian, E. & Deising, H.B. (2002) Chitin synthesis during *in planta* growth and asexual propagation of the cellulosic oomycete and obligate biotrophic grapevine pathogen *Plasmopara viticola. FEMS Microbiol. Lett.*, **208**, 169–173.

Whisson, S.C., van der Lee, T., Bryan, G.J., Waugh, R., Govers, F. & Birch, P.R.J. (2001) Physical mapping across an avirulence locus of *Phytophthora infestans* using a highly representative, large-insert bacterial artificial chromosome library. *Mol. Genet. Genomics*, **266**, 289–295.

Wilson, U.E. & Coffey, M.D. (1980) Cytological evaluation of general resistance to *Phytophthora infestans* in potato foliage. *Ann. Bot.*, **45**, 81–90.

Young, N.D. (2000) The genetic structure of resistance. *Curr. Opin. Plant Biol.*, **3**, 285–290.

Zentmyer, G.A. (1980) *Phytophthora cinnamomi* and the diseases it causes. Monograph No.**10**. USA: American Phytopathological Society, St Paul Minnesota, 96pp.

Zentmyer, G.A. (1983) The world of *Phytophthora.* In: *Phytophthora: Its Biology, Taxonomy, Ecology, and Pathology* (eds D.C. Erwin, S. Bartnicki-Garcia & P.H. Tsao), pp. 1–8. American Phytopathological Society Press, St Paul, MN.

Index